CYTOKINE INHIBITORS

CYTOKINE INHIBITORS

edited by
Gennaro Ciliberto
Rocco Savino
Istituto di Ricerche di Biologia Molecolare P. Angeletti
Rome, Italy

MARCEL DEKKER, INC. NEW YORK • BASEL

ISBN: 0-0434-7

This book is printed on acid-free paper.

Headquarters
Marcel Dekker, Inc.
270 Madison Avenue, New York, NY 10016
tel: 212-696-9000; fax: 212-685-4540

Eastern Hemisphere Distribution
Marcel Dekker AG
Hutgasse 4, Postfach 812, CH-4001 Basel, Switzerland
tel: 41-61-261-8482; fax: 41-61-261-8896

World Wide Web
http://www.dekker.com

The publisher offers discounts on this book when ordered in bulk quantities. For more information, write to Special Sales/Professional Marketing at the headquarters address above.

Current printing (last digit):
10 9 8 7 6 5 4 3 2 1

PRINTED IN THE UNITED STATES OF AMERICA

Preface

Cytokines play important physiological roles in cell-to-cell communication, development, and differentiation. It is well known that cell differentiation in the hemopoietic and immune systems depends on the complex interplay and sequential action of several members of this heterogeneous group of polypeptides, which have in common that they are produced in small quantities, often only upon exogenous stimuli, and in most cases act over a short distance in a paracrine or autocrine fashion. The differentiation-inducing activity of cytokines has opened up new avenues toward the therapy of hematological and immune disorders. Industrial investments on the part of biotech and pharmaceutical companies have been enormous over the past years, and many recombinant molecules are now marketed or used in advanced clinical trials for the treatment of primary or secondary anemia, granulocytopenias, thrombocytopenias and bone marrow transplantation protocols. Similarly, molecules acting as stimulators of immune functions, such as interferons, have found a wide market in the therapy of chronic hepatitis, cancer, and multiple sclerosis. It has to be stressed that the therapeutic utilization of these recombinant molecules has so far been only partially exploited due to a combination of factors, which include elevated production costs, limited half-life in vivo, and lack of efficient and safe systems for selective delivery of the desired therapeutic amount to the site of action.

Besides those cases in which exogenous cytokines can be used therapeutically either to substitute for or to potentiate physiological functions, there are contrasting cases in which it is desirable to block the activity of an endogenously overproduced cytokine. Research over the past ten years has clearly indicated that imbalanced cytokine production plays an important pathogenetic role in several inflammatory, chronic autoimmune diseases, and sometimes facilitates cancer growth. Thus, blocking the activity of a central molecule can result in the improvement of clinical and biochemical parameters.

While it is clear that blocking cytokine function is necessary to cure several pathological conditions, it is also evident that this is difficult to achieve. Naturally soluble cytokine antagonists, with the exception of a very limited number of cases, do not exist. Thus, ad hoc drug-discovery strategies have to be devised in order to obtain potent and specific cytokine blocking agents. Potency is an absolute requirement to counteract the activity of the endogenous cytokine, which is produced in microgram-per-day amounts in several pathologies and is fully active at picomolar levels. Specificity is also difficult to achieve, given the high degree of redundancy in the use of common receptors and/or signaling pathways. Finally, while several cytokine blocking agents in the form of recombinant proteins (monoclonal antibodies, soluble receptors, receptor antagonists) have been successfully generated over the years, they suffer from several major limitations, some of them identical to those encountered for wild-type cytokine delivery, i.e., the need for parenteral administration and the elevated costs of production. In addition, we face new problems, the most important being the potential immunogenicity of these nonnatural molecules. In order to overcome these hurdles, there is strong pressure to develop small-molecular-weight inhibitors.

Cytokine antagonists can be obtained at least at three general levels: 1) blockade of cytokine production, 2) inhibition of cytokine-induced assembly of a functionally competent receptor complex, 3) interference with cytokine signaling inside target cells. All these approaches have now become more feasible thanks to our greater understanding of both the intracellular signaling pathways and the structural determinants responsible for cytokine–receptor interaction. In many instances the 3-D structure of cytokines and their relevant receptors has been solved, as was that of intracellular tyrosine kinases. This knowledge is of enormous help in the molecular modeling and drug design of selective inhibitory compounds. Once a given target is identified, the search for its inhibitors can be undertaken using different strategies: random screening of libraries of compounds, molecular design, or a combination of the two. Furthermore, the technical improvements and the widespread diffusion of combinatorial chemistry are speeding up these steps enormously. We are thus experiencing a moment of high creativity and progress in this field, marked by frequent, important breakthroughs.

The aim of this book is to provide a comprehensive and updated overview of the field of cytokine inhibitors at the beginning of the third millennium. Several chapters illustrate cases of naturally produced or in vitro engineered polypeptide cytokine-blocking agents. In particular, some chapters

(Chapter 1, Dinarello and Fantuzzi; Chapter 2, van Deventer and van der Poll; Chapter 3, Bendtzen et al.; Chapter 9, Naka et al.; and Chapter 10, Starr) give clear examples of the homeostatic function of cytokine inhibition and indirectly substantiate a therapeutic approach oriented to potentiating a natural response of the organism, when required. In the group of ''first-generation'' recombinant cytokine inhibitors the most successful clinical examples are tumor necrosis factor antagonists (Chapter 4, Hitraya and Schaible), which have reached the market during the past year and promise to become prominent tools in the therapy of chronic inflammatory conditions over the next decade. One case is also shown of how the detailed knowledge of the interaction between a cytokine and a multimeric receptor complex has led to the rational design of potent and selective receptor antagonists (Chapter 8, Vernallis). Chapter 5, Murali et al.; Chapter 6, Moss et al.; Chapter 7, Proudfoot-Fichard et al.; Chapter 11, Lipson et al.; and Chapter 12, Gum and Young, illustrate various examples and approaches to the generation and clinical development of ''second-generation,'' small-molecular-weight, orally bioavailable cytokine inhibitors. In the future we can expect to see more and more examples of these inhibitors and eventually to find them on the shelves of pharmacies as novel and more potent anti-inflammatory, anticancer, and antiviral agents.

The editors wish to express their gratitude to Sara Razzicchia. Without her invaluable editorial assistance on this project the book might not have been born.

Gennaro Ciliberto
Rocco Savino

Contents

Contributors

M. Andersen, Ph.D. Department of Medicinal Chemistry, GlaxoWellcome, Inc., Research Triangle Park, North Carolina

R. C. Andrews, Ph.D. Department of Medicinal Chemistry, GlaxoWellcome, Inc., Research Triangle Park, North Carolina

R. Austin, Ph.D. Department of Medicinal Chemistry, GlaxoWellcome, Inc., Research Triangle Park, North Carolina

E. J. Beaudet, M.S. Department of Research BioMet, GlaxoWellcome, Inc., Research Triangle Park, North Carolina

J. D. Becherer, Ph.D. Department of Molecular Biochemistry, GlaxoWellcome, Inc., Research Triangle Park, North Carolina

Klaus Bendtzen, M.D. Institute for Inflammation Research, Rigshospitalet National University Hospital, Copenhagen, Denmark

Alan Berezov, Ph.D. Department of Pathology and Laboratory Medicine, University of Pennsylvania, Philadelphia, Pennsylvania

D. M. Bickett, M.S. Department of Molecular Biochemistry, GlaxoWellcome, Inc., Research Triangle Park, North Carolina

V. M. Boncek, M.S. Department of Molecular Pharmacology, GlaxoWellcome, Inc., Research Triangle Park, North Carolina

D. G. Bubacz, M.S. Department of Medicinal Chemistry, GlaxoWellcome, Inc., Research Triangle Park, North Carolina

J. H. Chan, Ph.D. Department of Medicinal Chemistry, GlaxoWellcome, Inc., Research Triangle Park, North Carolina

Gennaro Ciliberto, M.D. Department of Gene Therapy, Istituto di Ricerche di Biologia Molecolare P. Angeletti, Rome, Italy

J. G. Conway, Ph.D. Department of Molecular Pharmacology, GlaxoWellcome, Inc., Research Triangle Park, North Carolina

D. J. Cowan, M.S. Department of Medicinal Chemistry, GlaxoWellcome, Inc., Research Triangle Park, North Carolina

Charles A. Dinarello, M.D. Division of Infectious Diseases, Department of Medicine, University of Colorado Health Sciences Center, Denver, Colorado

Giamila Fantuzzi, Ph.D. Department of Medicine, University of Colorado Health Sciences Center, Denver, Colorado

M. D. Gaul, Ph.D. Department of Medicinal Chemistry, GlaxoWellcome, Inc., Research Triangle Park, North Carolina

K. Glennon, B.S. Department of Medicinal Chemistry, GlaxoWellcome, Inc., Research Triangle Park, North Carolina

Rebecca J. Gum, Ph.D. Department of Strategic and Exploratory Sciences, Abbott Laboratories, Abbott Park, Illinois

Y. Guo, M.S. Department of Medicinal Chemistry, GlaxoWellcome, Inc., Research Triangle Park, North Carolina

Morten Bagge Hansen, M.D. D.M.Sc. Institute for Inflammation Research, Rigshospitalet National University Hospital, Copenhagen, Denmark

Akihiro Hasegawa, Ph.D. Department of Pathology and Laboratory Medicine, University of Pennsylvania, Philadelphia, Pennsylvania

K. M. Hedeen, M.S. Department of Research BioMet, GlaxoWellcome, Inc., Research Triangle Park, North Carolina

Elena G. Hitraya, M.D., Ph.D. Medical Affairs Department—Immunology, Centocor, Inc., Malvern, Pennsylvania

Kiichi Kajino, M.D., Ph.D.* Second Department of Pathology, Shiga University of Medical Sciences, Otsu, Japan

I. Kaldor, M.S. Department of Medicinal Chemistry, GlaxoWellcome, Inc., Research Triangle Park, North Carolina

Tadamitsu Kishimoto, M.D., Ph.D. President, Osaka University, Suita City, Osaka, Japan

M. H. Lambert, Ph.D. Department of Structural Chemistry, GlaxoWellcome, Inc., Research Triangle Park, North Carolina

M. A. Leesnitzer, M.S. Department of Molecular Biochemistry, GlaxoWellcome, Inc., Research Triangle Park, North Carolina

Congxin Liang, Ph.D. Department of Chemistry, SUGEN, Inc., South San Francisco, California

Kenneth E. Lipson, Ph.D. Department of Drug Discovery and Cell Biology, SUGEN, Inc., South San Francisco, California

R. T. McConnell, M.S. Department of Molecular Biochemistry, GlaxoWellcome, Inc., Research Triangle Park, North Carolina

D. L. McDougald, M.S. Department of Medicinal Chemistry, GlaxoWellcome, Inc., Research Triangle Park, North Carolina

Gerald McMahon, Ph.D. Department of Drug Discovery, SUGEN, Inc., South San Francisco, California

* Formerly at University of Pennsylvania, Philadelphia, Pennsylvania

Christian Meyer, M.D. Institute for Inflammation Research, Rigshospitalet National University Hospital, Copenhagen, Denmark

J. L. Mitchell, B.S. Department of Molecular Biochemistry, GlaxoWellcome, Inc., Research Triangle Park, North Carolina

M. L. Moss, Ph.D. Department of Biochemistry and Biophysics, Lineberger Comprehensive Cancer Center, University of North Carolina at Chapel Hill, Chapel Hill, North Carolina

Ramachandran Murali, Ph.D. Department of Pathology and Laboratory Medicine, University of Pennsylvania, Philadelphia, Pennsylvania

D. L. Musso, M.S. Department of Medicinal Chemistry, GlaxoWellcome, Inc., Research Triangle Park, North Carolina

Tetsuji Naka, M.D. Department of Molecular Medicine, Graduate School of Medicine, Osaka University, Suita City, Osaka, Japan

Masashi Narazaki, M.D., Ph.D. Department of Molecular Medicine, Graduate School of Medicine, Osaka University, Suita City, Osaka, Japan

Amanda E. I. Proudfoot-Fichard, Ph.D. Department of Biotechnology, Serono Pharmaceutical Research Institute S.A., Geneva, Switzerland

M. H. Rabinowitz, Ph.D. Department of Medicinal Chemistry, GlaxoWellcome, Inc., Research Triangle Park, North Carolina

M. C. Rizzolio, M.S. Department of Pharmaceutics, GlaxoWellcome, Inc., Research Triangle Park, North Carolina

Christian Ross, M.D. Institute for Inflammation Research, Rigshospitalet National University Hospital, Copenhagen, Denmark

Rocco Savino, Ph.D. Department of Gene Therapy, Istituto di Ricerche di Biologia Molecolare P. Angeletti, Rome, Italy

L. T. Schaller, B.S. Department of Medicinal Chemistry, GlaxoWellcome, Inc., Research Triangle Park, North Carolina

Thomas F. Schaible, Ph.D. Department of Medical Affairs—Immunology, Centocor, Inc., Malvern, Pennsylvania

T. M. Seaton, B.S. Department of Molecular Biochemistry, GlaxoWellcome, Inc., Research Triangle Park, North Carolina

J. B. Stanford, M.S. Department of Medicinal Chemistry, GlaxoWellcome, Inc., Research Triangle Park, North Carolina

Robyn Starr, Ph.D. Division of Cancer and Haematology, The Walter and Eliza Hall Institute of Medical Research, Parkville, Victoria, Australia

Li Sun, Ph.D. Department of Chemistry, SUGEN, Inc., South San Francisco, California

Morten Svenson Institute for Inflammation Research, Rigshospitalet National University Hospital, Copenhagen, Denmark

Wataru Takasaki, Ph.D.* Drug Metabolism and Pharmacokinetics Research Laboratories, Sankyo Co., Ltd., Tokyo, Japan

T. K. Tippin, Ph.D. Department of Research BioMet, GlaxoWellcome, Inc., Research Triangle Park, North Carolina

Alexandra Trkola, Ph.D. Division of Infectious Diseases, Department of Internal Medicine, University Hospital Zurich, Zurich, Switzerland

Sander J. H. van Deventer, M.D. Department of Experimental Internal Medicine, Academic Medical Center, Amsterdam, The Netherlands

Tom van der Poll, M.D., Ph.D. Laboratory of Experimental Internal Medicine, Academic Medical Center, Amsterdam, The Netherlands

Ann B. Vernallis, Ph.D. School of Life and Health Sciences, Aston University, Birmingham, England

---

* Formerly at University of Pennsylvania, Philadelphia, Pennsylvania

J. R. Warner, B.S. Department of Molecular Biochemistry, GlaxoWellcome, Inc., Research Triangle Park, North Carolina

Timothy N. C. Wells Serono Pharmaceutical Research Institute, Geneva, Switzerland

L. G. Whitesell, Ph.D. Department of Pharmaceutics, GlaxoWellcome, Inc., Research Triangle Park, North Carolina

R. W. Wiethe, M.S. Department of Medicinal Chemistry, GlaxoWellcome, Inc., Research Triangle Park, North Carolina

Peter R. Young, Ph.D. Department of Cardiovascular Diseases, DuPont Pharmaceuticals, Wilmington, Delaware

1

Importance of the Interleukin-1β–Converting Enzyme in Disease Mediated by the Proinflammatory Cytokines IL-1β and IL-18

Charles A. Dinarello and Giamila Fantuzzi
University of Colorado Health Sciences Center, Denver, Colorado

INTRODUCTION

Cytokines are small, nonstructural proteins with molecular weights ranging from 8,000 to 40,000 D. Originally called lymphokines and monokines to indicate their cellular sources, it became clear that the term *cytokine* is the best description, since nearly all nucleated cells are capable of synthesizing these proteins and, in turn, respond to them. There is no amino acid sequence motif or three-dimensional structure that links cytokines; rather their biological activities allow them to be grouped into different classes. For the most part, cytokines are primarily involved in host responses to disease or infection. From gene deletion studies in mice and the use of neutralizing antibodies, their role in any homeostatic mechanism has been nonexistent or less than dramatic. For example, the interleukin-1β (IL-1β)– and IL-1β–converting enzyme (ICE)–deficient mouse has been bred in normal animal housing at several institutions without signs of increased resistance or development of spontaneous disease. In the case of the IL-10–deficient mouse, which develops spontaneous inflammatory bowel disease in normal housing, the disease does not develop in germ-free housing.

Many scientists have made the analogy of cytokines to hormones, but upon closer examination, this is not an accurate comparison. Why? First, hormones tend to be constitutively expressed by highly specialized tissues but cytokines are synthesized by nearly every cell. Whereas hormones are the primary synthetic product of a cell (e.g., insulin, thyroid, ACTH), cytokines account for a very small amount of the synthetic output of a cell. In addition, hormones are expressed in response to homeostatic control signals, many of which are part of a daily cycle. In contrast, most cytokine genes are not expressed (particularly at the translational level) unless specifically stimulated by noxious events. In fact, it has become clear that the protein kinases involved in triggering cytokine gene expression are activated by a variety of ''cell stressors.'' For example, ultraviolet light, heat shock, hyperosmolarity, or adherence to a foreign surface activate the mitogen-activated protein kinases (MAPK) which phosphorylate transcription factors for gene expression. Of course, infection and inflammatory products also use the MAPK pathway for initiating cytokine gene expression. One concludes then that cytokines themselves are produced in response to ''stress,'' whereas most hormones are produced by a daily intrinsic clock.

Based on their primary biological activities, cytokines are often grouped as lymphocyte growth factors, mesenchymal growth factors, interferons, chemokines, and colony-stimulating factors. Some have been given the name *interleukin* to indicate a product of a leukocyte and a target of a leukocyte. But the term *interleukin* does not correctly connote the true pleotropic nature of cytokines. Nevertheless, there are presently 18 cytokines called interleukins. By convention, a newly discovered cytokine can be called an interleukin when a biological activity is associated with a novel human DNA sequence and the gene product is expressed recombinantly and shown to possess the same property as the natural product. Other cytokines have retained their original biological description such as ''tumor necrosis factors''. Another way to look at some cytokines is their role in inflammation, and this is particularly relevant to the importance to pain. Hence, some cytokines clearly promote inflammation and are called proinflammatory cytokines, whereas other suppress the activity or production of proinflammatory cytokines and are called anti-inflammatory cytokines.

Cytokines are rather promiscuous molecules with many equally promiscuous cells as lovers. For example, IL-4 and IL-10 are potent activators of B lymphocytes. B lymphocytes are responsible for antibody formation in response to a foreign (or endogenous) antigen. Cytokines which ''help'' promote B-cell antibody formation are classified as T-lymphocyte helper cell of the type 2 class (Th2). The counterpart to T-lymphocyte helper cells of the type

2 class are cytokines that help the type 1 T lymphocytes (Th1). Th1 cytokines help the process of cellular immunity in which T lymphocytes attack and kill viral infected cells. Th1 cytokines include IL-2, IL-12, IL-18, and interferon-γ (IFN-γ). However, IL-4 and IL-10, although primarily considered to be Th2 cytokines, are also potent anti-inflammatory cytokines. They are anti-inflammatory cytokines by virtue of their ability to suppress genes for proinflammatory cytokines such as IL-1, tumor necrosis factor (TNF), and several chemokines.

IFN-γ is another example of the pleiotropic nature of cytokines. Although like IFN-α and IFN-β, IFN-γ possesses antiviral activity, IFN-γ is also an activator of the Th1 response which leads to cytotoxic T cells. IFN-γ is also considered to be a proinflammatory cytokine, because it augments TNF and IL-1 receptors, induces nitric oxide (NO), and upregulates endothelial adhesion molecules. However, listing IFN-γ as a purely proinflammatory cytokine should be done with an open mind in that depending on the biological process, IFN-γ may function as an anti-inflammatory cytokine. It should be pointed out that these same cytokines (IFN-γ, IL-6, IL-10, IL-4) are also major participants in directly controlling inflammation and hence serve as an example of the pleiotropic nature of cytokines.

Although a vast amount of data support the concept that TNF and IL-1 are highly proinflammatory molecules, why the interest in whether IFN-γ is proinflammatory or anti-inflammatory? IFN-γ production requires the activity of another cytokine; namely, IL-18. The biology of IL-18 and its role in inflammation are discussed below. However, the ability of IL-18 to induce IFN-γ is dependent on ICE (1,2). In addition to cleaving the IL-1β precursor into an active, proinflammatory cytokine, ICE also cleaves the inactive IL-18 precursor into an active, IFN-γ–inducing factor. Hence, inhibitors of ICE must be considered to affect the processing and activity of two distinct cytokines, IL-1β and IL-18. Moreover, the effect of inhibitors of ICE cannot be regarded as purely anti-inflammatory, since the production of IFN-γ will affect the immune (Th1) and anti-inflammatory properties of IFN-γ. Indeed, inhibitors of ICE reduce endotoxin-induced IFN-γ by a non–IL-1β mechanism (3), and mice deficient in ICE do not produce IFN-γ following administration of endotoxin, whereas mice deficient in IL-1β do.

The concept that inhibition of ICE may not be a purely anti-inflammatory strategy is best illustrated when cells are stimulated with IL-1β to produce prostaglandin E_2 (PGE_2) via the induction of cyclooxygenase-2 (COX-2) (reviewed in ref. 4). For the most part, pain is associated with inflammation. Nerve endings sense the swelling of local tissues which is a hallmark of inflammation. The other hallmark is erythema. The injection of IL-1β locally into humans results in pain, swelling, and erythema (4). IL-1β induces the genes coding for phospholipase A_2 type II, COX-2, and inducible nitric oxide synthase (iNOS). Figure 1 illustrates the basic events of this inflammatory process. Whether induced by infection, trauma, ischemia, immune-activated T cells, or toxins, IL-1 initiates the cascade of inflammatory mediators by targeting the endothelium.

However, IFN-γ is unique in that it suppresses IL-1β induction of PGE_2 (5). In fact, IFN-γ suppresses other IL-1–induced processes such as IL-1 induction of itself (6,7). IFN-γ also suppresses IL-1 induction of collagenases (8,9). Therefore, in blocking the processing of proIL-18 using inhibitors of ICE, one does not have the specificity of reducing the biological impact of a single cytokine.

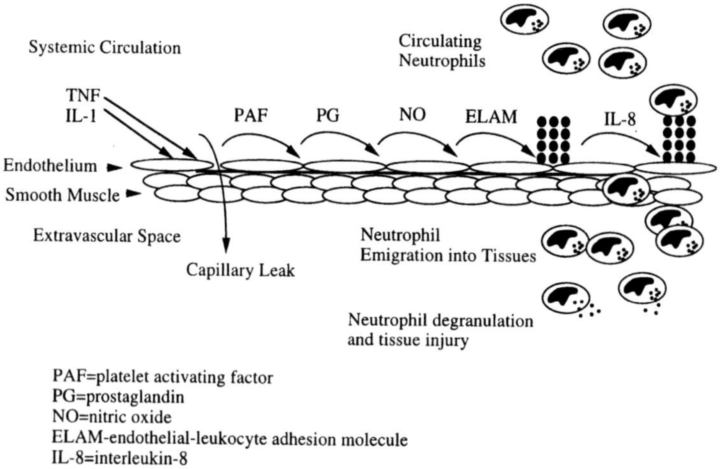

Figure 1 Proinflammatory effects of IL-1 and TNF are on the endothelium. TNF and IL-1 activate endothelial cells and trigger the cascade of proinflammatory small molecule mediators. Increased gene expression for phospholipase A_2 type II, cyclooxygenase type 2 (COX-2), and inducible nitric oxide synthase (iNOS) results in elevated production of their products, PAF, PGE_2, and NO. Alone or in combination, these mediators decrease the tone of vascular smooth muscle. IL-1 and TNF also cause increased capillary leak leading to swelling and pain. Pain threshold is lowered by PGE_2. The upregulation of endothelial leukocyte adhesion molecules results in adherence of circulating neutrophils to the endothelium, and increased production of chemokines such as IL-8 facilitates the emigration of neutrophils into the tissues. Chemokines also activate degranulation of neutrophils. Activated neutrophils lead to tissue destruction and more inflammation.

WHY INHIBIT IL-1?

Although the systemic effects of IL-1 have been studied in animals, there are now data on the effects and sensitivity to IL-1 in humans. The overwhelming conclusion of these studies is that IL-1 is a highly inflammatory molecule and reduction in its production or activity in a variety of disease states is likely a sensible therapeutic strategy. IL-1α or IL-1β has been injected in patients with various solid tumors or as part of a reconstitution strategy in bone marrow transplantation. Acute toxicities of either IL-1α or IL-1β were greater following intravenous compared to subcutaneous injection; subcutaneous injection was associated with significant local pain, erythema, and swelling (10,11). Chills and fever are observed in nearly all patients, even in the 1 ng/kg dose group (12). The febrile response increased in magnitude with increasing doses (13–17) and chills and fever were abated with indomethacin treatment (18). In patients receiving IL-1α (16,17) or IL-1β (13,14), nearly all subjects experienced significant hypotension at doses of 100 ng/kg or greater. Systolic blood pressure fell steadily and reached a nadir of 90 mm Hg or less 3–5 h after the infusion of IL-1. At doses of 300 ng/kg, most patients required intravenous pressors. By comparison, in a trial of 16 patients given IL-1β from 4 to 32 ng/kg subcutaneously, there was only one episode of hypotension at the highest dose level (10). These results suggest that the hypotension is probably due to induction of NO, and elevated levels of serum nitrate have been measured in patients with IL-1–induced hypotension (17).

Patients given 30–100 ng/kg of IL-1β had a sharp increase in cortisol levels 2–3 h after the injection. Similar increases were noted in patients given IL-1α. In 13 of 17 patients given IL-1β, there was a fall in serum glucose within the first hour of administration, and in 11 patients, glucose fell to 70 mg/100 mL or lower (14). In addition, there were increases in ACTH and thyroid-stimulating hormone but a decrease in testosterone (17). No changes were observed in coagulation parameters such as prothrombin time, partial thromboplastin, or fibrinogen degradation products. This latter finding is to be contrasted to TNF-α infusion into healthy humans which results in a distinct coagulopathy syndrome (19).

Not unexpectedly, IL-1 infusion into humans significantly increased circulating IL-6 levels in a dose-dependent fashion (17). At a dose of 30 ng/kg, mean IL-6 levels were 500 pg/mL 4 h after IL-1 (baseline < 50 pg/mL) and 8000 pg/mL after a dose of 300 ng/kg. In another study, infusion of 30 ng/kg of IL-1α induced elevated IL-6 levels within 2 h (20). These elevations in IL-6 are associated with a rise in C-reactive protein and a decrease in albumin.

BLOCKING IL-1 IN PATIENTS WITH RHEUMATOID ARTHRITIS WITH THE IL-1 RECEPTOR ANTAGONIST

The basic concept of inhibition of ICE to reduce the processing and secretion of IL-1β in disease states received a great deal of support following the publication of clinical trials of IL-1 receptor antagonist (IL-1Ra) in patients with rheumatoid arthritis and graft versus host disease. Also, the importance of endogenous IL-1Ra in patients with rheumatoid arthritis is supported by a study using the administration of soluble IL-1R type I to these patients. Since the soluble form of IL-1R type I binds IL-1Ra with a greater and near irreversible affinity than that of IL-1α or IL-1β, the use of the soluble IL-1R type in humans worsened disease in these patients (21). Although IL-1Ra was used in three trials to reduce 28-day mortality in sepsis, the overall success of any anticytokine-based therapy in this patient population precludes any conclusion whether the anticytokine is effective (22–24). In each of these three trials, there was clear evidence of improved outcome in subgroups treated with IL-1Ra compared to placebo-treated patients. However, in each trial, the entire group did not reach a statistically significant reduction in mortality. On the other hand, the use of IL-1Ra is patients with rheumatoid arthritis and graft versus host disease provides convincing evidence that blocking IL-1 reduces disease and, moreover, blocking IL-1 is safe. Of course, inhibition of IL-1β is not the sole effect of ICE inhibition, and the lack of specificity of this strategy in humans may yield different results when it is used in the same diseases.

IL-1Ra in Patients with Rheumatoid Arthritis

IL-1Ra was initially tested in a trial in 25 patients with rheumatoid arthritis. In the group receiving a single subcutaneous dose of 6 mg/kg, there was a fall in the mean number of tender joints ($P < .05$) (25). In patients receiving 4 mg/kg per day for 7 days, there was a reduction in the number of tender joints from 24 to 10, the erythrocyte sedimentation rate fell from 48 to 31, and C-reactive protein decreased from 2.9 to 1.9 μg/mL. In this group, the mean plasma concentration of IL-1Ra was 660 ± 240 ng/mL.

In an expanded double-blind trial, IL-1Ra was given to 175 patients (26). Patients were enrolled into the study with active disease and taking nonsteroidal anti-inflammatory drugs and/or up to 10 mg/day of prednisone. There was an initial phase of 3 weeks of either 20, 70, or 200 mg one, three, or seven times per week. Thereafter, patients received the same dose once weekly for 4 weeks. Placebo was given to patients once weekly for the entire 7-week study period. To maintain the blindness of the study, patients received daily injections of either IL-1Ra or placebo on the days IL-1Ra was not administered.

Four measurements of efficacy were used: number of swollen joints, number of painful joints, patient and physician assessment of disease severity. A reduction of 50% or greater in these scores from baseline was considered significant in the analysis. After 3 weeks, a statistically significant reduction in the total number of parameters was observed with the optimal improvement in patients receiving 70 mg per day. Daily dosing appeared more effective than weekly dosing when assessed by the number of swollen joints, the investigator and patient assessments of disease activity, pain score, and C-reactive protein levels.

A large double-blind, placebo-controlled multicenter trial of IL-1Ra in 472 patients with rheumatoid arthritis (RA) has been reported (27). The study comprised patients who had discontinued the use of disease-modifying agents such as gold and methotrexate 6 weeks prior to entry. Patients had active and severe RA (disease duration 8 years) and were recruited into a 24-week course of therapy into placebo or one of three IL-1Ra groups. Patients had stable disease of 2 or more years' duration and were controlled on nonsteroidal anti-inflammatory agents. Some were taking less than 10 mg of prednisone daily. There were three doses of IL-1Ra administered subcutaneously: 30, 75, and 150 mg/day for 24 weeks. At entry, age, sex, disease duration, and percentage of patients with rheumatoid factor and joint bone erosions were similar in each of the groups. After 24 weeks, 43% of the patients receiving 150 mg/day of IL-1Ra met the American College of Rheumatology criteria for response (the primary efficacy measure), 44% met the Paulus criteria, and statistically significant ($P = .048$) improvements were seen in the number of swollen joints, number of tender joints, investigator's assessment of disease activity, patient's assessment of disease activity, pain score on a visual analog scale, duration of morning stiffness, and Health Assessment score. In addition, there was a dose-dependent reduction in the C-reactive protein level and erythrocyte sedimentation rate.

Importantly, the rate of radiological progression in the patients receiving IL-1Ra was significantly less than in the placebo group at 24 weeks, as evidenced by the Larsen score and the erosive joint count. The reduction in new bone erosions was assessed by two radiologists who were blinded to the patient treatment as well as blinded to the chronology of the x-ray films. This finding suggests that IL-1Ra blocks the osteoclast-activating factor property of IL-1, as has been reported in myeloma cell cultures (28). This study confirmed both the efficacy and the safety of IL-1Ra in a large cohort of patients with active and severe rheumatoid arthritis.

The only side effect was the local skin rash which was observed primarily during the first 2 weeks of therapy. After 24 weeks, there were no statistically significant increases in infection in the IL-1Ra arm compared to placebo. When the trial was extended another 24 weeks, there was also no statistically significant increases in infection. Some patients have now been treated with IL-1Ra for over 3 years and there are no reports of increased infection or cancer. Another trial of IL-1Ra in patients with rheumatoid arthritis is being carried out in patients randomized to receive treatment with methotrexate or different doses of IL-1Ra plus methotrexate treatment.

Graft versus Host Disease

A Phase I/II trial of escalating doses of IL-1Ra in 17 patients with steroid-resistant graft versus host disease has been completed (29). IL-1Ra (400–3400 mg/day) was given as a continuous intravenous infusion every 24 h for 7 days. Using an organ-specific, acute disease scale, there was improvement in 16 of the 17 patients. Moreover, a decrease in the steady-state mRNA for TNF-α in peripheral blood mononuclear cells correlated with improvement ($P = .001$) (29). These studies in humans are similar to the use of IL-1Ra in animal models of graft versus host disease.

For clinical efficacy, IL-1Ra in patients with RA exhibits a dose-dependent response. Even the reduction of endotoxin-induced neutrophilia in healthy subjects is dose dependent. Animal studies support these clinical observations. The requirement for such high plasma levels of IL-1Ra is not completely understood, because IL-1Ra levels are already several logs higher than measurable IL-1 levels in the most severe cases of septic shock. Rapid renal clearance, binding to the soluble form of the type I receptor, and the effect of acidosis in the local or systemic situation may explain a need for these high levels.

IL-1β VERSUS IL-1α

Since the above studies provide evidence that blocking IL-1 in disease is a sensible strategy, the next question is how much of the effectiveness of IL-1Ra in animal models as well as in human disease is due to blocking IL-1β compared to blocking IL-1α? In humans, IL-1α does not circulate during various disease states compared to IL-1β. Unlike ICE, it is necessary to activate a nonspecific, membrane-bound protease, calpain, by calcium for the cleavage of proIL-1α (30). Calpain appears to be the protease that cleaves proIL-1α

Table 1 A Comparison of IL-1α and IL-1β

IL-1α	IL-1β
proIL-1α is active	proIL-1β is inactive
Active membrane is IL-1α	Membrane IL-1β not observed
Mature IL-1α does not circulate	mature IL-1β circulates
Nuclear localization	No nuclear localization
Intracellular role for proIL-1α	No intracellular role for proIL-1β
Neutralizing autoantibodies	Non-neutralizing autoantibodies
Calpain cleavage proIL-1α	ICE and PR-3 cleavage of proIL-1β
No disease link with calpain expression	Disease link with ICE expression
No correlation	Correlation with bone resorption
IL-1α KO mice normal	IL-1β KO resistant to disease
Anti–IL-1α ineffective in CIA	Anti–IL-1β effective in CIA
IL-1α expression in AML absent	IL-1β expression in AML present
No data	ICE inhibition ↓ brain ischemia
No data	ICE inhibition ↓ AML proliferation
No data	ICE anti-sense ↓ AML proliferation

and results in secretion. However, unlike IL-1β, proIL-1α is active, and hence the role of any protease in conferring biological activity is questionable.

Nevertheless, IL-1α is found in tissues, particularly intracellularly. It has been speculated that the biological role for IL-1α in disease is as a membrane-bound cytokine. Indeed, membrane-bound IL-1 activity has consistently been IL-1α, not IL-1β. IL-1β is secreted, whereas IL-1α remains biologically active as a cell-associated cytokine (31,32). Unlike IL-1β, normal skin and epithelial cells contain IL-1α. Also, autoantibodies to IL-1α are common and many are neutralizing antibodies (33).

IL-1α appears to be involved with nuclear location and has a role in intracellular events, whereas IL-1β has no role in such events. For example, proIL-1α has a strong nuclear localization sequence and, when overexpressed, increases maligant transformation (34,35). Antisense DNA to IL-1α reduces cell death in endothelial cells (36,37). Thus, IL-1α appears to act as an intracellular growth factor. Differences in IL-1α versus IL-1β are shown in Table 1.

INTERLEUKIN-1β

Gene Expression and Synthesis of IL-1β

In human monocytes, IL-1β mRNA levels rise rapidly within 15 min but, depending on the stimulus, start to fall after 4 h. The decrease is thought to be due to the synthesis of a transcriptional repressor and/or a decrease in mRNA half-life (38,39). Using IL-1 itself as a stimulant of its own gene expression, IL-1β mRNA levels are sustained for over 24 h (7,40). Raising cAMP levels in monocytes with histamine enhances IL-1β gene expression and protein synthesis (41). Retinoic acid induces IL-1β gene expression, but the primary precursor transcript fails to yield mature mRNA (39). Inhibition of translation by cycloheximide results in enhanced splicing of exons, excision of introns, and increased levels of mature mRNA (superinduction). Thus, synthesis of mature IL-1β mRNA requires an activation step to overcome an apparently intrinsic inhibition to process precursor mRNA.

A dissociation between transcription and translation is characteristic of IL-1β and also of TNF-α (42). Despite a vigorous signal for transcription by a variety of agents, including C5a, adherence, or even hyperosmolar NaCl (42–44), most of the IL-1β mRNA is degraded and no significant translation into proIL-1β takes place. Although the IL-1β mRNA assembles into large polyribosomes, there is little significant elongation of the peptide (45). However, adding bacterial endotoxin or IL-1 itself to cells with high levels of steady-state IL-1β mRNA results in augmented translation (42,44) in somewhat the same manner as the removal of cycloheximide following superinduction. One explanation is that stabilization of the AU-rich 3′ untranslated region takes place in cells stimulated with lipopolysaccharide (LPS). These AU-rich sequences are known to suppress normal hemoglobin synthesis. The stabilization of mRNA by microbial products may explain why low concentrations of LPS or a few bacteria or *Borrelia* organisms per cell induce the translation of large amounts of IL-1β (46).

Another explanation is that IL-1 stabilizes its own mRNA by preventing deadenylation as it does for the chemokine gro-α (47). Removal of IL-1 from cells after 2 h increases the shortening of polyadenylic acid (poly-A), and IL-1 apparently is an important regulator of *gro* synthesis, because it prevents deadenylation. In fact, of the several cytokines induced by IL-1, large amounts of the chemokine family are produced in response to low concentrations of IL-1. For example, 1 pM of IL-1 stimulates fibroblasts to synthesize 10 nM of IL-8 (48).

Following synthesis, proIL-1β remains primarily cytosolic until it is cleaved and transported out of the cell (Fig. 2). The IL-1β propiece (amino acids 1–116) is also myristoylated on lysine residues (34), but unlike IL-1α, proIL-1β has no known membrane form and proIL-1β is only marginally active (49). Some IL-1β is found in lysosomes (50) or associated with microtubules (51,52) and either localization may play a role in the secretion of IL-1β. In mononuclear phagocytes, a small amount of proIL-1β is secreted from

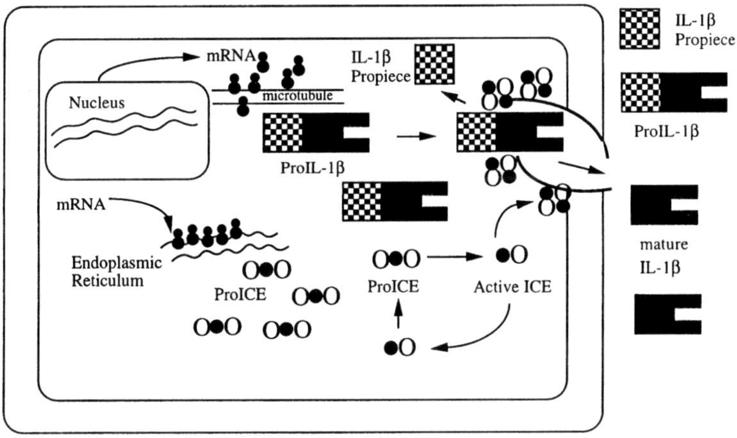

Figure 2 Human blood monocyte producing IL-1β. mRNA coding for proIL-1β is translated on polysomes in the cytosol and associated with microtubules (52). proIL-1β remains cytosolic until cleaved by ICE. ICE is translated in the endoplasmic reticulum as an inactive precursor (proICE) and requires two internal cleavage steps to form the enzymatically active heterodimer. Two heterodimers form a tetramer in association with two molecules of proIL-1β and cleavage occurs. Active ICE is found predominantly on the inner surface of the cell membrane (68). Following cleavage, 17-kD IL-1β is secreted into the extracellular compartment through a putative membrane channel. The 16 kD IL-1β "propiece" can be found both inside and outside the cell. A small amount of proIL-1β can be transported into the extracellular space from intact cells, presumably using the same channel; however, when ICE activity is inhibited, more proIL-1β is found in the extracellular compartment (65).

intact cells (53,54), but the pathway for this secretion remains unknown. On the other hand, release of mature IL-1β appears to be linked to processing at the aspartic acid–alanine (116–117) peptide cleavage by the IL-1β–converting enzyme (ICE) (55) (see below).

There are several sites in proIL-1β which are vulnerable to cleavage by enzymes in the vicinity of alanine 117. These are trypsin, elastase, chymotrypsin, a mast cell chymase, and a variety of proteases which are commonly found in inflammatory fluids (reviewed in ref. 56). The extent that these proteases play in the in vivo conversion of proIL-1β to mature forms is uncertain, but in each case, a biologically active IL-1β species is produced. For example, in ICE-deficient mice given a subcutaneous injection of turpentine, biologically active IL-1β is released (56). In the discussion on the soluble IL-1 receptor type II (below), the affinity of proIL-1β for this constitutively produced soluble receptor is high and may prevent haphazard cleavage of the precursor by these enzymes in inflammatory fluids.

ICE

As depicted in Figure 2, proIL-1β requires cleavage by ICE before the mature form is secreted. The 45-kD precursor of ICE requires two internal cleavages before becoming the enzymatically active heterodimer comprising a 10- and 20-kD chain. The active site cysteine is located on the 20-kD chain. ICE itself contributes to autoprocessing of the ICE precursor by undergoing oligomerization with itself or homologs of ICE (57,58). ICE is a member of a family of intracellular cysteine proteases called caspases, a term used to connote a cysteine proteases cleaving after an aspartic acid residue (59). ICE is caspase-1.

Two molecules of the ICE heterodimer form a tetramer with two molecules of proIL-1β for cleavage (57,60). The aspartic acid at position 116 of the proIL-1β is the so-called P1 recognition amino acid for ICE cleavage. ICE does not cleave the IL-1α precursor. Enzymes such as elastase (61) and granzyme A (62) cleave proIL-1β at amino acids 112 and 120, respectively, yielding biologically active IL-1β. The propiece of IL-1β can be found both inside and outside the cell (63). In addition, the propiece exhibits biological activity as a chemoattractant for fibroblasts via an IL-1R–mediated event (64).

In the presence of a tetrapeptide competitive substrate inhibitor of ICE, the generation and secretion of mature IL-1β is reduced and proIL-1β accumulates mostly inside but also outside the cell (65). This latter finding supports the concept that proIL-1β can be released from a cell independent of processing by ICE. Similar to that of thioredoxin (66) and basic fibroblast growth factor (FGF) (67), exocytosis has been proposed as a possible mechanism of proIL-1β release. A putative membrane "channel" where active ICE is localized has also been proposed. In this model, mature IL-1β is released through this channel (68). When ICE activity is blocked by a reversible competitive substrate inhibitor, greater amounts of proIL-1β are found in the supernatants (65,68) and, thus, the putative channel may provide a passive secretory pathway for both proIL-1β and mature IL-1β. Macrophages from ICE-deficient mice do not release mature IL-1β on stimulation in vitro (69,70).

PROTEINASE-3 AND EXTRACELLULAR CLEAVAGE OF proIL-1β

Neutrophil enzymes such as elastase and granzyme A (62) can cleave proIL-1β at sites close to alanine 117 and result in active IL-1β. In ICE-deficient mice injected subcutaneously with turpentine, neutrophils dominate in the inflammatory reaction and local as well as systemic effects in this model are unaffected in ICE-deficient mice (56). Therefore, there is likely a neutrophil enzyme that cuts proIL-1β in ICE-deficient mice which accounts for biologically active IL-1β. Elastase is a candidate but the neutrophil enzyme proteinase-3 (PR-3) cuts proIL-1β into an active, mature molecule (71). Moreover, specific inhibitors of PR-3, but not elastase, inhibit the processing of proIL-1β into an active cytokine (71). It is probable that enzymes such as PR-3 cleave extracellular proIL-1β, whereas ICE cleaves proIL-1β intracellularly. How much extracellular and how much intracellular cleavage of proIL-1β takes place in inflammation? That question will be answered when clinical trials of ICE inhibitors are initiated. At the present time, it seems certain that, in animal models of inflammatory disease, extracellular cleavage of proIL-1β takes place by PR-3.

Proteases present in the extracellular microenvironment are able to cleave cytokines already secreted into the extracellular space and generate biologically active molecules. In addition to elastase and PR-3, proIL-1β can be cleaved by cathepsin G, chymotrypsin, a mast cell chymase, and different matrix metalloproteinases to yield active IL-1β. These proteases are found in inflammatory fluids at sites of neutrophil, lymphocyte, or macrophage infiltration, and they can therefore participate in the generation of active IL-1β. Since proIL-1β can be actively released from live cells or passively from dead cells (72), it is likely that extracellular processing of proIL-1β by inflammatory proteases takes place. Processing by extracellular proteases is not limited to IL-1β or to cytokines lacking a signal peptide. The 26-kDa membrane-bound form of TNF-α can be processed in vitro to its 17-kDa soluble counterpart by the serine protease, PR-3 (71). PR-3 can also convert the 77–amino acid form of IL-8 into the 10-fold more potent 70–amino acid form (73). In addition, PR-3 activates TGF-β (transforming growth factor-β) from its latent form and a mast cell chymase cleaves the inactive, membrane-bound form of stem cell factor and releases the active molecule (74). The composition of the inflammatory microenvironment is therefore critical in determining which cytokines will be processed by which enzymes. Some cytokines will become active by interacting with extracellular proteases, whereas other enzymatic interactions may result in the loss of cytokine function. Therefore, the protease/protease inhibitor balance becomes an important factor in the regulation of cytokine activity and hence of the net result of inflammation.

We have used an in vitro model of inflammation by which IL-1β stimulates the neutrophil chemokine, IL-8, from monkey COS cells. These cells are highly responsive to IL-1β. The induction of IL-8 represents the role of IL-1β in inflammation. Using this model, proIL-1β was incubated with these cells

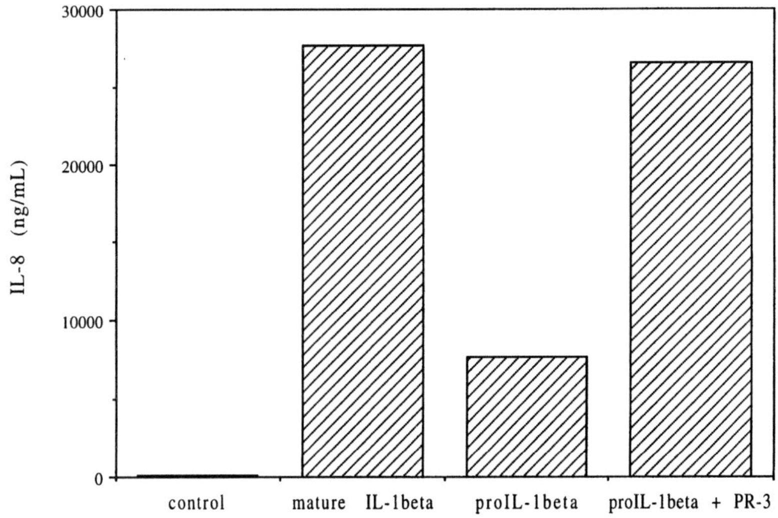

Figure 3 Cleavage of proIL-1β by PR-3. Mature IL-1β (10 ng/mL) was incubated with monkey COS cells and IL-8 was measured after 24 h. (121). proIL-1β (Cistron Biotechnology, Pine Brook, NJ) was incubated for 30 min at room temperature with and without PR-3 (2 μg/mL) and then added to COS cells for 24 h. The final concentration of proIL-1β during the COS cell incubation was 38 ng/mL.

and after 24 h little or no IL-8 was produced. On the other hand, when proIL-1β was cleaved with PR-3 for 30 min, proIL-1β had become active and induced IL-8. These data are shown in Figure 3.

ICE AND IL-18

An endotoxin-induced serum activity that induced IFN-γ from mouse spleen cells was described (75). This serum activity functioned not as a direct inducer of IFN-γ but rather as a costimulant together with IL-2 or mitogens. An attempt to purify the activity from postendotoxin mouse serum revealed an apparently homogeneous 50- to 55-kDa protein (76). Since other cytokines can act as costimulants for IFN-γ production, the failure of neutralizing antibodies to IL-1, IL-4, IL-5, IL-6, or TNF to neutralize the serum activity suggested it was a distinct factor. In 1995, the third report was published from the same investigators demonstrating that the endotoxin-induced costimulant for IFN-γ production was present in extracts of livers from mice preconditioned with *Propionibacterium acnes* (77). In this model, the hepatic macrophage population (Kupffer cells) expands, and in these mice, a low dose of bacterial lipopolysaccharide (LPS), which in nonpreconditioned mice is not lethal, becomes lethal. The factor, named IFN-γ–inducing factor (IGIF), was purified to homogeneity from 1200 g of *P. acnes*–treated mouse livers. Its molecular weight was 18–19 kDa and a N-terminal amino acid sequence was reported (78). Similar to the endotoxin-induced serum activity, IGIF did not induce IFN-γ by itself but functioned primarily as a costimulant with mitogens or IL-2. Degenerate oligonucleotides derived from amino acid sequences of purified IGIF were used to clone a murine IGIF cDNA (77). Recombinant IGIF did not induce IFN-γ by itself but only in the presence of a mitogen or IL-2. However, the coinduction of IFN-γ was independent of IL-12 induction of IFN-γ.

Neutralizing antibodies to mouse IGIF were shown to prevent the lethality of low-dose LPS in *P. acnes*–preconditioned mice (77). Others had reported the importance of IFN-γ as a mediator of LPS lethality in preconditioned mice. For example, neutralizing anti–IFN-γ antibodies protected mice against Shwartzman-like shock (79), and galactosamine-treated mice deficient in the IFN-γ receptor were resistant to LPS-induced death (80). Hence, it was not unexpected that neutralizing antibodies to murine IGIF protected *P. acnes*–preconditioned mice against lethal LPS. Antimurine IGIF treatment also protected surviving mice against severe hepatic cytotoxicty. After the murine form was cloned (77), the human cDNA sequence for IGIF was reported in 1996 (81). Recombinant human IGIF exhibited natural IGIF activity. Human recombinant IGIF was without direct IFN-γ–inducing activity on human T cells but acted as a costimulant for production of IFN-γ and other T helper cell-1 (Th1) cytokines (81). IGIF induced T cell and natural killer (NK) cell IFN-γ production independently of IL-12 (and vice versa) (77). To date, IGIF is thought of as primarily a costimulant for Th1 cytokine production (IFN-γ, IL-2, and granulocyte-macrophage colony-stimulating factor, GM-CSF) (82) and also as a costimulant for Fas ligand-mediated cytotoxicity of murine NK cell clones (83). In vivo, endogenous IGIF activity appears to account for IFN-γ production in *P. acnes* and LPS-mediated lethality (77).

Investigators working on other IFN-γ–inducing cytokines analyzed the computer-generated protein-folding pattern of murine IGIF and compared its pattern to those of others in the data bank. Using a validated compatibility relatedness program, the mature murine IGIF had the highest score with mature human IL-1β; furthermore, the IGIF amino acid sequence matched best with amino acids which form the all-β pleated sheet–folding pattern of human

IL-1β (84). A high degree of alignment was present in the sequences that comprise the 12 β sheets of the mature IL-1β structure. Using this alignment of conserved amino acids, there is a 19% positional identity of mature murine IGIF to mature human IL-1β and a 12% identity to human IL-1α. Using this same positional alignment, the identity of IL-1β to IL-1α is 23%. It was suggested that the name INF-γ–inducing factor be changed to interleukin-1γ (84). Does IGIF bind to IL-1 type I receptors? This would be an essential criterion for assigning the name IL-1γ, since the type I IL-1 receptor is the signaling receptor for the biological activity of IL-1. In the absence of evidence that IGIF binds to the IL-1 receptor type I (unpublished data), IL-18 rather than IL-1γ is a more appropriate name.

Similar to proIL-1β, precursor IL-18 (proIL-18) does not contain a signal peptide required for the removal of the precursor amino acids with subsequent secretion. The N-terminal amino acid sequence of the secreted form of murine IL-18 (78) was consistent with that following cleavage after an aspartic acid residue, a typical cleavage site for ICE. In fact, this analysis alerted investigators that the cleavage of proIL-18 at the aspartic acid site would likely require ICE (84). Therefore, it was not surprising that ICE cleaved proIL-18 and resulted in the mature and active protein (1,2).

IL-18 IS A MEMBER OF THE IL-1 FAMILY

Present Understanding of the IL-18 Receptor Complex

Torigoe et al. described the purification and characterization of a human IL-18 receptor using an antibody directed against a cell surface protein of the human Hodgkin's cell line L428 (85). The purified protein had a molecular weight between 60 and 100 kDa and its amino acid sequence revealed 100% identity to a previously known member of the IL-1 receptor family which had been termed the IL-1 receptor–related protein (IL-1Rrp) (86). IL-1Rrp had remained an orphan receptor until recently. When the IL-1Rrp was first reported, it was clear that it was indeed a member of the IL-1 receptor family, although it did not bind IL-1α, IL-1β, or IL-1Ra (86). This was an unexpected finding, because the IL-1Rrp was isolated from a library using degenerate oligonucleotides from conserved amino acids in the *extracellular* domains of the IL-1R type I.

Until the studies by Torigoe et al., the ligand for IL-1Rrp was unknown, but it would be likely that the ligand was structurally related to members of the IL-1 family. However, despite its relatedness, neither IL-1α nor IL-1β bind to IL-1Rrp (86). When the extracellular domain of the IL-1Rrp was expressed as a fusion protein linked to the Fc domain of IgG, the IL-1Rrp did not bind IL-1 on a Biacore chip (86). However, a chimeric molecule was constructed that comprised the extracellular domains of the IL-1 type I receptor linked to the transmembrane and cytosolic domain of the IL-1Rrp. When transfected into COS cells, this chimeric receptor did exhibit IL-1 functional activity on binding of IL-1. Therefore, although the IL-1Rrp is member of the IL-1 receptor family and its extracellular domains do not bind IL-1 itself, IL-1Rrp transmits a signal similar to that of the IL-1R.

IL-1Rrp Is the Binding Component of the IL-18 Receptor Complex

Torigoe et al. reported that IL-1Rrp is the functional component of IL-18 receptor primarily based on the observation that antibodies to IL-1Rrp prevent IL-18 activity (85). Yet, in terms of other cytokine receptors, IL-18 binding to the L428 cell line used in the purification process was low (dissociation constant [K_d] of 18.5 nM). As summarized by Boraschi, Thomassen reported a K_d of 25 nM for IL-18 (87). The Kd for IL-18 binding to IL-1Rrp expressed transiently on COS cells was 45 nM (85). These are unusually low affinities for a cytokine with a biological activity in the picomolar range. Nevertheless, IL-1Rrp is unquestionably essential for IL-18 activity and an integral part of the IL-18–signaling complex. Therefore, the term *IL-18Rα* replaces the term *IL-1Rrp*. But, a second binding chain in a putative IL-18 receptor complex likely exists in order to form a high-affinity complex which characterizes all cytokine receptor signaling.

Nakanishi's group has reported that there are both high– (0.4 nM) and low– (40 nM) affinity binding sites for IL-18 in murine primary T cells (88). Clearly, the IL-18Rα could account for the low-affinity IL-18 binding sites. Thus, a second chain resulting in a greater affinity would explain the higher affinity. This ''other'' receptor chain has been identified as the ''accessory protein-like (AcPL) receptor'' (89), and it is related to the IL-1 accessory protein identified by Greenfeder (90). Similar to the IL-1 accessory protein, the IL-18 AcPL does not bind to its ligand but rather binds to the complex formed by IL-18 with the IL-18Ra chain (89). This heterodimer is likely the high-affinity complex.

There is a comparable finding in the identification of the IL-1 receptor complex. IL-1 activity was thought to be due to a single chain receptor, the 80-kD IL-1R type I (91). The second chain was discovered in 1995; it was termed the IL-1 receptor accessory protein (IL-1R-AcP) (90). This latter receptor is essential to IL-1 signaling, although IL-1R-AcP does not bind to any

IL-1 ligand. Similar to many cytokine receptors, a heterocomplex of a ligand-binding chain (often termed the α chain of the receptor) plus an accessory chain (often termed the β chain) which transmits the signal. The importance of the IL-1R-AcP for IL-1 activity was shown using antibodies to IL-1R-AcP which is an essential part of the heterodimeric IL-1R complex.

Therefore, the IL-18Rα is unlikely the sole component of the IL-18 receptor–signaling apparatus and is certainly not the high-affinity ligand-binding receptor chain. It is likely that the IL-18 receptor complex comprises a ligand-binding chain of IL-1Rrp (IL-18Rα), with the second chain being the non–IL-18–binding AcPL (IL-18Rβ) (89). As described below, an IL-18–binding protein (IL-18BP) has been purified from human urine and would likely be the soluble (extracellular) form of the IL-18Rα chain (92). But a transmembrane form of the IL-18BP has not been isolated to date despite murine cDNA cloning and human genomic analyses. Thus, the IL-18BP is a third gene product and appears to be the "decoy" receptor of IL-18, similar to the IL-1R type II which is the decoy receptor for IL-1.

Isolation and Purification of an IL-18–Binding Protein From Human Urine

Using methods established for the purification and characterization of ligand-binding cytokine receptors in human urine (93–96), a 38-kD IL-18–binding protein was purified to homogeneity from urine using ligand affinity chromatography and the N-terminal amino acids determined (92). The amino acid sequence of the IL-18–binding protein (IL-18BP) is not that of a known cytokine receptor presently listed in the data banks. Although this binding protein likely represents the extracellular domain of the ligand-binding receptor chain, to date a transmembrane and cytosolic component to the IL-18BP has not been found.

Similar to neutralizing antibodies to IL-18 (3), the IL-18BP prevents LPS-induced IFN-γ production (92). Also, similar to anti–IL-18 antibodies, IL-18BP does not affect IFN-γ production following stimulation with mitogens such as concanavalin A. However, unlike antibodies to IL-18, the IL-18BP does not exhibit species specificity, and hence human IL-18BP neutralizes both human and murine IL-18 (92).

IL-18 Signal Transduction Is Similar to that of IL-1

It is clear that IL-1 and IL-18 do share signaling pathways, particularly in activating the processes by which nuclear factor-κB (NF-κB) translocates to the nucleus. In terms of IFN-γ production, IL-1α, IL-1β, and IL-18 each induce IFN-γ in the presence of IL-2 or IL-12 (97,98). IL-18 is 5–10 times more potent that either form of IL-1 (97), which suggests, expectedly, that signal transduction of IL-1 and IL-18 cannot be identical. However, the recruitment of interleukin-1 receptor–associated kinase (IRAK) to the IL-18 receptor complex places IL-18 signaling high in the signaling cascade that it shares with IL-1. For IL-1 signaling, one of the earliest events is the formation of the heterocomplex of the ligand IL-1 with the type I IL-1R and the IL-1R-AcP (90). This complex then recruits IRAK (99). The topic of IL-1 signal transduction related to NF-κB activation has been reviewed in detail (100,101). The IL-1R-AcP is essential for the recruitment and activation of IRAK (102,103). In fact, deletion of specific amino acids in the IL-1R-AcP cytoplasmic domain results in loss of IRAK association (102). Similar amino acids are present in the IL-18R-AcPL (89) (IL-18Rβ). In addition, an intracellular adapter molecule termed MyD88 appears to dock to the IL-1 complex allowing IRAK to become phosphorylated (99,104). Indeed, a null mutation of MyD88 results in loss of IL-18–induced IFN-γ production (105).

After phosporylation, IRAK then dissociates from the IL-1R complex and associates with TNF receptor–associated factor-6 (TRAF-6) (106). TRAF-6 then phosphorylates NIK (107) and NIK phosphorylates the inhibitory κB kinases (IKK-1 and IKK-2) (108). Once phosphorylated, IκB is rapidly degraded by a ubiquitin pathway liberating NF-κB, which translocates to the nucleus for gene transcription.

IRAK has been demonstrated to associate with the IL-18R complex (109,110). This was shown in IL-12–stimulated T cells followed by immunoprecipitation with anti–IL-18R or anti-IRAK (109). Furthermore, IL-18–triggered cells also recruited TRAF-6 (109). Like IL-1 signaling, MyD88 has a role in IL-18 signaling. MyD88-deficient mice do not produce acute phase proteins and have diminished cytokine responses. Recently, Th1-developing cells from MyD88-deficient mice were unresponsive to IL-18–induced activation of NF-κB and c-Jun N-terminal kinase (JNK) (105). Thus, MyD88 is an essential component in the signaling cascade that follows IL-1 receptor as well as IL-18 receptor binding. It appears that the cascade of sequential recruitment of MyD88, IRAK, and TRAF-6 followed by the activation of NIK and degradation of IκBK and release of NF-κB are nearly identical for IL-1 as well as for IL-18. Indeed, in cells transfected with IL-18Rα (IL-1Rrp) and then stimulated with IL-18, translocation of NF-κB is observed using electromobility shift assay (85). In U1 macrophages which already express the gene for IL-18Rα (IL-1Rrp), there is translocation of NF-κB and stimulation of the human immunodeficiency virus type 1 (HIV-1) production (111).

Figure 4 is a working model of the proposed IL-18 signaling resulting in translocation of NF-κB to the nucleus. However, not shown is the evidence that IL-18 signaling activates p38 mitogen-activated protein kinase (MAPK). In U1 cells stimulated with IL-18, IL-8 synthesis is reduced by nearly 90% in the presence of a specific p38 MAPK inhibitor (111). In murine T cells, IL-18 stimulated the phosphorylation of p42 MAPK and p561ck (112). Table 2 lists the biological effects shared by IL-1 and IL-18.

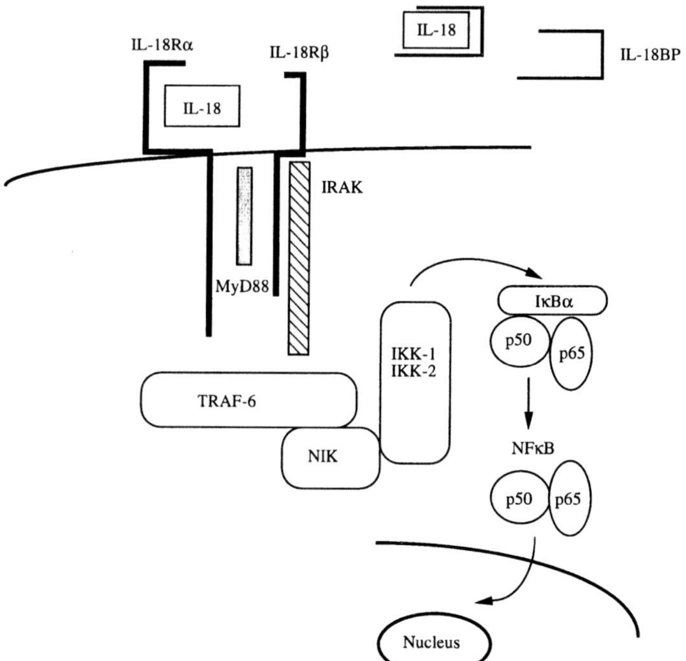

Figure 4 Model for IL-18 signal transduction. IL-18 binds to the IL-18Rα chain (85) (also known as IL-1Rrp [86]) and recruits the IL-18Rβ chain (also known as AcPL [89]). This heterodimeric complex recruits MyD88 and IRAK. Phosphorylated IRAK activates TRAF-6 and in turn NIK is phosphorylated. This results in activation of IKK and then phophorylation of IκB. Phosphorylated IκB degrades and releases the p50 and p65 components of NF-κB. Free NF-κB migrates through nuclear pores and translocates to nuclear DNA for gene transcription.

Table 2 A Comparison of Biological Properties of Two ICE-cleavable Cytokines, IL-18 and IL-1β

Property	IL-18	IL-1β
Precursor cleavage by ICE	+	+
Precursor cleavage by non-ICE	+	+
Induction of IFNγ	+	±
Synergy with IL-12 for IFNγ	+	+
↑ Expression in NOD	+	+
Induction of IL-8/IL-1β/TNFα	+	+
Increased HIV-1 Production	+	+
Activation of Th1 Responses	+	±
B-cell activation	+	+
Activation of Th2 responses	±	±
Adjuvant for tumor immunity	+	+
Increased protection in infection	+	+
Elevated levels in leukemias	+	+
↑ Expression in adhesion molecules	+	+

INHIBITION OF ICE AND EFFECT ON IL-18

Because inhibition of IL-1 activity with IL-1Ra in patients with rheumatoid arthritis has improved the clinical outcome (27), attention has focused on the effect of inhibiting IL-18, since processing and secretion of this cytokine involves ICE. Indeed, inhibitors of ICE have been used in the mouse collagen arthritis model to reduce disease (113). Two irreversible ICE inhibitors, VE-13,045 and VE-16,084, were injected into DBA/1J mice with collagen-induced arthritis. Pretreatment significantly delayed the onset of inflammation and was more effective than either indomethacin or methyl prednisolone. The ICE inhibitors were also effective the progression of arthritis when administered to mice with established disease. Although these studies are similar to those using anti–IL-1β in this model (114), they do not rule out a possible role for inhibiting the release of IL-18. However, the role of IFN-γ (as a downstream event in IL-18 biology) is less impressive in this model than IL-1β (115). Nevertheless, because IL-18 is a proinflammatory cytokine (116), inhibition of ICE may be more effective as an anti-inflammatory mechanism, because it would act to reduce the biological effect of two cytokines.

These findings are supported by studies where IL-18 was examined for its effects on articular chondrocytes. IL-18 mRNA was induced by IL-1β in chondrocytes. IL-18 stimulated gene expression and synthesis for inducible nitric oxide synthase, inducible cyclooxygenase (COX-2), IL-6, and stromelysin. Although the gene for COX-2 was elevated following exposure of chondrocytes to IL-18, elevations in PGE_2 were not reported (117). It is presently unclear whether blocking IL-18 will be as effective as blocking IL-1 in rheumatoid arthritis because of the importance of IL-18 to production of IFN-γ. Again, the role for IFN-γ in various animal models of rheumatoid arthritis is controversial (reviewed in ref. [115]).

Inhibition of ICE also reduces pancreatitis. IL-1β is produced in large amounts during acute pancreatitis and is believed to play a role in disease progression. Also, IFN-γ does play an agonist role in experimental pancreatitis. Pretreatment with ICE inhibitors reduced the severity and mortality of experimental pancreatitis in rats due to infusion of bile acid into the pancreatic duct (118). In ICE knock-out (KO) mice, lethal pancreatitis induced by feeding a choline-deficient, ethionine-supplemented diet was also reduced dramatically. However, it remains unclear how much of this process is due to reduced IL-18 activity.

CONCLUSIONS

The availability of orally active inhibitors of ICE will no doubt advance anticytokine-based therapies for a variety of inflammatory and autoimmune diseases. Initially, ICE inhibition was thought to be highly specific, because it blocked the secretion of IL-1β. Inhibition of IL-1β also appeared to be safe in terms of host defense perturbations. With the discovery that processing of proIL-18 is also accomplished by ICE, there was no longer the advantage of specificity. In terms of host defense effects of ICE inhibitors, the role of IFN-γ (as a product of IL-18 activity) must be considered, since a reduction in IFN-γ is thought to be undesirable in terms of defense against tuberculosis and some other infectious diseases. On the other hand, inhibition of the Th1 response by ICE inhibitors (due to reduced IFN-γ production) is a worthwhile strategy in many diseases. For example, graft versus host disease, organ rejection, type I diabetes, and other autoimmune diseases of the Th1 type. In many ways, a reduction in IFN-γ due to inhibition of ICE may be beneficial in multiple sclerosis where both inhibition of IL-1 (119) and IFN-γ result in reduced disease. Indeed, overexpression of IFN-γ in the central nervous system of mice results in a spontaneous development of a disease very similar to that of multiple sclerosis (120). The clinical challenge of studies with ICE inhibitors is not whether they will reduce inflammation due to IL-1β. Those disease have already been identified in humans. The challenge will be from the safety side, since ICE inhibitors reduce IFN-γ in several models of disease. Nevertheless, when compared to the well-established toxicities, side effects, and host defense abnormalities of chronic corticosteroid use, inhibitors of ICE offer a unique opportunity to reduce disease, because they encompass both anti-inflammatory and immunosuppressive properties without the metabolic consequences and global immunosuppression of corticosteroids.

ACKNOWLEDGMENT

This work was supported by NIH grant AI-15614.

REFERENCES

1. Y Gu, K Kuida, H Tsutsui, G Ku, K Hsiao, MA Fleming, N Hayashi, K Higashino, H Okamura, K Nakanishi, M Kurimoto, T Tanimoto, RA Flavell, V Sato, MW Harding, DL Livingston, MS-S Su. Activation of interferon-γ inducing factor mediated by interleukin-1β converting enzyme. Science 275:206–209, 1997.
2. T Ghayur, S Banerjee, M Hugunin, D Butler, L Herzog, A Carter, L Quintal, L Sekut, R Talanian, M Paskind, W Wong, R Kamen, D Tracey, H Allen. Caspase-1 processes IFN-gamma–inducing factor and regulates LPS-induced IFN-gamma production. Nature 386:619–623, 1997.
3. G Fantuzzi, AJ Puren, MW Harding, DJ Livingston, CA Dinarello. IL-18 regulation of IFN-γ production and cell proliferation as revealed in interleukin-1β converting enzyme-deficient mice. Blood 91:2118–2125, 1998.
4. CA Dinarello. Biological basis for interleukin-1 in disease. Blood 87:2095–2147, 1996.
5. JL Browning, A Ribolini. Interferon blocks interleukin 1–induced prostaglandin release from human peripheral monocytes. J Immunol 138:2857–2863, 1987.
6. P Ghezzi, CA Dinarello. IL-1 induces IL-1. III. Specific inhibition of IL-1 production by IFN-γ. J Immunol 140:4238–4244, 1988.
7. R Schindler, P Ghezzi, CA Dinarello. IL-1 induces IL-1. IV. IFN-γ suppresses IL-1 but not lipopolysaccharide-induced transcription of IL-1. J Immunol 144: 2216–2222, 1990.
8. HJ Andrews, RA Bunning, CA Dinarello, RG Russell. Modulation of human chondrocyte metabolism by recombinant human interferon gamma: in-vitro effects on basal and IL-1–stimulated proteinase production, cartilage degradation and DNA synthesis. Biochim Biophys Acta 1012:128–134, 1989.

9. HJ Andrews, RA Bunning, TA Plumpton, IM Clark, RG Russell, TE Cawston. Inhibition of interleukin-1–induced collagenase production in human articular chondrocytes in vitro by recombinant human interferon-gamma. Arthritis Rheum 33:1733–1738, 1990.
10. MJ Laughlin, G Kirkpatrick, N Sabiston, W Peters, J Kurtzberg. Hematopoietic recovery following high-dose combined alkylating-agent chemotherapy and autologous bone marrow support in patients in phase I clinical trials of colony stimulating factors: G-CSF, GM-CSF, IL-1, IL-2 and M-CSF. Ann Hematol 67:267–276, 1993.
11. T Kitamura, F Takaku. A preclinical and Phase I clinical trial of IL-1. Exp Med 7:170–177, 1989.
12. A Tewari, WC Buhles, Jr., HF Starnes, Jr. Preliminary report: effects of interleukin-1 on platelet counts. Lancet 336:712–714, 1990.
13. J Nemunaitis, FR Appelbaum, K Lilleby, WC Buhles, C Rosenfeld, ZR Zeigler, RK Shadduck, JW Singer, W Meyer, CD Buckner. Phase I study of recombinant interleukin-1β in patients undergoing autologous bone marrow transplantation for acute myelogenous leukemia. Blood 83:3473–3479, 1994.
14. J Crown, A Jakubowski, N Kemeny, M Gordon, C Gasparetto, G Wong, G Toner, B Meisenberg, J Botet, J Applewhite, S Sinha, M Moore, D Kelsen, W Buhles, J Gabrilove. A phase I trial of recombinant human interleukin-1β alone and in combination with myelosuppressive doses of 5-fluoruracil in patients with gastrointestinal cancer. Blood 78:1420–1427, 1991.
15. J Crown, A Jakubowski, J Gabrilove. Interleukin-1: biological effects in human hematopoiesis. Leuk Lymphoma 9:433–440, 1993.
16. JW Smith, D Longo, WG Alford, JE Janik, WH Sharfman, BL Gause, BD Curti, SP Creekmore, JT Holmlund, RG Fenton, M Sznol, LL Miller, M Shimzu, JJ Oppenheim, SJ Fiem, JC Hursey, GC Powers, WJ Urba. The effects of treatment with interleukin-1α on platelet recovery after high-dose carboplatin. N Engl J Med 328:756–761, 1993.
17. JW Smith, WJ Urba, BD Curti, LJ Elwood, RG Steis, JE Janik, WH Sharfman, LL Miller, RG Fenton, KC Conlon, J Rossio, W Kopp, M Shimuzut, JJ Oppenheim, D Longo. The toxic and hematologic effects of interleukin-1 alpha administered in a phase I trial to patients with advanced malignancies. J Clin Oncol 10:1141–1152, 1992.
18. T Iizumi, S Sato, T Iiyama, R Hata, H Amemiya, H Tomomasa, T Yazaki, T Umeda. Recombinant human interleukin-1 beta analogue as a regulator of hematopoiesis in patients receiving chemotherapy for urogenital cancers. Cancer 68:1520–1523, 1991.
19. T van der Poll, HR Bueller, H ten Cate, CH Wortel, KA Bauer, SJH van Deventer, CE Hack, HP Sauerwein, RD Rosenberg, JW ten Cate. Activation of coagulation after administration of tumor necrosis factor to normal subjects. N Engl J Med 322:1622–1627, 1990.
20. H Tilg, E Trehu, MB Atkins, CA Dinarello, JW Mier. Interleukin-6 (IL-6) as an anti-inflammatory cytokine: induction of circulating IL-1 receptor antag-

onist and soluble tumor necrosis factor receptor p55. Blood 83:113–118, 1994.

21. BE Drevlow, R Lovis, MA Haag, JM Sinacore, C Jacobs, C Blosche, A Landay, LW Moreland, RM Pope. Recombinant human interleukin-1 receptor type I in the treatment of patients with active rheumatoid arthritis. Arthritis Rheum 39: 257–265, 1996.
22. CJJ Fisher, GJ Slotman, SM Opal, J Pribble, RC Bone, G Emmanuel, D Ng, DC Bloedow, MA Catalano. Initial evaluation of human recombinant interleukin-1 receptor antagonist in the treatment of sepsis syndrome: a randomized, open-label, placebo-controlled multicenter trial. Crit Care Med 22:12–21, 1994.
23. CJJ Fisher, JF Dhainaut, SM Opal, JP Pribble, RA Balk, GJ Slotman, TJ Iberti, EC Rackow, MJ Shapiro, RL Greenman. Recombinant human interleukin-1 receptor antagonist in the treatment of patients with sepsis syndrome. Results from a randomized, double blind, placebo-controlled trial. JAMA 271:1836–1843, 1994.
24. SM Opal, CJJ Fisher, JF Dhainaut, J-L Vincent, R Brase, SF Lowry, JC Sadoff, GJ Slotman, H Levy, RA Balk, MP Shelly, JP Pribble, JF LaBrecque, J Lookabough, H Donovan, H Dublin, R Baughman, J Norman, E Demaria, K Matzel, E Abraham, M Seneff. Confirmatory interleukin-1 receptor antagonist trial in severe sepsis: a phase III, randomized, double-blind, placebo-controlled, multicenter trial. Crit Care Med 25:1115–1124, 1997.
25. ME Lebsack, CC Paul, DC Bloedow, FX Burch, MA Sack, W Chase, MA Catalano. Subcutaneous IL-1 receptor antagonist in patients with rheumatoid arthritis. Arthritis Rheum 34(suppl):S67, 1991.
26. GV Campion, ME Lebsack, J Lookabaugh, G Gordon, M Catalano. Dose-range and dose-frequency study of recombinant human interleukin-1 receptor antagonist in patients with rheumatoid arthritis. Arthritis Rheum 39:1092–1101, 1996.
27. B Bresnihan, JM Alvaro-Gracia, M Cobby, M Doherty, Z Domljan, P Emery, G Nuki, K Pavelka, R Rau, B Rozman, I Watt, B Williams, R Aitchison, D McCabe, P Musikic. Treatment of rheumatoid arthritis with recombinant human interleukin-1 receptor antagonist. Arthritis Rheum 41:2196–2204, 1998.
28. M Torcia, M Lucibello, E Vannier, S Fabiani, A Miliani, G Guidi, O Spada, SK Dower, JE Sims, AR Shaw, CA Dinarello, E Garaci, F Cozzolino. Modulation of osteoclast-activating factor activity of multiple myeloma bone marrow cells by different interleukin-1 inhibitors. Exp Hematol 24:868–874, 1996.
29. JH Antin, HJ Weinstein, EC Guinan, P McCarthy, BE Bierer, DG Gilliland, SK Parsons, KK Ballen, IJ Rimm, G Falzarano, JL Ferrara. Recombinant human interleukin-1 receptor antagonist in the treatment of steroid-resistant graft-versus-host disease. Blood 84:1342–1348, 1994.
30. Y Kobayashi, K Yamamoto, T Saido, H Kawasaki, JJ Oppenheim, K Matsushima. Identification of calcium-activated neutral protease as a processing enzyme of human interleukin 1 alpha. Proc Natl Acad Sci USA 87:5548–5552, 1990.

31. DT Brody, SK Durum. Membrane IL-1: IL-1α precursor binds to the plasma membrane via a lectin-like interaction. J Immunol 143:1183, 1989.
32. G Kaplanski, C Farnarier, S Kaplanski, R Porat, L Shapiro, P Bongrand, CA Dinarello. Interleukin-1 induces interleukin-8 from endothelial cells by a juxacrine mechanism. Blood 84:4242–4248, 1994.
33. M Svenson, MB Hensen, L Kayser, AK Rasmussen, CM Reimert, K Bendtzen. Effects of human anti-IL-1α autoantibodies on receptor binding and biological activities of IL-1. Cytokine 4:125–133, 1992.
34. FT Stevenson, SL Bursten, C Fanton, RM Locksley, DH Lovett. The 31-kDa precursor of interleukin-1α is myristoylated on specific lysines within the 16-kDa N-terminal propiece. Proc Natl Acad Sci USA 90:7245–7249, 1993.
35. FT Stevenson, J Turck, RM Locksley, DH Lovett. The N-terminal propiece of interleukin 1α is a transforming nuclear oncoprotein. Proc Natl Acad Sci 94: 508–513, 1997.
36. JAM Maier, M Statuto, G Ragnotti. Endogenous interleukin-1 alpha must be transported to the nucleus to exert its activity in human endothelial cells. Mol Cell Biol 14:1845–1851, 1994.
37. JAM Maier, P Voulalas, D Roeder, T Maciag. Extension of the life span of human endothelial cells by an interleukin-1α antisense oligomer. Science 249: 1570–1574, 1990.
38. MJ Fenton, MW Vermeulen, BD Clark, AC Webb, PE Auron. Human pro-IL-1 beta gene expression in monocytic cells is regulated by two distinct pathways. J Immunol 140:2267–2273, 1988.
39. N Jarrous, R Kaempfer. Induction of human interleukin-1 gene expression by retinoic acid and its regulation at processing of precursor transcripts. J Biol Chem 269:23141–23149, 1994.
40. E Serkkola, M Hurme. Synergism between protein-kinase C and cAMP-dependent pathways in the expression of the interleukin-1β gene is mediated via the activator-protein-1 (AP-1) enhancer activity. Eur J Biochem 213:243–249, 1993.
41. E Vannier, CA Dinarello. Histamine enhances interleukin (IL)–1–induced IL-1 gene expression and protein synthesis *via* H_2 receptors in peripheral blood mononuclear cells: comparison with IL-1 receptor antagonist. J Clin Invest 92: 281–287, 1993.
42. R Schindler, BD Clark, CA Dinarello. Dissociation between interleukin-1β mRNA and protein synthesis in human peripheral blood mononuclear cells. J Biol Chem 265:10232–10237, 1990.
43. L Shapiro, CA Dinarello. Osmotic regulation of cytokine synthesis in vitro. Proc Natl Acad Sci USA 92:12230–12234, 1995.
44. R Schindler, JA Gelfand, CA Dinarello. Recombinant C5a stimulates transcription rather than translation of IL-1 and TNF; cytokine synthesis induced by LPS, IL-1 or PMA. Blood 76:1631–1638, 1990.
45. RL Kaspar, L Gehrke. Peripheral blood mononuclear cells stimulated with C5a or lipopolysaccharide to synthesize equivalent levels of IL-1β mRNA show

unequal IL-1β protein accumulation but similar polyribosome profiles. J Immunol 153:277–286, 1994.
46. LC Miller, S Isa, E Vannier, K Georgilis, AC Steere, CA Dinarello. Live *Borrelia burgdorferi* preferentially activate IL-1β gene expression and protein synthesis over the interleukin-1 receptor antagonist. J Clin Invest 90:906–912, 1992.
47. MY Stoeckle, L Guan. High-resolution analysis of gro-α mRNA poly (A) shortening: regulation by interleukin-1β. Nucl Acid Res 21:1613–1617, 1993.
48. L Shapiro, N Panayotatos, SN Meydani, D Wu, CA Dinarello. Ciliary neurotrophic factor combined with soluble receptor inhibits synthesis of proinflammatory cytokines and prostaglandin E_2 in vitro. Exp Cell Res 215:51–56, 1994.
49. SA Jobling, PE Auron, G Gurka, AC Webb, B McDonald, LJ Rosenwasser, L Gehrke. Biological activity and receptor binding of human prointerleukin-1β and subpeptides. J Biol Chem 263:16372, 1988.
50. O Bakouche, DC Brown, LB Lachman. Subcellular localization of human monocyte interleukin 1: evidence for an inactive precursor molecule and a possible mechanism for IL 1 release. J Immunol 138:4249–4255, 1987.
51. FT Stevenson, F Torrano, RM Locksley, DH Lovett. Interleukin-1: the patterns of translation and intracellular distribution support alternative secretory mechanisms. J Cell Physiol 152:223–231, 1992.
52. A Rubartelli, F Cozzolino, M Talio, R Sitia. A novel secretory pathway for interleukin-1 beta, a protein lacking a signal sequence. EMBO J 9:1503–1510, 1990.
53. PE Auron, SJ Warner, AC Webb, JG Cannon, HA Bernheim, KJ McAdam, LJ Rosenwasser, G LoPreste, SF Mucci, CA Dinarello. Studies on the molecular nature of human interleukin 1. J Immunol 138:1447–1456, 1987.
54. HU Beuscher, C Guenther, M Roellinghoff. IL-1β is secreted by activated murine macrophages as biologically inactive precursor. J Immunol 144:2179–2183, 1990.
55. RA Black, SR Kronheim, M Cantrell, MC Deeley, CJ March, KS Prickett, J Wignall, PJ Conlon, D Cosman, TP Hopp. Generation of biologically active interleukin-1 beta by proteolytic cleavage of the inactive precursor. J Biol Chem 263:9437–9442, 1988.
56. G Fantuzzi, G Ku, MW Harding, DL Livingston, JD Sipe, K Kuida, RA Flavell, CA Dinarello. Response to local inflammation of IL-1β converting enzyme-deficient mice. J Immunol 158:1818–1824, 1997.
57. KP Wilson, JA Black, JA Thomson, EE Kim, JP Griffith, MA Navia, MA Murcko, SP Chambers, RA Aldape, SA Raybuck, DJ Livingston. Structure and mechanism of interleukin-1β converting enzyme. Nature 370:270–275, 1994.
58. Y Gu, J Wu, C Faucheu, J-L Lalanne, A Diu, DL Livingston, MS-S Su. Interleukin-1β converting enzyme requires oligomerization for activity of processed forms in vivo. EMBO J 14:1923–1931, 1995.

59. ES Alnemri, DJ Livingston, DW Nicholson, G Salvesen, NA Thornberry, WW Wong, J Yuan. Human ICE/CED-3 protease nomenclature. Cell 87:123, 1996.
60. NP Walker, RV Talanian, KD Brady, LC Dang, NJ Bump, CR Ferenz, S Franklin, T Ghayur, MC Hackett, LD Hammill. Crystal structure of the cysteine protease interleukin-1 beta-converting enzyme: a (p20/p10)2 homodimer. Cell 78: 343–352, 1994.
61. CA Dinarello, JG Cannon, JW Mier, HA Bernheim, G LoPreste, DL Lynn, RN Love, AC Webb, PE Auron, RC Reuben, A Rich, SM Wolff, SD Putney. Multiple biological activities of human recombinant interleukin 1. J Clin Invest 77: 1734–1739, 1986.
62. M Irmler, S Hertig, HR MacDonald, R Sadoul, JD Becherer, A Proudfoot, R Solari, J Tschopp. Granzyme A is an interleukin-1β–converting enzyme. J Exp Med 181:1917–1922, 1995.
63. GC Higgins, JL Foster, AE Postlethwaite. Interleukin-1 beta propeptide is detected intracellularly and extracellularly when human monocytes are stimulated with LPS in vitro. J Exp Med 180:607–614, 1994.
64. GC Higgins, JL Foster, AE Postlethwaite. Synthesis and biological activity of human interleukin-1β propiece in vitro. Arthrititis Rheum 39:S153, 1993.
65. NA Thornberry, HG Bull, JR Calaycay, KT Chapman, AD Howard, MJ Kostura, DK Miller, SM Molineaux, JR Weidner, J Aunins, JA Schmidt, M Tocci. A novel heterodimeric cysteine protease is required for interleukin-1 beta processing in monocytes. Nature 356:768–774, 1992.
66. A Rubartelli, A Bajetto, G Allavena, E Wollman, R Sitia. Secretion of thioredoxin by normal and neoplastic cells through a leaderless secretory pathway. J Biol Chem 267:24161–24164, 1992.
67. P Mignatti, DB Rifkin. Release of basic fibroblast growth factor, an angiogenic factor devoid of secretory signal sequence: a trivial phenomenon or a novel secretion mechanism? J Cell Biochem 47:201–217, 1991.
68. II Singer, S Scott, J Chin, EK Bayne, G Limjuco, J Weidner, DK Miller, K Chapman, MJ Kostura. The interleukin-1 beta–converting enzyme (ICE) is localized on the external cell surface membranes and in the cytoplasmic ground substance of human monocytes by immuno-electron microscopy. J Exp Med 182:1447–1459, 1995.
69. P Li, H Allen, S Banerjee, S Franklin, L Herzog, C Johnston, J McDowell, M Paskind, L Rodman, J Salfeld, E Towne, D Tracey, S Wardwell, F-Y Wei, W Wong, R Kamen, T Seshadri. Mice deficient in interleukin-1 converting enzyme (ICE) are defective in producton of mature interleukin-1β and resistant to endotoxic shock. Cell 80:401–411, 1995.
70. K Kuida, JA Lippke, G Ku, MW Harding, DJ Livingston, MS-S Su, RA Flavell. Altered cytokine export and apoptosis in mice deficient in interleukin-1β converting enzyme. Science 267:2000–2003, 1995.
71. C Coeshott, C Ohnemus, A Pilyavskaya, S Ross, M Wieczorek, H Kroona, AH Leimer, J Cheronis. Converting enzyme–independent release of TNFα and IL-1β from a stimulated human monocytic cell line in the presence of activated

neutrophils or purified proteinase-3. Proc Natl Acad Sci USA 96:6261–6266, 1999.
72. DJ Hazuda, JC Lee, PR Young. The kinetics of interleukin 1 secretion from activated monocytes. Differences between interleukin 1 alpha and interleukin 1 beta. J Biol Chem 263:8473–8479, 1988.
73. M Padrines, M Wolf, A Walz, M Baggiolini. Interleukin-8 processing by neutrophil elastase, cathepsin G and proteinase-3. FEBS Lett 352:231–235, 1994.
74. BJ Longley, L Tyrrell, Y Ma, DA Williams, R Halaban, K Langley, HS Lu, NM Schehter. Chymase clavage of stem cell factor yields a bioactive, soluble product. Proc Natl Acad Sci USA 94:9017–9021, 1997.
75. K Nakamura, H Okamura, M Wada, K Nagata, T Tamura. Endotoxin-induced serum factor that stimulates gamma inteferon production. Infect Immun 57: 590–595, 1989.
76. K Nakamura, H Okamura, K Nagata, T Komatsu, T Tamura. Purification of a factor which provides a costimulatory signal for gamma interferon production. Infect Immun 61:64–70, 1993.
77. H Okamura, H Tsutsui, T Komatsu, M Yutsudo, A Hakura, T Tanimoto, K Torigoe, T Okura, Y Nukada, K Hattori, K Akita, M Namba, F Tanabe, K Konishi, S Fukuda, M Kurimoto. Cloning of a new cytokine that induces interferon-γ. Nature 378:88–91, 1995.
78. H Okamura, K Nagata, T Komatsu, T Tanimoto, Y Nukata, F Tanabe, K Akita, K Torigoe, T Okura, S Fukuda, M Kurimoto. A novel costimulatory factor for gamma interferon induction found in the livers of mice causes endotoxic shock. Infect Immun 63:3966–3972, 1995.
79. H Heremans, J van Damme, C Dillen, R Dikman, A Billiau. Interferon-γ, a mediator of lethal lipopolysaccharide-induced Shwartzman-like shock in mice. J Exp Med 171:1853–1861, 1990.
80. BD Car, VM Eng, B Schn yder, L Ozmen, S Huang, P Gallay, D Heumann, M Aguet, B Ryffel. Interferon γ receptor deficient mice are resistant to endotoxic shock. J Exp Med 179:1437–1444, 1994.
81. S Ushio, M Namba, T Okura, K Hattori, Y Nukada, K Akita, F Tanabe, K Konishi, M Mcallef, M Fujii, K Torigoe, T Tanimoto, S Fukuda, M Ikeda, H Okamura, M Kurimoto. Cloning of the cDNA for human IFN-γ–inducing factor, expression in Escherichia coli, and studies on the biologic activities of the protein. J Immunol 156:4274–4279, 1996.
82. K Kohno, J Kataoka, T Ohtsuki, Y Suemoto, I Okamoto, M Usui, M Ikeda, M Kurimoto. IFN-γ–inducing factor (IGIF) is a co-stimulatory factor on the activation of Th1 but not Th2 cells and exerts its effect independently of IL-12. J Immunol 158:1541–1550, 1997.
83. H Tsutsui, K Nakanishi, K Matsui, K Higashino, H Okamura, Y Miyazawa, K Kaneda. IFN-γ–inducing factor up-regulates Fas ligand–mediated cytotoxic activity of murine natural killer cell clones. J Immunol 157:3967–3973, 1996.
84. JF Bazan, JC Timans, RA Kaselein. A newly defined interleukin-1? Nature 379:591, 1996.

85. K Torigoe, S Ushio, T Okura, S Kobayashi, M Taniai, T Kunikate, T Murakami, O Sanou, H Kojima, M Fuji, T Ohta, M Ikeda, H Ikegami, M Kurimoto. Purification and characterization of the human interleukin-18 receptor. J Biol Chem 272:25737–25742, 1997.
86. P Parnet, KE Garka, TP Bonnert, SK Dower, JE Sims. IL-1Rrp is a novel receptor-like molecule similar to the type I interleukin-1 receptor and its homologues T1/ST2 and IL-1R AcP. J Biol Chem 271:3967–3970, 1996.
87. D Boraschi, MG Cifone, W Falk, H-D Flad, A Tagliabue, MU Martin. Cytokines in inflammation. Eur Cytokine Netw 9:205–212, 1998.
88. T Yoshimoto, K Takeda, T Tanaka, K Ohkusu, S Kashiwamura, H Okamura, S Akira, K Nakanishi. IL-12 upregulates IL-18 receptor expression on T cells, Th1 cells and B cells: synergism with IL-18 for IFNγ production. J Immunol 161:3400–3407, 1998.
89. TL Born, E Thomassen, TA Bird, JE Sims. Cloning of a novel receptor subunit, AcPL, required for interleukin-18 signaling. J Biol Chem 273:29445–29450, 1998.
90. SA Greenfeder, P Nunes, L Kwee, M Labow, RA Chizzonite, G Ju. Molecular cloning and characterization of a second subunit of the interleukin-1 receptor complex. J Biol Chem 270:13757–13765, 1995.
91. JE Sims, MA Gayle, JL Slack, MR Alderson, TA Bird, JG Giri, F Colotta, F Re, A Mantovani, K Shanebeck, KH Grabstein, SK Dower. Interleukin-1 signaling occurs exclusively via the type I receptor. Proc Natl Acad Sci USA 90:6155–6159, 1993.
92. D Novick, S-H Kim, G Fantuzzi, L Reznikov, CA Dinarello, M Rubinstein. Interleukin-18 binding protein: a novel modulator of the Th1 cytokine response. Immunity 10:127–136, 1999.
93. D Novick, H Engelmann, D Wallach, M Rubinstein. Soluble cytokine receptors are present in normal human urine. J Exp Med 170:1409–1414, 1989.
94. H Engelmann, D Novick, D Wallach. Two tumor necrosis factor–binding proteins purified from human urine. Evidence for immunological cross-reactivity with cell surface tumor necrosis factor receptors. J Biol Chem 265:1531–1536, 1990.
95. D Novick, B Cohen, M Rubinstein. The human interferon α/β receptor: characterization and molecular cloning. Cell 77:391–400, 1994.
96. D Novick, H Engelmann, D Wallach, O Leitner, M Revel, M Rubinstein. Purification of soluble cytokine receptors from normal human urine by ligand-affinity and immunoaffinity chromatography. J Chromatogr 510:331–337, 1990.
97. CA Hunter, J Timans, P Pisacane, S Menon, G Cai, W Walker, M Aste-Amezaga, R Chizzonite, JF Bazan, RA Kastelein. Comparison of the effects of interleukin-1α, interleukin-1β and interferon-γ inducing factor on the production of interferon-γ by natural killer. Eur J Immunol 27:2787–2792, 1997.
98. CA Hunter, R Chizzonite, JS Remington. IL-1β is required for IL-12 to induce the production of IFN-γ by NK cells. J Immunol 155:4347–4354, 1995.
99. GE Croston, Z Cao, DV Goeddel. NFkB activation by interleukin-1 requires

an IL-1 receptor–associated protein kinase activity. J Biol Chem 270:16514–16517, 1995.
100. MU Martin, W Falk. The interleukin-1 receptor complex and interleukin-1 signal transduction. Eur Cytokine Netw 8:5–17, 1997.
101. LAJ O'Neill, C Greene. Signal transduction pathways activated by the IL-1 receptor family: ancient signaling machinery in mammals, insects, and plants. J Leuk Biol 63:650–657, 1998.
102. H Wesche, C Korherr, M Kracht, W Falk, K Resch, MU Martin. The interleukin-1 receptor accessory protein is essential for IL-1-induced activation of interleukin-1 receptor–associated kinase (IRAK) and stress-activated protein kinases (SAP kinases). J Biol Chem 272:7727–7731, 1997.
103. J Huang, X Gao, S Li, Z Cao. Recruitment of IRAK to the interleukin 1 receptor complex requires interleukin 1 receptor accessory protein. Proc Natl Acad Sci USA 94:12829–12832, 1997.
104. Z Cao. Signal transduction of interleukin-1 (abst). Eur Cytokine Netw 9: 378(abs), 1998.
105. O Adachi, T Kawai, K Takeda, M Matsumoto, H Tsutsui, M Sakagami, K Nakanishi, S Akira. Targeted disruption of the MyD88 gene results in loss of IL-1- and IL-18–mediated function. Immunity 9:143–150, 1998.
106. Z Cao, J Xiong, M Takeuchi, T Kurama, DV Goeddel. Interleukin-1 receptor activating kinase. Nature 383:443–446, 1996.
107. NL Malinin, MP Boldin, AV Kovalenko, D Wallach. MAP3K-related kinase involved in NF-kappaB induction by TNF, CD95 and IL-1. Nature 385:540–544, 1997.
108. JA DiDonato, M Hayakawa, DM Rothwarf, E Zandi, M Karin. A cytokine-responsive I kappaB kinase that activates the transcription factor NF-kappaB. Nature 388:548–554, 1997.
109. H Kojima, M Takeuchi, T Ohta, Y Nishida, N Arai, M Ikeda, H Ikegami, M Kurimoto. Interleukin-18 activates the IRAK-TRAF6 pathway in mouse EL-4 cells. Biochem Biophys Res Commun 244:183–186, 1998.
110. D Robinson, K Shibuya, A Mui, F Zonin, E Murphy, T Sana, SB Hartley, S Menon, R Kastelein, F Bazan, A O'Garra. IGIF does not drive Th1 development but synergizes with IL-12 for interferon-γ production and activates IRAK and NFκB. Immunity 7:571–581, 1997.
111. L Shapiro, AJ Puren, HA Barton, D Novick, RL Peskind, MS-S Su, Y Gu, CA Dinarello. Interleukin-18 stimulates HIV type 1 in monocytic cells. Proc Natl Acad Sci USA 95:12550–12555, 1998.
112. K Tsuji-Takayama, S Matsumoto, K Koide, M Takeuchi, M Ikedo, T Ohta, M Kurimoto. Interleukin-18 induces activation and association of p56lck and MAPK in a mirine TH1 clone. Biochem Biophys Res Commun 237:126–130, 1997.
113. G Ku, T Faust, LL Lauffer, DJ Livingston, MW Harding. Interleukin-1 beta converting enzyme inhibition blocks progression of type II collagen-induced arthritis in mice. Cytokine 8:377–386, 1996.

114. WB van den Berg, LAB Joosten, M Helsen, FAJ Van de Loo. Amelioration of established murine collagen-induced arthritis with anti–IL-1 treatment. Clin Exp Immunol 95:237–243, 1994.
115. G Froyen, F Billiau. Potential therapeutic use of antibodies directed towards HuIFNγ. Biotherapy 10:49–57, 1997.
116. CA Dinarello. IL-18: A Th1-inducing, proinflammatory cytokine and new member of the IL-1 family. J Allergy Clin Immunol 103:11–24, 1999.
117. T Olee, S Hashimoto, J Quach, M Lotz. IL-18 is produced by articular chondrocytes and induces proinflammatory and catabolic responses. J Immunol 162: 1096–1100, 1999.
118. J Norman, J Yang, G Fink, G Carter, G Ku, W Denham, D Livingston. Severity and mortality of experimental pancreatitis are dependent on interleukin-1 converting enzyme (ICE). J Interferon Cytokine Res 17:113–118, 1997.
119. V Badovinac, M Mostarica-Stojkovic, CA Dinarello, S Stosic-Grujicic. Interleukin-1 receptor antagonist suppresses experimental autoimmune encephalomyelitis (EAE) in rats by influencing the activation and proliferation of encephalitogenic cells. J Neuroimmunol 85:87–95, 1998.
120. MS Horwitz, CF Evans, DB McGavern, M Rodriguez, MB Oldstone. Primary demyelination in transgenic mice expressing interferon-γ. Nat Med 3:1037–1041, 1997.
121. AJ Puren, G Fantuzzi, Y Gu, MS-S Su, CA Dinarello. Interleukin-18 (IFN-γ-inducing factor) induces IL-1β and IL-8 via TNFα production from non-CD14+ human blood mononuclear cells. J Clin Invest 101:711–724, 1998.

2
Interleukin-10: A Cytokine with Multiple Anti-Inflammatory and Some Proinflammatory Activities

Sander J. H. van Deventer and Tom van der Poll
Academic Medical Center, Amsterdam, The Netherlands

INTRODUCTION

Interleukin-10 (IL-10) was first identified as a protein produced by mouse T helper type 2 (Th2) lymphocytes that suppressed the production of cytokines by Th1 lymphocytes (1,2). Sequence analysis of the mouse and human cDNAs revealed extensive homology, and the IL-10 family also includes a protein encoded by the Epstein-Barr virus (BCRF1), now know as viral IL-10 (vIL-10) (3). The gene for human IL-10 (hIL-10) is located on chromosome 1, and it encodes an 18.5-kD protein (4). Two intramolecular disulfide bonds importantly influence the IL-10 protein structure, and secreted soluble IL-10 is a homodimer (5). The crystal structure of hIL-10 revealed a tight dimer composed of two intercalating subunits forming a V-shaped structure with an overall topology reminiscent of IFN-γ (6). Interestingly, the IL-10 receptor (IL-10R) is a member of the IFN-γ receptor family, and it is striking that two cytokines that have such antagonistic biological effects share structural similarities. The cytoplasmic domain of the IL-10R contains two motifs that have been previously recognized as signal-transducing units of the IL-6 receptor and are responsible for signal transducer of activated T cells (STAT3) activation (7). Indeed, engagement of IL-10R causes activation of STAT3 and STAT1 but not of STAT5. A more detailed analysis revealed that only STAT3

Table 1 Anti-Inflammatory Effects of IL-10

Cell type or disease model	Effect
Monocyte	Downregulates production of: IL-1α, IL-1β, IL-6, IL-8, IL-12, TNFα, G-CSF, and GM-CSF
Monocyte	Decreases HLA class II expression
Monocyte	Decreases tissue factor expression
T lymphocyte	Lowers proliferation
T lymphocyte	Decreases production of IL-2 and IL-5
Neutrophil	Decreases production of IL-8
Mucosal immune system	Maintains normal homeostasis by downregulation of IFN-γ, IL-2, and TNF-α production by lamina propria cells
T-lymphocyte transfer colitis	Recombinant IL-10 as well as IL-10–producing Tr1 cells are protective
Collagen-induced arthritis	Inhibits expression of proinflammatory cytokines, ameliorates disease
Experimental allergic encephalitis	Beneficial effect of IL-10 administration (depends on genetic background) and of IL-10 transgenic T lymphocytes
Endotoxemia	Decreases release of proinflammatory cytokines in mouse, baboon, and human volunteers. Decreases severity of Shwarztman reaction
Cecal ligation and puncture in mice	IL-10 is protective
Experimental pancreatitis	IL-10 is protective
Experimental hepatitis (galactosamine or ConA induced)	IL-10 is protective

is directly recruited to ligand-activated IL-10R through intracellular phosphotyrosine containing domains (8). Comparison of the different binding affinities of hIL-10 and vIL-10 with the observed biological effects in hIL-10R and mIL-10R transfected cells indicated that IL-10 signaling involves more proteins than the IL-10R, which was therefore named IL-10Rα (9).

In addition to T lymphocytes, various other cell types, including natural killer (NK) cells (10,11), monocytes-macrophages (12), mesangial cells (13), B lymphocytes (14,15), and various tumor cells can produce IL-10 (16), and in many inflammatory conditions, including rheumatoid arthritis and septic shock, monocytes-macrophages are considered to be the main source for IL-10.

The cellular targets of IL-10 include T lymphocytes, B lymphocytes, monocytes, macrophages, dendritic cells, and epithelial cells. These effects may be exerted during cell-to-cell contact (i.e., in the case of regulatory T lymphocytes or macrophage-lymphocyte interactions) or at long-distance (i.e., during systemic inflammatory responses in sepsis). Hence, IL-10 can have autocrine, paracrine, and endocrine activities. Many of the biological effects of IL-10 result in inhibition of T lymphocyte and monocyte-initiated inflammatory responses (Table 1). For example, IL-10 decreases the production of IL-1α, IL-1β, IL-6, IL-8, IL-12, tumor necrosis factor (TNF-α), granulocyte colony-stimulating factor (G-CSF), and granulocyte-macrophage CSF (GM-CSF) by human monocytes in vitro (17) as well as the production of proinflammatory cytokines by polymorphonuclear neutrophils (18). IL-10 also decreases the expression of tissue factor, the initiator of the extrinsic pathway of coagulation activation, by monocytes (19). The immunosuppressive activity of IL-10 is a result of several effects. IL-10 impairs the antigen-presenting capacity of monocytes and macrophages by decreasing expression of major histocompatibility complex (MHC) class II molecules as well as costimulatory molecules and by decreasing release of IL-12 (20–22). In addition, IL-10 interferes with the production of IL-2 and IL-5 by T lymphocytes (23,24). Chronic activation of $CD4^+$ T lymphocytes in the presence of IL-10 gives rise to T-cell clones that have low proliferative capabilities, which produce low levels of IL-2 and IL-4, but high levels of IL-10, and therefore are distinct from Th1 and Th2 clones. These T-cell clones, designated T-regulatory-1 (Tr1), suppress proliferation of $CD4^+$ lymphocytes in an antigen-specific manner, and they have been shown to be able to prevent Th1-mediated inflammatory responses (25).

Although in many pathological conditions the main activity of IL-10 is to suppress inflammation, it should be noted that IL-10 has several potential proinflammatory activities. IL-10 causes proliferation of $CD8^+$ lymphocytes, and, in synergism with IL-2, it can stimulate NK cells to proliferate and produce cytokines (26). In addition, IL-10 causes proliferation of B lymphocytes (this is one of the important activities of Epstein-Barr–derived vIL-10) and induces these cells to produce immunoglobulins. For a more detailed review of the cellular effects of IL-10, the reader is referred to recent excellent reviews (27–30).

REGULATION OF IL-10 PRODUCTION

The production of IL-10 by T lymphocytes and monocytes is regulated at multiple levels. IL-10 is a Th2 cytokine, and the events that lead to Th1/Th2 differentiation importantly influence the production of IL-10 by T lymphocytes. For a detailed discussion of these events, the reader is referred to recent comprehensive reviews (31,32). Following CD3/CD28 stimulation, T lymphocytes secrete IL-10 at a later time point than interferon-γ (IFN-γ) and IL-2 (33). IL-2 and other cytokines that bind ''common gamma chain'' receptors (IL-4, IL-7, and IL-15) boost stimulation of IL-10 production through CD3 engagement (33). INF-α is a potent inhibitor of IL-5 production by $CD4^+$ T lymphocytes but upregulates IL-10 (as well as IFN-γ) release (34,35). The production of IL-10 by T lymphocytes is inhibited by cyclosporin and FK506 (tacrolimus) but not by rapamycin (33). Lipopolysaccharide (LPS)–induced IL-10 release by monocytes is not inhibited by cyclosporin, and is even enhanced by such treatment in vivo (36).

LPS-induced production of IL-10 by monocytes in vitro is dependent on production of IL-1 and TNF-α and is signaled through the p38 (but not the p42) mitogen-activated protein kinase (37). In monocytes, the production of IL-10 is regulated at transcriptional and posttranscriptional levels: In the presence of the protein synthesis inhibitor cycloheximide (prevents degradation of instabile mRNA), IL-10 is superinduced (12). In humans and chimpanzees, both LPS and TNF-α induce IL-10, but LPS is a more potent inducer of IL-10 than TNF-α, and induction of IL-10 in primate endotoxemia could only be partially prevented by administration of TNF-α–neutralizing antibodies (38). In contrast, platelet-activating factor and IL-6 do not affect IL-10 release in primates (39).

Adenosine has potent anti-inflammatory effects, and the activities of the immune-modulating drugs methotrexate and sulphasalazine have been related to adenosine release. In part, the anti-inflammatory activities of adenosine may be related to its ability to induce IL-10 production by monocytes (TNF-α or LPS-stimulated) (40). Methylprednisone increased the production of IL-10 by mouse peritoneal macrophages and, depending on the dose, differentially influenced the production of TNF-α and IL-10 in mouse endotoxemia: At a low concentration, TNF-α, but not IL-10, production was inhibited, whereas at a high concentration, the production of both cytokines was decreased (41). In humans, hypercortisolemia markedly increases LPS-induced IL-10 production in vivo, but timing is important: When hypercortisolemia preceded the endotoxin challenge by more than 6 hs, no effect was observed (42). Epinephrine importantly affects immune responses by reducing TNF-α and increasing IL-10 production by human monocytes in vitro and in vivo (43). Another inhibitor of LPS-induced TNF-α release, chlorpromazine, was also shown to increase IL-10 production in LPS-challenged mice (44).

In conclusion, IL-10 seems to be differentially regulated in different cell types (in particular lymphocytes and monocytes) and, at different concentrations corticosteroids have opposite effects. Because many cytokines and hormones that are induced in the course of inflammation influence IL-10 release, it is hard to translate various in vitro findings relating to IL-10 prodcution to the in vivo situation.

IL-10 IN ANIMAL MODELS

The effects of IL-10 have been investigated in numerous animal studies, of which a selection is discussed.

Immune-Mediated Inflammation

The most impressive illustration of the regulatory function of IL-10 is provided by inactivation of the IL-10 gene in mice (IL-10 −/− mice). These mice are fertile, and they have offspring that initially develop normally with normal number of leukocytes and antibody concentrations. However, at the age of 3 months, most IL-10 −/− mice suffer from an inflammatory bowel disease involving the small and large bowels, which eventually is complicated by the development of adenocarcinomas and leads to death because of anemia and cachexia. The underlying defect seems to be increased production of proinflammatory cytokines (IL-2, IFN-γ, TNF-α, IL-1α) by lamina propria lymphocytes (45). The expression of disease is dependent on the presence of bacteria within the gut lumen and is mediated by $CD4^+$ T cells, and it therefore seems to be antigen dependent (46). In addition, genetic factors are important, and certain mouse strains seem to be relatively resistant to the effects of deletion of IL-10. These data indicate that IL-10 has an important role in maintaining the immunological homeostasis within the gut-associated lymphoid tissue, and that in the absence of IL-10, the antigenic pressure within the gut lumen is sufficient to induce T-lymphocyte activation. The occurrence of disease in IL-10 −/− mice can be prevented by early administration of IL-10, but at later stages, the disease is refractory to such treatment and becomes IL-12 dependent (47).

Another example of the regulatory role of IL-10 is given by findings in another model of colitis, which results from transfer of naive $CD4^+$ lymphocytes from BALB/C mice into SCID mice. Transfer of all $CD4^+$ lymphocytes from healthy BALB/C mice into SCID mice results in repopulation of various organs, including the gut lamina propria with $CD4^+$ cells. However, when the $CD4^+$ $CD45RB^{high}$ population is transferred, transmural inflammation of the mucosa of the stomach and large bowel ensues that is characterized by production of high levels of TNF-α and IFN-γ (48–50). The disease does not occur when $CD4^+$ $CD45RB^{low}$ cells are cotransferred, suggesting the presence of regulatory T lymphocytes within this latter population. Indeed, it has been recently reported that IL-10–producing Tr1 cells were capable of preventing $CD4^+$ $CD45RB^{high}$-induced colitis (25,51). Interestingly, similar findings have been reported using $CD4^+$ lymphocytes obtained from mice that were made transgenic using a construct consisting of the IL-10 gene under the control of the IL-2 promoter (52). As a result, these mice express IL-10 only when T lymphocytes become activated. The finding that transgenic T lymphocytes were able to prevent $CD4^+$ $CD45RB^{high}$-induced colitis in SCID mice suggests that T lymphocytes can be ''engineered'' to become Tr1-like cells.

IL-10 also decreases inflammation in other models of intestinal inflammation. Immune complex–induced colitis in rabbits is a result of complement and monocyte activation within the colonic mucosa by immune complexes. In this acute, T-lymphocyte–independent model of inflammation, IL-10 administration reduced inflammation as well as the production of tissue myeloperoxidase and leukotriene B_4 (53).

IL-10 administration had a modest effect on the course of collagen-induced arthritis, but in combination with IL-4, both early and delayed treatment significantly reduced IL-1β and TNF-α production in the synovia and ameliorated the arthritis. Conversely, anti–IL-10 antibodies worsened disease (54).

Experimental allergic encephalitis (EAE) is caused by $CD4^+$ basic myelin protein-specific T lymphocytes, which secrete Th1 cytokines. In rats, hIL-10 interfered with TNF-α production of these pathogenic T-cell clones, and when administered in the initiation phase, interfered with the induction of the disease (55). These data were not confirmed by a subsequent study in mice in which IL-10 completely failed to improve the severity of disease (56). It is possible that the route of IL-10 administration is important: Intranasal application in Lewis rats was very effective in mitigating disease and preventing the development of protracted relapses (57). In SJL mice, IL-10 had modest effects and delayed onset of disease when administered prophylactically (58). Interestingly, in this latter model, instead of the expected synergism, IL-4 antagonized the effects of IL-10. As in experimental colitis in mice, a more potent approach may be adoptive transfer of T lymphocytes engineered to produce IL-10 when activated. (59).

In autoimmune diabetes, IL-10 has a complex role. Unexpectedly, IL-10 has been implicated in the pathogenesis of pancreatic insulinitis, as it has been reported that pancreatic expression of IL-10 is sufficient to induce diabetes in NOD mice, which lack most known diabetes susceptibility loci (60,61). Similarly, transgenic mice expressing IL-10 within the pancreas were much more susceptible to streptozotocin-induced diabetes (62). On the other hand, the insulinitis that was caused in NOD mice by adoptive transfer of pancreatic islet-specific Th1 clones occurred less frequently when these clones were transduced with a retroviral construct encoding IL-10 (63). Similarly, treatment of NOD mice with a long-acting IL-10/Fc fusion protein was highly protective against the development of diabetes (64). Hence, the role of IL-10 in the pathogenesis of autoimmune pancreatic insulinitis in complex, and expression of IL-10 by different cells leads to contrasting outcomes.

Microbial Pathogens and Endotoxin

IL-10 has an important regulatory function in endotoxin-induced systemic inflammatory responses. IL-10 −/− mice are exceedingly susceptible to endotoxin and to the induction of the Shwartzman reaction (65). Similarly, neutralization of endogenous IL-10 by administration of antibodies renders normal mice more susceptible to the effects of endotoxin, increases the amount of TNF released, and greatly increases mortality. Conversely, administration of IL-10 dose dependently reduced the mortality of endotoxemia owing to a reduction of TNF production (66). Administration of recombinant hIL-10 to baboons strongly reduced the release of proinflammatory cytokines in baboons challenged with a sublethal dose of endotoxin. However, in this model, IL-10 administration did not prevent activation of several more peripheral inflammatory pathways, such as activation of the coagulation and fibrinolytic systems, and the effect on granulocyte degranulation was only modest (67). In a localized model of gram-negative infection induced by ligation and puncture of the cecum in C57Bl/6 mice, IL-10 was rapidly (within 1 h) induced in various distant organs, and serum levels of IL-10 became increased 2 hs after the insult (68). In this model, administration of IL-10–neutralizing antibodies caused a significant increase of mortality. It should be noted that IL-10 is not protective in all models of bacterial infection. For example, in murine pneumococcal pneumonia, administration of IL-10 impairs the host's ability to clear bacteria and results in an increase of bacteria in blood and greater mortality (69). Conversely, in this model, neutralization of IL-10 lowers bacterial counts and improves survival. In conclusion, in inflammatory responses induced by endotoxin, gram-negative or gram-positive bacteria, IL-10 generally downregulates the release of proinflammatory cytokines. In severe generalized inflammatory responses, this effect may be beneficial and result in a reduction of mortality. However, in some models, bacterial clearance is dependent on the activities of proinflammatory cytokines (IFN-γ, IL-1β, TNF-α), and, in these circumstances, IL-10 may increase mortality as a consequence of overwhelming bacteremia.

In a T-lymphocyte–dependent model of generalized inflammation induced by injection of staphylococcal enterotoxin B (SEB), IL-10 was rapidly produced by $CD4^+$ T lymphocytes. This endogenous release of IL-10 counteracted the release of IFN-γ and IL-2 (but not of TNF-α), and neutralizing IL-10 antibodies caused increased (IFN-γ-dependent) mortality (70).

Depending on the route of priming (intranasal or intraperitoneal), mice develop predominantly Th1- or Th2-mediated inflammatory responses to pulmonary challenge with *Aspergillus fumigatus* antigens. In IL-10 −/− mice, such allergic responses are exaggerated and associated with significantly higher concentrations of IL-4, IL-5, and IFN-γ (71). Hence, IL-10 can suppress both Th1 and Th2 responses to *Aspergillus* antigens.

The role of IL-10 in Chagas' disease is complex: Natural resistance of mice against infection with *Trypanosoma cruzei* is known to be associated with high levels of IFN-γ and a low production rate of IL-10 (72). Indeed, infection of IL-10 −/− mice with *T. cruzei* resulted in a higher production of TNF-α, IFN-γ, and IL-12 that was associated with a lower parasite burden (73). In this model, this resulted in an earlier mortality of IL-10 −/− mice that could be reversed by administration of anti-CD4 antibodies. Hence, it seemed that the early mortality due to *T. cruzei* infection was caused by cytokines derived from $CD4^+$ T lymphocytes (e.g., IL-12–induced IFN-γ). Apparently in contrast with these findings, it was demonstrated that IL-10–deficient macrophages had a defect in intracellular killing of *T. cruzei* because of insufficient induction of nitrous oxide, which could be corrected by IL-10 supplementation (74). A potential explanation for these findings is that, although IL-10 can upregulate NO production in macrophages, in vivo this effect is offset by an important counterregulatory effect on IFN-γ production (another potent inducer of NO synthesis). In *Plasmodium berghei* induced cerebral malaria in mice, which is known to be dependent on TNF-α production, IL-10 had a protective effect (75).

In intestinal-derived candidiasis in mice, administration of IL-4 and IL-10 caused a Th2 shift in the Peyer's patches, and this was associated with a greatly increased susceptibility and fatal outcome of disease (76). On the other hand, exogenous IL-10 did not influence outcome when administered to mice that had established cellular (Th1-mediated) immunity to *Candida* and were rechallenged with the organism.

Noninfectious Inflammatory Disease

In experimental pancreatitis in mice, IL-10 administration reduced the severity of pancreatitis, as assessed by serum amylase concentrations and histology

(77). Similar results were obtained in cerulin-induced pancreatitis in rats (78). Cold liver ischemia and reperfusion in mice is associated with induction of TNF-α as well as IL-10 mRNA, which are both dependent on reactive oxygen intermediates and can be reduced by administration of N-acetylcysteine (79). IL-10 also has a protective effect in galactosamine/endotoxin–induced hepatotoxicity (80,81) and in concanavalin A (ConA)-induced liver injury (82) by reducing monocyte- and lymphocyte-derived cytokines, respectively.

IL-10 PRODUCTION IN INFLAMMATORY DISEASES IN HUMANS

The production of IL-10 is increased in most human inflammatory studied, and even in "prototype" Th1-mediated diseases such as rheumatoid arthritis and Crohn's disease, the local IL-10 production is elevated (83,84). Hence, although the disease phenotype of Crohn's disease and the inflammatory bowel disease in IL-10 −/− mice may be similar, their pathogeneses are clearly different. Nonetheless, the increased IL-10 production in these human diseases represents an important counterregulatory mechanism, because neutralization of IL-10 in synovial cultures results in a two- to three-fold increase of TNF production (84).

The role of IL-10 in systemic inflammatory responses in humans has been recently investigated in detail, and some of the results are startling. In healthy individuals, serum IL-10 levels are below the detection limit of most assays (5–10 pg/mL), but a substantial fraction (57%) of patients with gram-positive and gram-negative sepsis do have detectable circulating IL-10 (range 12–2740 pg/mL), and serum concentrations are higher in patients with septic shock (85). Indeed, another study reported that IL-10 serum levels in septic patients were related to the severity of sepsis and organ failure (86). In severe meningococcal septic shock in children, very high IL-10 serum concentrations (median 6021 pg/mL; range 137–24,600 pg/mL) were reported, and nonsurvivors had higher levels than survivors (87). These findings were initially interpreted to reflect an anti-inflammatory response that was secondary to release of cytokines such as TNF-α, but recent data have cast doubt on this view. In a study of 190 first-degree family members of 61 patients with meningococcal septic shock, low TNF production in family members was associated with increased risk for mortality (OR 8.9, Cl [confidence interval] 1.8–45.0), whereas *high* IL-10 production showed an even stronger association (OR 19.5 Cl 2.3–165.0) (88). Because the capacity to produce these cytokines shows significant heritability, these data suggest that patients who are prone to have less rapid proinflammatory responses are at risk for an adverse outcome of meningococcal septic shock. A possible explanation for this finding would be that in the early stages of disease these patients clear bacteria less efficiently, as indeed was noted in animal experiments. This phenomenon may not be restricted to meningococcal septic shock: In a study in 464 febrile patients, those with a higher serum IL-10 concentration at admission were at a higher risk for mortality (89). If these data are confirmed, they would indicate that, in humans, an anti-inflammatory cytokine profile is a risk factor for an adverse outcome of sepsis. However, it remains to be determined whether such a profile is an *independent* predictor of outcome (i.e., not affected by clinical sepsis scores), whether such a profile is stable in time, and whether these data can be generalized for all bacterial infections. Nonetheless, these combined data make IL-10 a less attractive candidate for treatment of septic shock.

Transplantation of HLA-mismatched stem cells in patients with severe combined immune deficiency can result in successful engraftment of donor T lymphocytes; surprisingly without graft versus host disease. In these patients, the rate of endogenous IL-10 production by accessory cells may be an important factor in the induction of transplant tolerance, as it was demonstrated that both host monocytes and donor T lymphocytes produce very high amounts of IL-10 (90). In combination with animal data, these findings indicate that IL-10 is involved in tolerance induction (91–94).

EFFECTS OF IL-10 IN HUMAN VOLUNTEERS

The effects of administration of recombinant human IL-10 to human volunteers have been investigated in several studies. The first study reported the effects of intravenous IL-10 administration on the production of proinflammatory cytokines in blood that was stimulated ex vivo. Intravenous IL-10 administration (at doses ranging from 1 to 25 μg/kg) caused transient neutrophilia and monocytosis, but T-lymphocyte counts decreased (95). T-lymphocyte proliferative responses were reduced by IL-10, and, as expected, IL-10 dose dependently reduced LPS-induced production of TNF-α and IL-1β. However, no effects were observed on the induction of interleukin-1 receptor antagonist and the soluble p55 TNF receptor (95). At the IL-10 doses used in this study, no side effects were observed, but a subsequent pharmacokinetic study reported flu-like symptoms at doses above 25 μg/kg (96). In a placebo-controlled, crossover study in endotoxin-challenged volunteers (*Escherichia coli* endotoxin, 4 ng/kg), rhuIL-10, when injected simultaneously with the endotoxin (at a dose of 25 μg/kg), reduced the TNF, IL-6, and IL-8 peaks, and mitigated the febrile response (97). Moreover, dynamic scintigraphy indicated that IL-10 prevented the endotoxin-induced accumulation of neutrophils in the lungs. Postponed IL-10 treatment of endotoxin-challenged volunteers (30 mins after the endotoxin challenge) had no significant effects of TNF-α release but reduced IL-6 production. IL-10 treatment also significantly reduced endotoxin-induced intravascular thrombin formation and mitigated both fibrinolytic and antifibrinolytic responses (98). In conclusion, these data indicate that, very similar to earlier findings in mice, IL-10 was well tolerated and effectively inhibited production of proinflammatory cytokines by monocytes and interfered with T-lymphocyte proliferative responses and cytokine production.

IL-10 AS A THERAPEUTIC AGENT

The efficacy of recombinant human IL-10 has been tested in several diseases and conditions, including Crohn's disease, rheumatoid arthritis, and OKT3-induced systemic inflammatory responses.

In a controlled randomized clinical trial in 46 steroid-refractory patients with long-standing Crohn's disease, IL-10 was administered for 7 consecutive days at various doses (ranging from 0.5 to 25.0 μg/kg) as a daily intravenous bolus injection. The treatment was well tolerated, and the number of adverse events in the placebo and IL-10–treated groups were comparable. Although the study was not designed to detect clinical efficacy, more IL-10–treated (50%) than placebo-treated patients (23%) had a clinical response to treatment (99). In a subsequent trial, IL-10 was administered to patients with mild to moderate Crohn's disease as a daily subcutaneous injection for 28 days. In this study, patients treated with a low dose of IL-10 (5 μg/kg) had a significantly better outcome than placebo-treated patients, but the effect was lost at higher doses, resulting in a "bell-shaped" dose-response curve. The cause of this effect has not been completely elucidated, but preliminary data seem to indicate that Crohn's disease patients experience proinflammatory effects at IL-10 doses above 10 μg/kg.

OKT3 is an antibody that is directed against CD3 and is used to prevent allograft rejection following renal transplantation. Infusion of OKT3 is associated with substantial toxicity, which is caused by a antibody-induced release of proinflammatory cytokines. IL-10 is known to inhibit OKT3-induced cytokine responses in vitro and this provided a good rationale to study the efficacy of IL-10 for reducing OKT3-related systemic cytokine release in renal transplant patients (100). Indeed, IL-10 (at the 0.1, 1.0, and 10.0 μg/kg doses) significantly decreased TNF release and lowered IL-2 and IFN-γ concentrations (101). Antibodies against OKT3 were not induced in any of the patients treated with placebo or the lower two IL-10 doses, but this occurred in all three patients treated with 10 μg/kg IL-10 and resulted in neutralization of OKT3 and graft rejection in one of these patients. Hence, IL-10 was able to downregulate OKT3-induced T-lymphocyte activation, but boosted humoral immune responses, probably by its known effects on B lymphocytes. Obviously, further development of IL-10 for this indication has stopped.

CONCLUSIONS

Interleukin-10 is cytokine with important immune-modulatory effects that are critically involved in controlling systemic and localized inflammatory responses. Functional inactivation of IL-10 in mice leads to chronic inflammatory bowel disease, anemia, cachexia, the development of adenocarcinomas, and death. According to their genetic backgrounds, mice differ in the production of IL-10, and these differences have been related to varying immune responses in experimental inflammatory disease. Because no humans with IL-10 deficiency are known, it is likely that the activities of IL-10 are critical for a normal function of the immune system and for survival. Two studies in humans with bacterial infections have related increased IL-10 production with adverse outcome. Several lines of evidence indicate that downregulation of proinflammatory responses, in particular the production of IFN-γ and TNF-α by IL-10, may interfere with effective clearance of microbial pathogens.

Administration of IL-10 to human volunteers was well tolerated and had potent anti-inflammatory effects. The therapeutic efficacy of IL-10 treatment is currently investigated in several clinical trials, and early data from these trials indicate that (at higher doses) such treatment can have proinflammatory effects by induction of antibodies or proinflammatory cytokines. In the future, problems may be circumvented by IL-10 delivery in a cell-dependent manner. Indeed, in animal models, ''IL-10 engineered'' T lymphocytes effectively downregulated T-lymphocyte–dependent inflammation.

REFERENCES

1. T Suda, I MacNeil, M Fischer, KW Moore, A Zlotnik. Identification of a novel thymocyte growth factor derived from B cell lymphomas. Advances in Exp Med Biol 292:115–120, 1991.

2. H Yssel, R De Waal Malefyt, MG Roncarolo, et al. IL-10 is produced by subsets of human CD4+ T cell clones and peripheral blood T cells. J Immunol 149: 2378–2384, 1992.
3. KW Moore, P Vieira, DF Fiorentino, ML Trounstine, TA Khan, TR Mosmann. Homology of cytokine synthesis inhibitory factor (IL-10) to the Epstein-Barr virus gene BCRFI (published erratum appears in Science 1990 Oct 26; 250[4980]:494). Science 248:1230–1234, 1990.
4. JM Kim, CI Brannan, NG Copeland, NA Jenkins, TA Khan, KW Moore. Structure of the mouse IL-10 gene and chromosomal localization of the mouse and human genes. J Immunol 148:3618–3623, 1992.
5. WT Windsor, R Syto, A Tsarbopoulos, et al. Disulfide bond assignments and secondary structure analysis of human and murine interleukin 10. Biochemistry 32:8807–8815, 1993.
6. A Zdanov, C Schalk-Hihi, A Gustchina, M Tsang, J Weatherbee, A Wlodawer. Crystal structure of interleukin-10 reveals the functional dimer with an unexpected topological similarity to interferon gamma. Structure 3:591–601, 1995.
7. CF Lai, J Ripperger, KK Morella, et al. Receptors for interleukin (IL)–10 and IL-6–type cytokines use similar signaling mechanisms for inducing transcription through IL-6 response elements. J Biol Chem 271:13968–13975, 1996.
8. RM Weber-Nordt, JK Riley, AC Greenlund, KW Moore, JE Darnell, RD Schreiber. Stat3 recruitment by two distinct ligand-induced, tyrosine-phosphorylated docking sites in the interleukin-10 receptor intracellular domain. J Biol Chem 271:27954–27961, 1996.
9. L Ying, R de Waal Malefyt, F Briere, et al. The EBV IL-10 homologue is a selective agonist with impaired binding to the IL-10 receptor. J Immunol 158: 604–613, 1997.
10. ML Boulland, V Meignin, K Leroyviard, et al. Human interleukin-10 expression in T natural killer-cell lymphomas—association with anaplastic large cell lymphomas and nasal natural killer-cell lymphomas. Am J Pathol 153:1229–1237, 1998.
11. PT Mehrotra, RP Donnelly, S Wong, et al. Production of Il-10 by human natural killer cells stimulated with Il-2 and/or Il-12. J Immunol 160:2637–2644, 1998.
12. P Stordeur, L Schandene, P Durez, C Gerard, M Goldman, T Velu. Spontaneous and cycloheximide-induced interleukin-10 mRNA expression in human mononuclear cells. Mol Immunol 32:233–239, 1995.
13. B Fouqueray, V Boutard, C Philippe, et al. Mesangial cell–derived interleukin-10 modulates mesangial cell response to lipopolysaccharide. Am J Pathol 147: 176–182, 1995.
14. A O'Garra, M Howard. IL-10 production by CD5 B cells. Ann NY Acad Sci 651:182–199, 1992.
15. A O'Garra, R Chang, N Go, R Hastings, G Haughton, M Howard. Ly-1 B (B-1) cells are the main source of B cell–derived interleukin 10. Eur J Immunol 22:711–717, 1992.
16. GA Gastl, JS Abrams, DM Nanus, et al. Interleukin-10 production by human

carcinoma cell lines and its relationship to interleukin-6 expression. Int J Cancer 55:96–101, 1993.
17. R de Waal Malefyt, J Abrams, B Bennett, CG Figdor, JE de Vries. Interleukin 10 (IL-10) inhibits cytokine synthesis by human monocytes: an autoregulatory role of IL-10 produced by monocytes. J Exp Med 174:1209–1220, 1991.
18. MA Cassatella, L Meda, S Bonora, M Ceska, G Constantin. Interleukin 10 (IL-10) inhibits the release of proinflammatory cytokines from human polymorphonuclear leukocytes. Evidence for an autocrine role of tumor necrosis factor and IL-1 beta in mediating the production of IL-8 triggered by lipopolysaccharide. J Exptal Med 178:2207–2211, 1993.
19. O Pradier, C Gerard, A Delvaux, et al. Interleukin-10 inhibits the induction of monocyte procoagulant activity by bacterial lipopolysaccharide. Eur J Immunol 23:2700–2703, 1993.
20. R de Waal Malefyt, J Haanen, H Spits, et al. Interleukin 10 (IL-10) and viral IL-10 strongly reduce antigen-specific human T cell proliferation by diminishing the antigen-presenting capacity of monocytes via downregulation of class II major histocompatibility complex expression. J Exp Med 174:915–924, 1991.
21. F Willems, A Marchant, JP Delville, et al. Interleukin-10 inhibits B7 and intercellular adhesion molecule-1 expression on human monocytes. Eur J Immunol 24:1007–1009, 1994.
22. B Koppelman, JJ Neefjes, JE de Vries, R de Waal Malefyt. Interleukin-10 down-regulates MHC class II alphabeta peptide complexes at the plasma membrane of monocytes by affecting arrival and recycling. Immunity 7:861–871, 1997.
23. R de Waal Malefyt, H Yssel, JE de Vries. Direct effects of IL-10 on subsets of human CD4+ T cell clones and resting T cells. Specific inhibition of IL-2 production and proliferation. J Immunol 150:4754–4765, 1993.
24. L Schandene, C Alonso-Vega, F Willems, et al. B7/CD28–dependent IL-5 production by human resting T cells is inhibited by IL-10. J Immunol 152:4368–4374, 1994.
25. H Groux, A O'Garra, M Bigler, et al. A CD4+ T-cell subset inhibits antigen-specific T-cell responses and prevents colitis. Nature 389:737–742, 1997.
26. H Groux, M Bigler, JE de Vries, MG Roncarolo. Inhibitory and stimulatory effects of IL-10 on human CD8+ T cells. J Immunol 160:3188–3193, 1998.
27. M Goldman, A Marchant, L Schandene. Endogenous interleukin-10 in inflammatory disorders: regulatory roles and pharmacological modulation. Ann NY Acad Sci 796:282–293, 1996.
28. KW Moore, A O'Garra, R de Waal Malefyt, P Vieira, TR Mosmann. Interleukin-10. Ann Rev Immunol 11:165–190, 1993.
29. MA Cassatella. The neutrophil: one of the cellular targets of interleukin-10. Int J Clin Lab Res 28:148–161, 1998.
30. P Stordeur, M Goldman. Interleukin-10 as a regulatory cytokine induced by cellular stress: molecular aspects. Int Rev Immunol 16:501–522, 1998.

31. M Rincon, RA Flavell. T-cell subsets: transcriptional control in the Th 1/Th2 decision. Curr Biol 7:R729–732, 1997.
32. TR Mosmann, S Sad. The expanding universe of T-cell subsets: Th1, Th2 and more (see Comments). Immunol Today 17:138–146, 1996.
33. SB Cohen, SL Parry, M Feldmann, B Foxwell. Autocrine and paracrine regulation of human T cell IL-10 production. J Immunol 158:5596–5602, 1997.
34. L Schandene, GF Del Prete, E Cogan, et al. Recombinant interferon-alpha selectively inhibits the production of interleukin-5 by human CD4+ T cells. J Clin Invest 97:309–315, 1996.
35. L Schandene, E Cogan, A Crusiaux, M Goldman. Interferon-alpha upregulates both interleukin-10 and interferon-gamma production by human CD4+ T cells (letter). Blood 89:1110–1111, 1997.
36. P Durez, D Abramowicz, C Gerard, et al. In vivo induction of interleukin 10 by anti-CD3 monoclonal antibody or bacterial lipopolysaccharide: differential modulation by cyclosporin A. J Exp Med 177:551–555, 1993.
37. AD Foey, SL Parry, LM Williams, M Feldmann, BM Foxwell, FM Brennan. Regulation of monocyte IL-10 synthesis by endogenous IL-1 and TNF-alpha: role of the p38 and p42/44 mitogen-activated protein kinases. J Immunol 160: 920–928, 1998.
38. T van der Poll, J Jansen, M Levi, H ten Cate, JW ten Cate, SJ van Deventer. Regulation of interleukin 10 release by tumor necrosis factor in humans and chimpanzees. J Exp Med 180:1985–1988, 1994.
39. T van der Poll, J Jansen, M Levi, et al. Interleukin 10 release during endotoxaemia in chimpanzees:role of platelet-activating factor and interleukin 6. Scand J Immunol 43:122–125, 1996.
40. O Le Moine, P Stordeur, L Schandene, et al. Adenosine enhances IL-10 secretion by human monocytes. J Immunol 156:4408–4414, 1996.
41. A Marchant, Z Amraoui, C Gueydan, et al. Methylprednisolone differentially regulates IL-10 and tumour necrosis factor (TNF) production during murine endotoxaemia. Clin Exp Immunol 106:91–96, 1996.
42. T van der Poll, AE Barber, SM Coyle, SF Lowry. Hypercortisolemia increases plasma interleukin-10 concentrations during human endotoxemia—a clinical research center study. J Clin Endocrinol Metab 81:3604–3606, 1996.
43. T van der Poll, SM Coyle, K Barbosa, CC Braxton, SF Lowry. Epinephrine inhibits tumor necrosis factor-alpha and potentiates interleukin 10 production during human endotoxemia. J Clin Invest 97:713–719, 1996.
44. M Mengozzi, G Fantuzzi, R Faggioni, et al. Chlorpromazine specifically inhibits peripheral and brain TNF production, and up-regulates IL-10 production, in mice. Immunol 82:207–210, 1994.
45. DJ Berg, N Davidson, R Kuhn, et al. Enterocolitis and colon cancer in interleukin-10-deficient mice are associated with aberrant cytokine production and CD4(+) TH1-like responses. J Clin Invest 98:1010–1020, 1996.
46. RK Sellon, S Tonkonogy, M Schultz, et al. Resident enteric bacteria are neces-

sary for development of spontaneous colitis and immune system activation in interleukin-10–deficient mice. Infect Immun 66:5224–5231, 1998.
47. NJ Davidson, SA Hudak, RE Lesley, S Menon, MW Leach, DM Rennick. Il-12, but not Ifn-gamma, plays a major role in sustaining the chronic phase of colitis in Il-10–deficient mice. J Immunol 161:3143–3149, 1988.
48. MW Leach, AG Bean, S Mauze, RL Coffman, F Powrie. Inflammatory bowel disease in C.B-17 scid mice reconstituted with the CD45RBhigh subset of CD4+ T cells. Am J Pathol 148:1503–1515, 1996.
49. F Powrie, MW Leach. Genetic and spontaneous models of inflammatory bowel disease in rodents: evidence for abnormalities in mucosal immune regulation. Ther Immunol 2:155–123, 1995.
50. F Powrie. T cells in inflammatory bowel disease: protective and pathogenic roles. Immunity 3:171–174, 1995.
51. C Asseman, F Powrie. Interleukin 10 is a growth factor for a population of regulatory T cells. Gut 42:157–158, 1998.
52. A Hagenbaugh, S Sharma, SM Dubinett, et al. Altered immune responses in interleukin 10 transgenic mice. J Exp Med 185:2101–2110, 1997.
53. TA Grool, H Vandullemen, J Meenan, et al. Anti-inflammatory effect of interleukin-10 in rabbit immune complex–induced colitis. Scand J Gastroenterol 33:754–758, 1998.
54. LA Joosten, E Lubberts, P Durez, et al. Role of interleukin-4 and interleukin-10 in murine collagen-induced arthritis. Protective effect of interleukin-4 and interleukin-10 treatment on cartilage destruction. Arthritis Rheum 40:249–260, 1997.
55. O Rott, B Fleischer, E Cash. Interleukin-10 prevents experimental allergic encephalomyelitis in rats. Eur J Immunol 24:1434–1440, 1994.
56. B Cannella, YL Gao, C Brosnan, CS Raine. IL-10 fails to abrogate experimental autoimmune encephalomyelitis. J of Neurosci Res 45:735–746, 1996.
57. BG Xiao, XF Bai, GX Zhang, H Link. Suppression of acute and protracted-relapsing experimental allergic encephalomyelitis by nasal administration of low-dose IL-10 in rats. J Neuroimmunol 84:230–237, 1998.
58. L Nagelkerken, B Blauw, M Tielemans. IL-4 abrogates the inhibitory effect of IL-10 on the development of experimental allergic encephalomyelitis in SJL mice (published erratum appears in Int Immunol 1997 9(11):1773). Int Immunol 9:1243–1251, 1997.
59. PM Mathisen, M Yu, JM Johnson, JA Drazba, VK Tuohy. Treatment of experimental autoimmune encephalomyelitis with genetically modified memory T cells. J Exp Med 186:159–164, 1997.
60. MS Lee, R Mueller, LS Wicker, LB Peterson, N Sarvetnick. IL-10 is necessary and sufficient for autoimmune diabetes in conjunction with NOD MHC homozygosity. J Exp Med 183:2663–2668, 1996.
61. MS Lee, LS Wicker, LB Peterson, N Sarvetnick. Pancreatic IL-10 induces diabetes in NOD. B6 Idd3 Idd10 mice. Autoimmunity 26:215–221, 1997.
62. R Mueller, MS Lee, SP Sawyer, N Sarvetnick. Transgenic expression of

interleukin 10 in the pancreas renders resistant mice susceptible to low dose streptozotocin-induced diabetes. J Autoimmunity 9:151–158, 1996.
63. M Moritani, K Yoshimoto, S Ii, et al. Prevention of adoptively transferred diabetes in nonobese diabetic mice with IL-10-transduced islet-specific Th1 lymphocytes. A gene therapy model for autoimmune diabetes. J Clin Invest 98: 1851–1859, 1996.
64. XX Zheng, AW Steele, WW Hancock, et al. A noncytolytic IL-10/Fc fusion protein prevents diabetes, blocks autoimmunity, and promotes suppressor phenomena in NOD mice. J Immunol 158:4507–4513, 1997.
65. DJ Berg, R Kuhn, K Rajewsky, et al. Interleukin-10 is a central regulator of the response to LPS in murine models of endotoxic shock and the Shwartzman reaction but not endotoxin tolerance. J Clin Invest 96:2339–2347, 1995.
66. C Gerard, C Bruyns, A Marchant, et al. Interleukin 10 reduces the release of tumor necrosis factor and prevents lethality in experimental endotoxemia. J Exp Med 177:547–550, 1993.
67. T van der Poll, PM Jansen, WJ Montegut, et al. Effects of IL-10 on systemic inflammatory responses during sublethal primate endotoxemia. J Immunol 158: 1971–1975, 1997.
68. T van der Poll, A Marchant, WA Buurman, et al. Endogenous IL-10 protects mice from death during septic peritonitis. J Immunol 155:5397–5401, 1995.
69. T van der Poll, A Marchant, CV Keogh, M Goldman, SF Lowry. Interleukin-10 impairs host defense in murine pneumococcal pneumonia. J Infect Dis 174: 994–1000, 1996.
70. S Florquin, Z Amraoui, D Abramowicz, M Goldman. Systemic release and protective role of IL-10 in staphylococcal enterotoxin B–induced shock in mice. J Immunol 153:2618–2623, 1994.
71. G Grunig, DB Corry, MW Leach, BW Seymour, VP Kurup, DM Rennick. Interleukin-10 is a natural suppressor of cytokine production and inflammation in a murine model of allergic bronchopulmonary aspergillosis. J Exp Med 185: 1089–1099, 1997.
72. P Minoprio, MC el Cheikh, E Murphy, et al. Xid-associated resistance to experimental Chagas' disease is IFN-gamma dependent. J Immunol 151:4200–4208, 1993.
73. CA Hunter, LA Ellis-Neyes, T Slifer, et al. IL-10 is required to prevent immune hyperactivity during infection with Trypanosoma cruzi. J Immunol 158:3311–3316, 1997.
74. F Jacobs, D Chaussabel, C Truyens, et al. IL-10 up-regulates nitric oxide (NO) synthesis by lipopolysaccharide (LPS)–activated macrophages: improved control of Trypanosoma cruzi infection. Clin Exp Immunol 113:59–64, 1998.
75. S Kossodo, C Monso, P Juillard, T Velu, M Goldman, GE Grau. Interleukin-10 modulates susceptibility in experimental cerebral malaria. Immunology 91: 536–540, 1997.
76. L Tonnetti, R Spaccapelo, E Cenci, et al. Interleukin-4 and -10 exacerbate candidiasis in mice. Euro J Immunol 25:1559–1565, 1995.

77. AM Kusske, AJ Rongione, SW Ashley, DW McFadden, HA Reber. Interleukin-10 prevents death in lethal necrotizing pancreatitis in mice (see Comments). Surgery 120:284–288, 1996; discussion 289.
78. AJ Rongione, AM Kusske, K Kwan, SW Ashley, HA Reber, DW McFadden. Interleukin 10 reduces the severity of acute pancreatitis in rats. Gastroenterology 112:960–967, 1997.
79. O Le Moine, H Louis, P Stordeur, JM Collet, M Goldman, J Deviere. Role of reactive oxygen intermediates in interleukin 10 release after cold liver ischemia and reperfusion in mice. Gastroenterology 113:1701–1706, 1997.
80. H Louis, O Le Moine, MO Peny, et al. Hepatoprotective role of interleukin 10 in galactosamine/lipopolysaccharide mouse liver injury. Gastroenterology 112: 935–942, 1997.
81. L Santucci, S Fiorucci, M Chiorean, et al. Interleukin 10 reduces lethality and hepatic injury induced by lipopolysaccharide in galactosamine-sensitized mice. Gastroenterology 111:736–744, 1996.
82. H Louis, O Le Moine, MO Peny, et al. Production and role of interleukin-10 in concanavalin A-induced hepatitis in mice. Hepatology 25:1382–1389, 1997.
83. SB Cohen, PD Katsikis, CQ Chu, et al. High level of interleukin-10 production by the activated T cell population within the rheumatoid synovial membrane. Arthritis Rheum 38:946–952, 1995.
84. PD Katsikis, CQ Chu, FM Brennan, RN Maini, M Feldmann. Immunoregulatory role of interleukin 10 in rheumatoid arthritis. J Exp Med 179:1517–1527, 1994.
85. A Marchant, J Deviere, B Byl, D De Groote, JL Vincent, M Goldman. Interleukin-10 production during septicaemia. Lancet 343:707–708, 1994.
86. G Friedman, S Jankowski, A Marchant, M Goldman, RJ Kahn, JL Vincent. Blood interleukin 10 levels parallel the severity of septic shock. J Crit Care 12:183–187, 1997.
87. B Derkx, A Marchant, M Goldman, R Bijlmer, S van Deventer. High levels of interleukin-10 during the initial phase of fulminant meningococcal septic shock. J Infect Dis 171:229–232, 1995.
88. RG Westendorp, JA Langermans, TW Huizinga, et al. Genetic influence on cytokine production and fatal meningococcal disease (published erratum appears in Lancet 1997 Mar 1;349(9052):656) (see Comments). Lancet 349:170–173, 1997.
89. JT van Dissel, P van Langevelde, RG Westendorp, K Kwappenberg, M Frolich. Anti-inflammatory cytokine profile and mortality in febrile patients (see Comments). Lancet 351:950–953, 1998.
90. R Bacchetta, M Bigler, JL Touraine, et al. High levels of interleukin 10 production in vivo are associated with tolerance in SCID patients transplanted with HLA mismatched hematopoietic stem cells. J Exp Med 179:493–502, 1994.
91. V Donckier, D Abramowicz, P Durez, C Gerard, T Velu, M Goldman. Defective production of IFN-gamma in neonatally tolerant mice: involvement of IL-10. Transplant Proc 26:237, 1994.

92. D Abramowicz, P Durez, C Gerard, et al. Neonatal induction of transplantation tolerance in mice is associated with in vivo expression of IL-4 and -10 mRNAs. Transplant Proc 25:312–313, 1993.
93. H Groux, M Rouleau, R Bacchetta, MG Roncarolo. T-cell subsets and their cytokine profiles in transplantation and tolerance. Ann NY Acad Sci 770:141–148, 1995.
94. MT Bejarano, R de Waal Malefyt, JS Abrams, et al. Interleukin 10 inhibits allogeneic proliferative and cytotoxic T cell responses generated in primary mixed lymphocyte cultures. Int Immunol 4:1389–1397, 1992.
95. AE Chernoff, EV Granowitz, L Shapiro, et al. A randomized, controlled trial of IL-10 in humans. Inhibition of inflammatory cytokine production and immune responses. J Immunol 154:5492–5499, 1995.
96. RD Huhn, E Radwanski, SM O'Connell, et al. Pharmacokinetics and immunomodulatory properties of intravenously administered recombinant human interleukin-10 in healthy volunteers. Blood 87:699–705, 1996.
97. D Pajkrt, L Camoglio, MC Tiel-van Buul, et al. Attenuation of proinflammatory response by recombinant human IL-10 in human endotoxemia: effect of timing of recombinant human IL-10 administration. J Immunol 158:3971–3977, 1997.
98. D Pajkrt, T van der Poll, M Levi, et al. Interleukin-10 inhibits activation of coagulation and fibrinolysis during human endotoxemia. Blood 89:2701–2705, 1997.
99. SJ van Deventer, CO Elson, RN Fedorak. Multiple doses of intravenous interleukin 10 in steroid-refractory Crohn's disease. Crohn's Disease Study Group. Gastroenterology 113:383–389, 1997.
100. L Schandene, C Gerard, A Crusiaux, D Abramowicz, T Velu, M Goldman. Interleukin-10 inhibits OKT3-induced cytokine release: in vitro comparison with pentoxifylline. Transplantation Proc 25:55–56, 1993.
101. KM Wissing, E Morelon, C Legendre, et al. A pilot trial of recombinant human interleukin-10 in kidney transplant recipients receiving OKT3 induction therapy. Transplantation 64:999–1006, 1997.

3

Natural and Induced Anticytokine Antibodies in Humans

Klaus Bendtzen, Christian Ross, Christian Meyer, Morten Bagge Hansen, and Morten Svenson
Institute for Inflammation Research, Rigshospitalet National University Hospital, Copenhagen, Denmark

INTRODUCTION

Cytokines are polypeptide or glycopeptide signaling molecules that act at extremely low concentrations as regulators of cell growth and essential mediators of inflammation and immune reactions. The production and functions of cytokines are tightly regulated by the cytokines themselves and by several other factors. Most cytokines act locally, but some of the clinically most important cytokines also act systemically as pleiotropic hormones with overlapping and potentially dangerous functions (immunoinflammatory cytokines).

It is known that blood from both healthy individuals and patients suffering from various immunoinflammatory diseases contain antibodies (Ab) to certain cytokines and that Ab to these mediators may be induced in patients treated with recombinant human cytokines (1–8) (Table 1). The immunological basis for induction of Ab to cytokines is not clear. In the case of therapy-induced anticytokine Ab, the pharmaceutical formulation, route of administration, dosage and duration of therapy, and the immunological status of the patient all appear to be important (9). The majority of anticytokine Ab detected in humans are of the IgG type and are found with highly different prevalences. The latter may relate to the different binding characteristics and sensitivities of the methods used for Ab detection. Indeed, the lack of standardization in

Table 1 Cytokines to Which Natural Antibodies Have Been Repeatedly Found

Cytokine	MW	Main producers	Major functions
IFN-α	16–27 × 2	Virus-infected cells MØ, T and B-cells	Activates: MØ, NK cells, B cells MHC I and MHC II modulation Inhibits: proliferation of many cells (antitumor effect) Antiviral activity
IFN-β	20 × 2	Virus-infected cells fibroblasts	As IFN-α
IFN-γ	17–25 × 2	T cells (Th0, Th1) endothelial cells, smooth muscle cells	Activates: MØ, $CD8^+$ T cells, B cells, fibroblasts, and smooth muscle cells (vasoconstriction) MHC II (and MHC I) expression (many cells) General inhibitor of cell growth Antiangiogenic Antiviral activity (weak)
IL-2	15–20	T cells (Th0, Th1)	Promotes: T- and B-cell growth and differentiation NK cell growth and activity
IL-1α	17	Constitutive: keratinocytes, epithelial cells, granulosa cells of ovary, hypothalamic cells Nonconstitutive: MØ, B cells, NK cells, neutrophils, dendritic cells, Langerhans' cells, Kupffer cells, and many other cells	Activates: T, B, and NK-cells, polymorphonuclear cells, endothelial cells, smooth muscle cells, nerve cells, adipocytes, hepatocytes, chondrocytes, osteoclasts, fibroblasts, thyrocytes, pancreatic β cells (low conc.) Cytotoxic to: melanocytes, pancreatic β cells (high conc.) General in vivo effects: fever, anorexia, slow-wave sleep, acute-phase protein induction, leukocytosis, radioprotection
IL-6	21–28	MØ, T and B cells, fibroblasts, keratinocytes, hepatocytes, endothelial cells, neuronal cells, astrocytes, thyrocytes, pancreatic islet cells, neoplastic cells	As IL-1 (few exceptions) Stimulates: B cell Ig secretion, hepatocytes (acute-phase proteins), megakaryocytes Shortens: G0 in hematopoietic progenitor cells

IL-10	19 × 2	MØ, T cells (Th0, Th2), B cells, mast cells, keratinocytes, epidermal cells, Epstein-Barr virus-infected cells (vIL-10)	Coactivates: Tc cells, thymocytes, B cells, mast cells Inhibits: IFN-γ, IL-2, IL-3, GM-CSF, LT-α production by Th1 cells MØ and NK-cell functions: MHC II, IL-1, IL-6, TNF-α, nitric oxide, parasite killing
TNF-α	17 (× 1–3)	MØ, T cells (Th1), NK cells, neutrophils, keratinocytes, astrocytes, endothelial cells, smooth muscle cells	Activates: T and B cells, neutrophils, eosinophils, endothelial cells, fibroblasts, chondrocytes, osteoclasts, nerve cells Induces: IL-1, IL-6 in many cells Cytotoxic to: transformed and virus-infected cells
NGF	14 (× 2)	MØ, neurons, astrocytes, Schwann cells, smooth muscle cells, fibroblasts, epithelial cells	Activates: sympathetic and sensory neurons B cells, basophils
IL-8 (C-X-C)	6–8 (× 2)	MØ, T cells, neutrophils, fibroblasts, chondrocytes, synovial cells, keratinocytes, endothelial cells, epithelial cells, hepatocytes, neoplastic cells	Chemotactic for: neutrophils (activates), T cells, basophils (activates), keratinocytes (activates), melanoma cells Angiogenic
MCP-1 (C-C)	8–10	MØ, B cells, fibroblasts, endothelial cells, smooth muscle cells, neoplastic cells	Chemotactic for: MØ (regulates adhesion molecules), lymphocytes
MIP-1α/β (C-C)	8	MØ, T cells, B cells, mast cells, fibroblasts	Chemotactic for: MØ, $CD4^+$ T cells (MIP-1β), $CD8^+$ T cells, B cells, NK cells, granulocytes Activate: endothelial cells (β_1-integrin expression)
GM-CSF	14–22	MØ, T cells, granulocytes, fibroblasts, endothelial cells	Activates and differentiates: multipotential progenitor cells, MØ, T cells, neutrophils and precursors, eosinophils

detecting anticytokine Ab is a major concern because of the often insufficient characterization of anticytokine Ab provided in the literature.

The cytokine-induced Ab usually disappear after termination of therapy. In contrast, although the transient appearance of naturally occurring Ab to some cytokines have been described, long-lasting and stable levels is the rule in the case of Ab to interferon-α (IFN-α), IFN-β, interleukin-1α (IL-1α), IL-6, and granulocyte-macrophage colony-stimulating factor (GM-CSF) (8).

The biological significance of the naturally occurring Ab is poorly understood. Animal studies show that efficient cytokine neutralization can be obtained by administering relatively high doses of monoclonal Ab (mAb) to cytokines, and high levels of therapy-induced Ab have been shown to abrogate the effect of the administered cytokine.

An in vivo agonistic rather than antagonistic effect has been observed in situations where in vitro neutralizing mAb are present at low levels, particularly if the cytokine is bound in small complexes (10). Hence, accelerated clearance of a cytokine has been observed if the mediator form complexes with more than two IgG molecules (11). It is likely, therefore, that an agonistic effect may also be obtained in vivo with low-avidity, polyclonal Ab to selective cytokines. Since both nonneutralizing and neutralizing anticytokine Ab have been detected in vitro, determination of the in vitro biological effects as well as binding characteristics are important when trying to predict their in vivo functions.

The most extensively studied anticytokine Ab are those that in normal individuals bind IFN-α, IL-1α, and IL-6. These IgG Ab bind selectively and with high avidity to both recombinant and native forms of the cytokines and neutralize their activities in vitro and in vivo (see below).

NATURAL ANTIBODIES TO CYTOKINES

It is unknown why and how Ab are induced to some cytokines and not to others in apparently healthy individuals. However, cytokine-reactive B cells seem to occur frequently. As T-cell help is thought to be essential for the production of high-affinity IgG Ab, termination of T-cell tolerance is likely to be involved in the course of events leading to Ab induction. One possibility is that activation of cross-reactive B and T cells is initiated if a native cytokine as a hapten binds to a carrier encoded by microorganisms (8). For example, poxviruses and herpesviruses encode IFN receptors, as well as TNF-, IL-1, and IL-6 receptors (12). Many of these are truncated and lack the transmembrane domains; the receptors are therefore secreted from the infected cells. Some resemble their host counterparts, whereas others have functional similarity but little homology, which favors their role as immunological carriers. A chronic or repeated infection or a permanent break in T-cell tolerance to epitopes on the native cytokine would explain a continued production of these Ab.

IFN-α/β

IFN-α is a group of at least 16 subtypes of 16- to 27-kD glycoproteins. They are produced by virus-infected leukocytes and interfere with the replication of many viruses (see Table 1). Produced by antigen- or mitogen-activated monocytes-macrophages (MØ) and T and B lymphocytes, the IFN-α group of cytokines also has important immunoregulatory functions along with antiproliferative effects on many cell types. IFN-α was the first cytokine used effectively in clinical trials with cancer patients and is approved for the treatment of several human malignancies.

Unlike IFN-α, IFN-β is a single type of protein produced primarily by fibroblasts (see Table 1). It shares a receptor component with IFN-α and has similar bioactivities.

Antibodies to IFN-α and IFN-β

Natural Ab against IFN in humans were first reported in 1981 as a case of neutralizing IFN-α Ab in a patient with varicella-zoster (13). Later, IFN-α Ab were found in patients with other viral infections, with neoplastic and autoimmune diseases, and in a patient with chronic graft versus host disease (6,7,13–24). The prevalence of IgG anti–IFN-α Ab in these patients is about 10%.

Characterization. Demonstration of natural Ab to IFN species in sera of healthy individuals has met with considerable difficulties. Even though human sera are known to suppress the antiviral activity of IFN in highly sensitive assays and recombinant IFN species may bind to IgG using an immunoblotting technique, specific Ab to IFN-α, IFN-β, and IFN-γ are usually not the cause of this antiviral effect (25,26). For example, sera from only 7 of 599 (1%) healthy individuals were found to be immunoreactive to lymphocytic IFN (lyIFN) in direct enzyme-linked immunosorbent assay (ELISA) and only one neutralized IFN activity (27). Other investigations have failed to detect IFN-α Ab in groups of 60 up to 350 healthy donors (reviewed in ref. 18). IFN-α2a Ab have been described in less than 10% of sera of pregnant women and in cord blood samples of their fetuses (18).

The existence of these Ab in healthy individuals is verified by the finding of specific and high-avidity IFNα Ab (K_d 10^{-10}–10^{-12} M) in pharmaceutically prepared normal human IgG pooled from more than 1000 donors (28) (Table 2). These Ab suppressed the antiviral effect of IFN-α through saturable binding to the Fab fragments of the IgG molecules.

In some cases, Ab that bind IFN-β have been found in patients with infectious diseases (29). Low levels of IgG and IgM capable of neutralizing both IFN-α, IFN-β, and IFN-γ in vitro have also been detected in blood of healthy individuals (29). However, as in the cases of IFN-α Ab in patients, these Ab were not demonstrated to bind in a specific manner or with appreciable affinity. Since IgG-induced neutralization of IFN-α or IFN-β in biological assays may not necessarily implicate the presence of IFN Ab, we have recently investigated the presence of Ab to radiolabeled IFN-α2a and IFN-β in plasma samples of 4000 and 2500 blood donors, respectively. Only two donors were IFN-α Ab–positive and only one was IFN-β Ab–positive. Interestingly, these donors showed remarkably high binding capacities, suggesting that the IFN Ab in pooled human IgG preparations are contributed by only a few but highly positive donors.

IFN-γ

IFN-γ is produced primarily by T helper type 1 (Th1) T cells and natural killer (NK) cells (see Table 1). It has no homology with the various IFN-α family proteins or IFN-β. Mature IFN-γ is a noncovalently linked homodimer, and human IFN-γ is biologically active only in human and primate cells. IFN-γ has antiviral activity but also antiproliferative, immunoregulatory, and proinflammatory activities. IFN-γ upregulates the expression of MHC class I and class II molecules, Fc receptor, and leukocyte adhesion molecules. It induces the production of cytokines, activates MØ, and influences isotype switching and secretion of immunoglobulins by B cells.

Antibodies to IFN-γ

Natural Ab to IFN-γ were first reported in 1989 in human immunodeficiency virus (HIV)–infected individuals, correlating with the stage of the HIV infection (30). These Ab bound to IFN-γ in a solid phase radioimmunoassay and were of the IgG isotype. However, when the Ab were affinity purified, none of them impaired the antiviral activity of IFN-γ, whereas IFN-γ–induced HLA-DR expression was inhibited, apparently without interfering with the binding to cellular IFN-γ receptors (31).

Table 2 Characteristics of Antibodies Binding Native Cytokines in Healthy Adults

	IFNα Ab	IL-1α Ab	IL-6 Ab	GM-CSF Ab
Frequency in pharmaceutical IgG[a]	80%	100%	100%	100%
Detectable binding capacity[b]	80%	>90%	80%	>90%
Frequency in normal sera[c]	<1–10%[d]	30–75%	10–20%	<1–10%
Increased frequency with age	No	Yes	No	?
Increased frequency in males	No	Yes	No	?
Predominant Ig class	IgG1	IgG 4,2,1	IgG1	ND
Block receptor binding	Yes	Yes	Yes/no	Yes
Block bioactivity in vitro	Yes	Yes	Yes	ND
Ligand binding				
K_d	$<10^{-10}$ M	$<10^{-11}$ M	$<10^{-10}$ M	$<10^{-10}$ M
Bind with Fab	Yes	Yes	Yes	Yes
Miscellaneous	Crossreacts with IFN-β	A cloned Ab is IgG4 κ		

ND, not determined.

[a] Nordimmun (Novo-Nordisk), Sandoglobulin (Sandoz), and Gammagard (Baxter).

[b] 200 pg radiolabeled cytokine to 10 mg/mL of IgG.

[c] Detected by double antibody precipitation and protein G binding of untreated serum or human IgG preparations (preexisting cytokine-Ig complexes may go undetected).

[d] Ab to IFN-α, although measurable in human immunoglobulin preparations, are difficult to detect in normal sera. This is most likely because they are present in serum in a saturated form, complexed with their respective cytokines, or because of the presence of other inhibitory factors in serum. The Ab are polyclonally derived, because they are recognized in individual sera by antibodies to both immunoglobulin light chains.

IFN-γ Ab have also been reported in humans with other viral infections, in Africans suffering from trypanosomiasis, and at lower levels in healthy individuals (31). IFN-γ Ab were recently detected in cerebrospinal fluids from patients with multiple sclerosis and Guillain-Barré syndrome (32,33). The specificity of the reaction between IFN-γ and Ab was confirmed by selective and saturable binding to Fab fragments generated from purified immunoglobulins. Since IFN-γ is thought to play a pathogenic role in multiple sclerosis, these Ab may be of clinical significance, particularly if they are shown to bind with high affinities to native IFN-γ and to be present in sufficient quantity to react with IFN-γ molecules in the brain. In patients with Guillain-Barré syndrome, recovery was associated with increased levels of neutralizing IFN-γ Ab (33).

IL-2

IL-2 is produced primarily by Th-cells and is an important autocrine and paracrine growth factor for T and B cells as well as NK cells (see Table 1). IL-2 also appears to influence the survival and/or growth of MØ, neutrophils, γδ-T cells, and neurons and glial cells.

Antibodies to IL-2

Natural Ab against IL-2 in humans were first reported in 1988 (34). The study was initiated because homology between the carboxy terminal six amino acids, LERILL, of the HIV envelope protein gp41 and the human IL-2 sequence, LEHLLL, suggested that gp41 Ab induced in HIV-infected individuals might cross react with IL-2.

Characterization. When measured by ELISA using the HIV sequence LERILL conjugated to bovine serum albumin, the above study showed that nearly 90% of individuals seropositive for HIV expressed anti-LERILL IgG and IgM Ab, which could be displaced by recombinant IL-2 (34). There was a positive correlation between Ab titers assayed in ELISA by binding to fixed IL-2 and LERILL, respectively. It has not been reported whether these Ab neutralized IL-2 bioactivity. However, $F(ab)_2$ fragments generated from IL-2 affinity-purified Ab in pooled normal immunoglobulin were later shown to inhibit the lymphocyte costimulatory effect of IL-2. The specificity of the $F(ab)_2$ fragments was shown by reaction with IL-2 but not IFN-γ in Western blot analysis and the blocking of binding to fixed IL-2 by soluble IL-2 in ELISA (35,36).

IL-2 IgG Ab have also been found by direct ELISA of sera from multiple sclerosis patients (37). A positive correlation between inhibition of proliferation of the IL-2–dependent cell line CTLL-2 and the IL-2 Ab activity of the 150-kD fraction of gel-filtrated individual sera was taken as indication of the presence of neutralizing anti–IL-2 Ab in these individuals. Unfortunately, a possible interference from heat-sensitive nonimmunoglobulin factors was not investigated (9,38).

We have recently tested for IL-2 Ab in sera from 16 healthy individuals and 55 HIV positive patients, including patients with acquired immunodeficiency syndrome (AIDS) (C. Meyer, unpublished data). None expressed saturable binding of radiolabeled IL-2 to IgG or to other serum components, and none interfered with IL-2 receptor binding or IL-2–induced proliferation of antigen-stimulated human T cells. Furthermore, all of several preparations of pooled normal human IgG were negative for IL-2 Ab (39). Other investigators have also failed to detect IL-2 Ab in the sera of cancer patients prior to IL-2 therapy (9,38,40,41). The reason for this discrepancy is not clear but could be related to the types of assays used (see below).

IL-1

IL-1α is a multifunctional cytokine synthesized by a number of cell types (see Table 1). It is part of the IL-1 family of cytokines: IL-1α, IL-1β, and IL-1 receptor antagonist (IL-1Ra). The first two are highly inflammatory cytokines which affect nearly every cell type in the body, whereas IL-1Ra functions as a specific receptor antagonist. IL-1α, in contrast to IL-1β and IL-1Ra, is primarily associated with the cell that produces it, for example MØ and keratinocytes. IL-1α is usually absent or present in the circulation only at low concentrations. During infection and inflammation, however, substantial amounts of IL-1α may be found in the blood. IL-1α is also found in a biologically active form on the surface of several cells, for example, on MØ and B cells; that is, "professional" antigen-presenting cells.

IL-1α, like IL-1β, is highly inflammatory when given to humans, and it has been implicated in the pathogenesis of autoimmune diseases such as rheumatoid arthritis (RA), in insulin-dependent diabetes mellitus, and in complications of infections and trauma.

Antibodies to IL-1α

Natural Ab against IL-1α in humans were first found in 1988 by demonstration of direct binding of IgG from healthy individuals to radiolabeled human recombinant IL-1α and by IgG-mediated competitive interference with IL-1α binding to cellular IL-1 receptors (42). At the same time, IgG from patients with RA were found to interfere in vitro with the biological effect of IL-1α, and sera from these patients interfered with the effect of both IL-1α and IL-1β (43). Subsequently, sera of healthy individuals and of patients with various immunoinflammatory disorders as well as pharmaceutical preparations of human IgG were found to contain Ab that bind to IL-1α with high avidity (44–55).

Characterization. The IL-1α Ab are dominated by the IgG isotype, mainly of the IgG4 class, and are the main binding molecules for IL-1α in the circulation (44,49,55) (see Table 2). They bind with high avidities ($K_d < 10^{-10}$ M). Size exclusion chromatography of IgG or Fab fragments in complex with IL-1α has shown a predominance of two binding sites on IL-1α (44,55). However, IgG4 molecules, normally constituting only about 2% of total IgG in plasma, are functionally monovalent, and complexes of this IgG class bind weakly to type 2 and type 3 Fc receptors and do not activate complement (56–59). Accordingly, a reduced clearance of IL-1α in complex with human IL-1α Ab was seen in rats (49), and low amounts of IL-1α in complex with IgG have been reported in 25% of IL-1α Ab–positive individuals (54,60). Thus, even though carrier functions cannot be excluded, the high binding avidity, comparable to the affinity of the IL-1 receptors, and the high stability of the complexes suggest that IL-1α Ab function as a specific IL-1α antagonist in vivo (47). The IL-1α Ab detected in direct ligand binding radioimmunoassay neutralize receptor binding and the in vitro biological activity of natural IL-1α (45,48,51,52,55,61). They also neutralize the activity of precursor and membrane-associated IL-1α (52). IL-1α Ab of low receptor blocking activity have been reported in some individuals; their inhibitory effect appears to depend on the target cell (49,52).

An IL-1α Ab has recently been cloned (62). This monoclonal Ab is an IgG4/κ Ab reacting with IL-1α but not IL-1β, IL-1Ra, or several other cytokines. It binds with high affinity ($K_d \approx 10^{-10}$ M), and the presence of somatic mutations in the variable regions suggests an antigen-driven affinity maturation.

Prolonged antigenic challenge results in predominance of IgG4 Ab (63). It is likely therefore that some individuals may be efficiently primed to IL-1α early in life, and that the increased levels of IL-1α Ab with age is a consequence of persistent antigenic stimulation, perhaps by endogenous IL-1α.

Clinical Importance. IL-1α Ab have been detected in 1–60% of healthy adults (46,49,51,54,61,64). The differences may reflect the sensitivity of different assays. Analysis of sera from 466 Danish blood donors aged 18–65 years revealed a doubling in incidence from 20 to 40% to 40 to 60% among the oldest donors. These individuals also had higher levels of IL-1α Ab. Although males were more frequently positive, there were no differences in the binding avidities or IgG subclass distribution between the two sexes. The IL-1α Ab–positive individuals generally stayed positive at the same or slightly increased levels, but a few initially negative individuals became highly positive over periods of up to 3 years (22,54). Similar observations have been made in patients with chronic renal and joint diseases (50,65). In such patients, IL-1α Ab have been reported with increased, reduced, and similar prevalences to healthy controls. For example, a normal prevalence of IL-1α Ab appeared in patients with chronic liver diseases, without correlation to etiology or inflammatory activity of the disease or risk of recurrent infections (64,66). In chronic renal failure patients with or without renal replacement therapy, there was no clear association with kidney or renal graft disease (50). However, the relative number of IL-1α Ab positives was increased in hemodialysis patients: 33% versus 6–9% in the other groups of renal patients, as were the serum levels of IL-1α. Increased frequencies have also been reported in patients with Graves' disease, Sézary's syndrome, and Schnitzler's syndrome, whereas reduced frequencies have been observed in atopic patients and in patients suffering from Crohn's disease of the gut (3,49,67). In Sézary's syndrome, the leukemic stage of chronic T-cell lymphoma, IL-1α Ab have been proposed to promote dissemination of the disease by neutralizing the induction of adhesion molecules in the skin.

A frequency of IL-1α Ab below that of healthy controls has been detected in patients with systemic lupus erythematosus (SLE), systemic sclerosis, and chronic hepatitis B infection (46,51,65,67). In contrast, patients with RA usually present with higher prevalences (22,43,51,61,65). It has been suggested that the levels of these Ab fluctuate in parallel with RA activity (51); this however was not found in other studies (22,65). Common to these studies was the fact that patients most positive for IL-1α Ab were the ones with milder disease manifestations and low titers of rheumatoid factor, and that a relative higher number of positive patients suffered from less severe forms of polyarthritis; for example, primary Sjögren's syndrome or self-limiting inflammatory arthritis (61,65). A negative relationship between the presence of IL-1α Ab and HLA-DR4 was suggested from these studies (61). Association of IL-1α Ab positivity to a particular major histocompatibility complex (MHC) type was not apparent in healthy individuals (46).

Recently, a 3-year study of unselected patients with chronic polyarthritis revealed that both incidence and levels of IL-1α Ab were higher in a subset characterized by an increased proportion of patients with primary Sjögren's syndrome or self-limiting inflammatory arthritis, diseases with a much better prognosis than RA (65). The relative risk of developing RA was 12 in the absence of high Ab levels and 18.2 when associated with the presence of HLA-DR4. If confirmed, detection of these Ab would be the best-known predictor of nonerosive disease in patients with polyarthritis and might serve to select patients for less aggressive forms of antiarthritic therapies.

Since IL-1 causes arthritis in experimental animals, IL-1α Ab might also prove to be useful for the treatment of RA patients. In this regard, however, one must realize that the above finding is no proof of involvement of IL-1α in the pathogenesis of erosive disease.

Absence of Antibodies to IL-1β and IL-1 Receptor Antagonist

Natural Ab against IL-1β in humans have been reported by some investigators but not by others (42,44–49). Of the three plasma samples found to be positive for IL-1β Ab, two had Ab cross reacting with IL-1α (46). In the blood, the predominant IL-1β–binding protein therefore seems to be the soluble form of the IL-1 type 2 receptor (68).

Natural Ab against IL-1Ra have so far not been detected in healthy individuals (39,69). Hence, the most important serum protein binding IL-1ra appears to be the soluble form of the IL-1 type 1 receptor (69,70).

Tumor Necrosis Factor α and Lymphotoxin α

Tumor necrosis factor-α (TNF-α) is synthesized as a type II membrane protein, which can be cleaved to produce a soluble form. In crystal form and at high concentrations, soluble TNF-α is a trimer. At low concentrations, however, it functions also as a monomer. TNF-β, also termed lymphotoxin (LT), is expressed as LT-α and LT-β in certain cytotoxic T cells. LT-α is produced as a soluble cytokine with a signal sequence, whereas the latter is a type II membrane protein. LT-α and LT-β associate on the cell surface at high concentrations as trimers and at low concentrations as monomers or dimers.

Antibodies to TNF-α and LT-α

Natural Ab against TNF-α, detected by immunometric assay and Western blotting, were first reported in 1989 (71,72). They were found in 40% of healthy adult donors and in more than 60% of patients with bacterial infections and various rheumatic diseases. The specificity of the assays was shown by the prevention of binding of serum immunoglobulin after preincubation with recombinant human TNF-α. Both IgG and, with a stronger activity, IgM Ab were found, particularly in patients with sepsis. Up to 50% inhibition of in vitro TNF-α bioactivity was obtained with affinity-purified TNF-α Ab that consisted primarily of the IgM isotype (73). These Ab could identify three sequences from randomly displayed nonapeptides in a phage library, and these peptides together inhibited by 50% the binding of the purified TNF-α Ab in direct ELISA. Phages displaying the nonapeptides induced Ab in mice that reacted with immobilized TNF-α. Ab reacting with human recombinant TNF-α and correlating with disease activity have been reported in cerebrospinal fluids and blood of patients with multiple sclerosis (32). There Ab may be of clinical significance, because TNF-α is thought to be pathogenically involved in multiple sclerosis. Whereas the above studies, and those of others (74,75), suggest a relative high prevalence of TNF-α Ab in healthy individuals and in patients with infection and inflammation, several normal immunoglobulin preparations and a number of patients were found to be Ab negative with Western blotting when nonfat dry milk was included to reduce nonspecific protein interaction (76). We have also failed to detect significant binding to radiolabeled TNF-α in preparations of normal human immunoglobulin (53). Natural Ab against LT-α, detected by immunometric assay, were first reported in 1989 in normal human sera (72). To our knowledge, these findings have not been confirmed.

Characterization. A TNF-α IgM Ab with strong reactivity in direct ELISA and Western blotting has been cloned (77,78). This monoclonal Ab failed to neutralize TNF-α cytotoxicity and bound only weakly to soluble TNF-α, and it failed to compete with the binding of three neutralizing mouse monoclonal Ab to fixed TNF-α. In contrast, the Ab bound to cell surface TNF-α on human and chimpanzee as well as murine cells and apparently also to receptor-bound TNF-α. Because the cloned Ab did not react with other antigens commonly recognized by polyreactive natural IgM Ab, it was proposed that this Ab recognizes a linear epitope shared by TNF-α from the different species, and that this epitope is strongly expressed by TNF-α fixed to a solid phase and as a precursor in the cell membrane (77,78).

It appears from the above that human TNF-α Ab have low binding to, and limited influence on, the activity of soluble bioactive TNF-α. It is unknown whether binding of such Ab to membrane-associated TNF-α is cytotoxic to the target cells or can influence the processing of membrane TNF-α.

IL-6

IL-6 is a multifunctional cytokine produced by many cell types (see Table 1). It participates in hematopoiesis, in terminal differentiation of activated B cells into Ab-producing cells, and it is of central importance in host defense, acute phase reactions, and immune responses. IL-6 is involved in the pathogenesis of many diseases, including multiple myeloma, Castleman's disease, glomerulonephritis, autoimmune diseases, and certain neurological disorders. In addition, patients with certain leukemias and autoimmune disorders have improved after administration of heterologous in vitro neutralizing IL-6 Ab.

Antibodies to IL-6

Natural Ab against IL-6 in humans were first observed in 1991, because they, in analogy with IL-1α Ab, interfered with IL-6 measurements by ELISA (79). Since then, the presence of naturally occurring IL-6 Ab and similar Ab in patients with immunoinflammatory and fibrotic diseases has been confirmed (53,80–84).

Characterization. Saturable IgG binding of radiolabeled IL-6 in solution was observed in 61 of 467 (13%) sera of healthy Danish blood donors (54) (see Table 2). IL-6 binding Ab are present at low levels, usually less than 10^{-10} M, but about 3% of normal individuals have IL-6 Ab at levels exceeding 10^{-10} M; in one case, even at the micromolar level (79,81). These relatively few donors appear healthy and contribute the bulk of Ab–IL-6 activity found in pooled human IgG preparations (M. B. Hansen et al., unpublished results).

Generally, the Ab bind IL-6, including native IL-6, with high affinity ($K_d \approx 10^{-10}$ M) (54,81). Not surprisingly, a small proportion (0.1–5.0%) of the Ab are circulating in complex with IL-6 and possibly also with the soluble IL-6 receptor CD126 (83,85). The Ab response to IL-6 is polyclonal, since both κ and λ light chains are involved. The IgG subclass is almost entirely restricted to IgG1, and several epitopes on the IL-6 molecule can be attached by different Ab. Therefore, relatively large immune complexes can be made in vitro when the Ab are in excess (79,84).

Clinical Importance. There has been some discrepancy as to whether natural IL-6 Ab neutralize IL-6 in vitro (81,83,86). In this regard, it is important to consider that most IL-6 bioassays are performed with highly sensitive murine cells that may respond to other IL-6–related factors (82,84). In addition, the avidity of polyclonal Ab may decline in diluted samples and dissociation of IL-6 from preexisting immune complexes may occur. Thus, bioassay of the IL-6 Ab is likely to result in underestimation of the neutralizing ability of undiluted samples.

Interference with IL-6 binding to cells of human origin seems to be an appropriate way to judge IL-6 Ab neutralization. Using U-937 (monocytoid) and U-266 (plasmacytoid) cells, individual Ab-positive samples interfered in a competitive and concentration-dependent manner with cellular high-affinity binding of IL-6 (81,84). Since the high-affinity binding of IL-6 to cells involves formation of a hexameric complex between IL-6, CD126, and the signal-transducing unit gp130 (CD130) (87), there are two possibilities for biological neutralization by IL-6 Ab. Ab bind to sites on IL-6 directly involved in the binding to CD126 molecules, or Ab bind to sites on IL-6 not involved in CD126 binding but interfering with the assembly of the complex with CD130 (81,84). The latter implicates that IL-6–containing immune complexes can bind to CD126-positive cells. Such Ab could express higher bioinhibitory activity than fully blocking Ab, because they at the same time bind IL-6 and form complexes competing with receptor binding of free IL-6 (83).

Soluble forms of CD126 and CD130 occur naturally in human sera and in concentrations far exceeding those of IL-6 Ab (88). However, the binding to these receptors is of low affinity and accounts for less than 30% of IL-6 binding in normal plasma (81,85). Significant binding of IL-6 has only been observed in samples from IL-6 Ab–positive individuals. In most cases, the high-avidity binding of IL-6 to Ab is almost irreversible, because 30–50% of the complexes are intact after 20 h (84). This is compatible with the observation of small, bioinactive IL-6–containing immune complexes in the circulation (81,89). Thus, in systemic sclerosis patients, the highest serum IL-6 bioactivity was found in IL-6 Ab–positive individuals who presented with complexes containing IL-6, IgG, and soluble CD126 (83).

Various clinical trials together with animal experiments with IL-6 Ab have given insight into the possible in vivo functions of IL-6 Ab. Depending on their concentration, avidities, epitope recognition pattern, and subclass, IL-6 Ab may suppress or prolong the in vivo activity of IL-6 (1,11,90–94). Indeed, IL-6 Ab are inhibitors in vivo only when they are present in molar excess and able to form immune complexes that can be cleared via Fc receptors (10).

Because the concentration range in which IL-6 is bioactive is 2–3 logs lower than that of a typical natural IL-6 Ab and owing to the high affinity of IL-6 Ab and their ability to form large complexes with IL-6, the Ab are likely to function as inhibitors in vivo. This is further expected because of the predominant IgG1 subclass distribution making the complexes prone to effective clearance by macrophages. In this regard, IL-6 has been detected coupled to complement products in normal blood (95).

The reason for the apparent increased prevalence of IL-6 Ab among patients with inflammatory diseases is unknown, but conditions associated with increased IL-6 production and/or augmented B-cell activation, for example, systemic sclerosis and RA, may predispose to production of IL-6 Ab (22,80,83).

Nerve Growth Factor

Nerve growth factor (NGF) is a neurotrophic peptide essential for survival and differentiation of sensory and sympathetic nerves in both the peripheral and central nervous systems.

Antibodies to NGF

Natural Ab against NGF in human and rabbit sera were first reported in 1991 (96). The Ab were found in higher titers in patients with herpes virus infections, with SLE, insulin-dependent diabetes mellitus, autoimmune thyroiditis, RA, and Alzheimer's disease (97–99). Recently, NGF Ab were also found in synovial fluids from patients with spondylarthropaties (100). NGF Ab of all Ig classes have been found, they were detected by ELISA but have also been found to bind to NGF-Sepharose columns and, in some cases, they have been shown to neutralize the bioactivity of NGF in vitro.

Chemokines

IL-8 is produced by many cell types (see Table 1). It is a potent chemoattractant for, and activator of, neutrophils, basophils, T cells, and eosinophils. It is angiogenic both in vivo and in vitro, and it enhances the adherence of neutrophils to endothelial cells and subendothelial matrix proteins. IL-8 is a member of the α-chemokine subfamily (C-X-C).

Monocyte chemotactic protein-1 (MCP-1) is a member of the β-chemokine subfamily (C-C) (see Table 1). Other members within this family include macrophage inflammatory protein-1α (MIP-1α), MIP-1β, and RANTES (acronym for *r*egulated upon *a*ctivation, *n*ormal *T*-cell *e*xpressed and presumably *s*ecreted). MCP-1 is produced primarily by fibroblasts, MØ, smooth muscle cells, endothelial cells and various tumor cells; MIP-1α and RANTES are produced primarily by T cells. The β-chemokines are MØ chemoattractants, but they also attract basophils (MCP-1 and RANTES), eosinophils (MIP-1 and RANTES), $CD8^+$ T cells, B cells (MIP-1α), and both $CD4^+$ and $CD8^+$ T cells with the naive and memory phenotypes (RANTES). C-C chemokine receptors are important ports of entry for the HIV.

Antibodies to Chemokines

Since the first reports in 1992, several papers have described natural Ab against chemokines in healthy and diseased individuals (101–111). These investigations have mainly been based on detection and quantitation of the Ab in direct

ELISA, and in the case of Ab against IL-8 and MCP-1, also by presumed specific ELISAs for circulating levels of the cytokine bound in Ig complexes.

Characterization. Ab IL-8 have been found to belong to all major isotypes of antibodies, with a predominance of IgG, and containing both κ and λ light chains indicating a polyclonal origin for these Ab (102,107). In most cases, the Ab were found by ELISA, and the possible contribution of heterophilic or nonspecific Ab has generally not been addressed (see below). However, in one study using ELISA, the Ab IL-8 were also found to bind to radiolabeled IL-8 (108).

Increased ELISA reactivity for IL-8 IgG Ab has been reported in patients with RA, systemic sclerosis and related disorders, heparin-associated thrombocytopenia, and adult respiratory distress syndrome (101,104,106,107). In patients with IgA nephropathy, only IL-8 Ab of the IgA class was increased (109). IgA Ab have also been found at increased levels in the bronchial mucosa in patients with asthma and in the gastric mucosa in patients with *Helicobacter pylori* infection (103,111). A correlation between the levels of anti–IL-8 Ab, levels of IL-8, and disease activity was reported in some of the studies (101,103,107,109,111).

The Ab IL-8 found in healthy individuals were predominantly of the IgG3 and IgG4 classes, but IgG2 Ab have also been found (102,107). Concentrations up to several nanomolars of IL-8–IgG complexes have been reported (102,107). Using molecular size separation, the IL-8–IgG complexes coelute with monomeric IgG and dissociate poorly even in 9 M of urea at pH 2 (102,107). Estimated by assuming equilibrium between free IL-8, Ab, and the IL-8–Ab complexes, a K_d below 10^{-11} M has been estimated (102). This is comparable with a K_d of 0.5 to 1×10^{-11} M obtained from binding analysis using radiolabeled IL-8 and affinity-purified free IL-8 IgG Ab, but not to the K_d of nearly 10^{-6} M when estimated by competition with soluble IL-8 in direct ELISA (107,108).

Functional analysis of IL-8 IgG Ab has shown either no inhibition or up to 50% inhibition of binding of radiolabeled IL-8 to human neutrophils, and the chemotactic activity of IL-8 was reported to be inhibited (102,107).

IL-8 Ab have been detected by ELISA in pooled human IgG preparations (102,107). In contrast, we have only been able to find low and nonsaturable binding to radiolabeled homodimeric IL-8 in several preparations of pharmaceutical IgG (unpublished results). It is possible therefore that IL-8 Ab detected by ELISA do not have the potential to bind native dimeric IL-8 in solution and perhaps not in a specific manner, at least in healthy individuals. IgG Ab against MCP-1 were reported in healthy individuals both in free form and complexed to MCP-1 (105).

Recently, IgG against MIP-1α and RANTES were detected by direct ELISA in sera from asymptomatic HIV-positive patients, including those with AIDS, as well as in healthy controls (110). The Ab were found at highest levels in asymptomatic HIV-positive patients. Because affinity-purified Ab against the V3-loop of the viral envelope glycoprotein gp120 reacted with the solid-phase fixed chemokines, it was suggested that the higher levels of MIP-1α and RANTES Ab in HIV-positive but asymptomatic patients might be due to a cross reaction with anti–V3-loop Ab raised against gp120 of macrophage-tropic HIV-1 strains. Using direct radioligand binding assays, we have generally failed to detect MIP-1 Ab or RANTES Ab in 505 HIV-positive patients with or without AIDS. We also found no differences from healthy controls, because RANTES Ab were undetectable in all of 2000 sera tested, and MIP-1 IgG Ab were found in only four HIV-positive patients (<1%) and in nine controls (<0.5%) (C. Meyer, unpublished data).

IL-10

IL-10 is a 35- to 40-kD homodimeric protein; it has profound effects on cells involved in the immune response (see Table 1). IL-10 is produced primarily by MØ and T lymphocytes, Th2 cells in particular, and the cytokine is a potent suppressor of MØ, chiefly because it counteracts the stimulatory functions of IFN-γ; for example, induction of MHC class II expression and cytokine synthesis. IL-10 therefore inhibits antigen presentation and indirectly T-cell functions. The cytokines, whose production are most affected by IL-10, are those originating from Th1 cells. Interestingly, two herpesviruses have acquired an IL-10 gene, an equine herpesvirus and the Epstein-Barr virus (EBV). Thus, analysis of the coding sequence of the IL-10 gene has revealed that it is highly homologous to the EBV open reading frame BCRF1, and that the EBV-derived polypeptide, viral IL-10, has the same biological activities as IL-10. IL-10 is expressed by some neoplastic cells, including primary B-cell tumors as well as basal cell and bronchogenic carcinomas. IL-10 suppresses the functions of cytotoxic T cells and NK cells, and it is a potent inhibitor of tumor cytotoxicity by human MØ.

Antibodies to IL-10

Pharmaceutic preparations of human IgG have been shown to bind homodimeric but only weakly to monomeric IL-10 (67,112). The binding was saturable and only at a low level of activity (39). However, none of 50 individual sera was positive in a direct radioligand binding assay. This suggests that

IL-10 Ab are rarely present in healthy individuals and/or that these Ab are inaccessible for detection in serum, because they are blocked by IL-10 or other serum factors. This is supported by the finding of low-level and neutralizing IL-10 Ab in 7 of 1860 sera collected from healthy individuals and patients suffering from various, mostly autoimmune, diseases (113). Using a radiolabeled IL-10 binding assay, three of the seven cross bound to viral IL-10, the isoform expressed by EBV, suggesting that in some cases cross reactive IL-10 Ab may be induced by infection with this virus. Purification of the Ab from one serum revealed a single IgG1, lambda Ab, that neutralized IL-10 bioactivity. The Ab levels were constant for more than a year. Recently, Ab to viral IL-10 were detected by ELISA in patients with EBV infections, including chronic infectious mononucleosis, nasopharyngeal carcinoma, and EBV-associated lymphoproliferative disease. In one patient with chronic infectious mononucleosis, the Ab also neutralized natural IL-10 bioactivity in vitro (114).

Colony-stimulating Factors

Granulocyte colony-stimulating factor (G-CSF) is a pleiotropic cytokine produced mainly by MØ. In addition, various carcinoma cell lines and myeloblastic leukemia cells can express G-CSF constitutively. G-CSF stimulates growth, differentiation, and functions of cells from the neutrophil lineage. It also has blast cell growth factor activity. G-CSF plays important roles in defense against infection, inflammation, and tissue repair.

GM-CSF is also a pleiotropic cytokine produced by a number of different cell types (see Table 1). Besides granulocyte-macrophage progenitors, GM-CSF is also a growth factor for erythroid, megakaryocyte, and eosinophil progenitors. GM-CSF also modulates the function of differentiated white blood cells. For example, GM-CSF stimulates macrophages for antimicrobial and antitumor effects. The cytokine further enhances healing and repair by its actions on endothelial cells, fibroblasts, and epidermal cells.

GM-CSF is the pivotal mediator of the maturation and function of dendritic cells, the most important cell type for the induction of primary T-cell immune responses. GM-CSF may enhance Ab-dependent cellular cytotoxicity in several cell types, and the generation and cytotoxicity of NK cells. Recently, GM-CSF was found to be essential in an RA model using GM-CSF knock out mice. Human GM-CSF is species specific.

Antibodies to Colony-stimulating Factors

Low levels of Ab against G-CSF of both the IgG and IgM isotype have been detected by ELISA and immunoblots in nearly 10% of sera from healthy adults

(115). Recently, however, several preparations of pooled normal IgG were all judged to be negative when assayed by binding to radiolabeled G-CSF (39).

Using ELISA with fixed recombinant GM-CSF, low levels of Ab against GM-CSF were found in less then 2% of healthy individuals (116) (see Table 2). In a similar study, more than 10% had low-level GM-CSF Ab (117). In a biological assay, 2 of 115 (2%) patients with autoimmune diseases were positive for neutralizing Ab (118). Also, 4 of 1238 (0.3%) blood donors were positive in a direct radioligand binding assay, and the Ab bound GM-CSF with high avidities (K_d 10^{-11}–10^{-10} M) and binding capacities of 5–320 μmol/mol IgG (39). Since recombinant human GM-CSF was used in the assays, it has been questioned whether similar reactivity can be found to endogenous GM-CSF. We found that although there was no difference between endogenous GM-CSF and *Escherichia coli* CSF and *E. coli*–derived recombinant human GM-CSF, this was the case when using recombinant protein expressed in Chinese hamster ovary cells or yeast cells (39). Thus, detection of natural anticytokine Ab may depend not only on the assay system but also on the form of the recombinant cytokine.

Other Cytokines and Soluble Cytokine Receptors

IgG Ab to IL-12 have recently been detected by ELISA in sera of myasthenia gravis patients (24). The Ab were absent in healthy individuals but present in nearly 10% of patients without thymoma and in over 50% of those with thymoma. In a few patients, the levels of these Ab increased considerably around the time of tumor recurrence/metastasis, suggesting that their production is provoked or enhanced by the tumor (N. Willcox, personal communication). There was a positive linear correlation between the levels of IL-12 Ab and their in vitro neutralization of IL-12 bioactivity.

We have previously reported preliminary findings of IgG and IgM Ab to leukemia-inhibitory factor (LIF) in sera of 80% of healthy individuals (67). These Ab bound in a saturable manner to a glycosylated form of yeast recombinant human LIF. However, they also bound to fragments generated by pronase treatment, suggesting binding to the carbohydrate moiety of this recombinant form of LIF (Fig. 1).

There are a few reports about Ab to other cytokines and soluble cytokine receptors in both healthy and diseased individuals. For example, sera of AIDS patients contained IgE Ab to TNF-α, LT-α, IFN-γ, and IL-4 when detected by ELISA (119). These Ab specifically sensitized normal basophils to release histamine on challenge with the respective cytokine. Also, Ab to the soluble TNF receptor p75 were detected by ELISA in a patient with SLE (120).

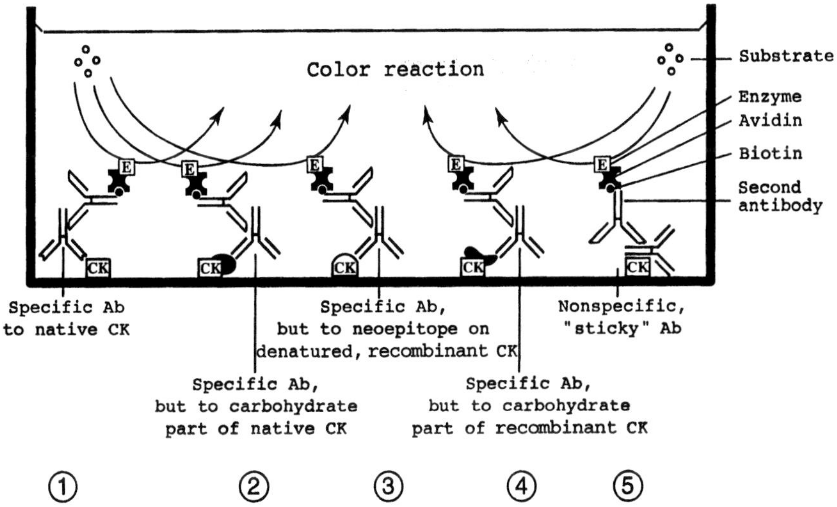

Figure 1 Specific and nonspecific reactions in ELISA for cytokine antibodies. Only cases (1) and (2) detect specific Ab to the native cytokine. These cases should be discriminated from cases (3), (4), and (5) by competition experiments using excess (nondenatured) cytokine in solution. However, although Ab binding is displaceable in both cases with the natural cytokine, a nonglycosylated, nondenatured, recombinant cytokine would displace binding only in case (1). In practice, the native cytokine used to capture Ab in case (2) will rarely be available, and any carbohydrate-specific Ab will therefore go unnoticed. The binding in case (4) is displaceable only by the glycosylated recombinant cytokine and, possibly, by relevant saccharides. In case (5), it will be difficult or impossible to displace the binding with soluble cytokine. These principles also apply to the detection of Ab to soluble peptides by many other methods, particularly those involving solid-phase binding techniques. CK, cytokine; E, enzyme.

Cytokine Antibodies in Normal Human Immunoglobulin

Pooled human IgG preparations are used for treatment of an increasing number of infectious and immunoinflammatory diseases (121). These preparations usually contain IgG from several thousand individuals and hence a broad range of binding specificities and idiotypes. Since, as mentioned above, Ab reacting with native cytokines are found in apparently healthy individuals, it is not surprising that they are also found in pooled human IgG preparations; in some cases at levels which interfere with selective cytokines during high-dose IgG therapy (28,39,47,52,122) (see Table 2). Hence, IgG from different manufacturers has been demonstrated to contain high-avidity Ab to IL-1α, IL-6, IL-10, IFN-α, and GM-CSF. They are polyclonal containing both κ and λ light chains, and the IgG1 subclass predominates, together with IgG4 and IgG2 in the case of IL-1α Ab.

ANTIBODIES INDUCED BY CYTOKINE THERAPY

Induction of anticytokine Ab has been reported in patients treated with human recombinant cytokines such as IFN-α, IFN-β, IFN-γ, IL-2, and GM-CSF (3,6,7,123–127). ELISA and Western blotting have been used most frequently for the detection of cytokine-binding Ab, but this does not necessarily predict a neutralizing capacity of these Ab in vitro or in vivo (see below). Furthermore, these Ab are in many cases not autoreactive and therefore cannot be considered as autoantibodies, because they fail to react with the native cytokines. For example, patients with response failure to human recombinant IFN-α due to Ab development may respond to natural IFN-α (6). On the other hand, patients with preexisting Ab, at least to IFN-α, appear to be particularly at risk of developing high levels of neutralizing Ab and response failure (C. Ross et al. unpublished data).

Interferons

Since the introduction of pharmaceutical preparations of recombinant IFN-α, IFN-β, and IFN-γ, it is being increasingly recognized that antibodies to these cytokines develop in many patients, particularly during prolonged therapy (6). There is some controversy, however, as to whether the development of such antibodies always leads to response failure. Some investigators distinguish between "binding" and "neutralizing" Ab and argue that only the latter have clinical importance. In this regard, it is appropriate to realize that immunohistochemical binding analyses and biological assays often vary significantly in sensitivity, specificity, and precision. For example, misinterpretations of antiviral neutralization bioassays are likely if they are carried out without controls for sample toxicity and endogenous antiviral activity. In our experience, high-titered "binding" IFN Ab always neutralize their respective cytokine in vitro. It is therefore advisable to monitor patients on IFN therapy for IFN Ab using validated analyses. Should Ab develop, determinations of their specificity might allow for rescue treatment using alternative IFN preparations.

Antibodies to Recombinant and Native IFN-α

Because of its antiviral, antitumor, and immunoregulatory effects, IFN-α has been given to patients with a variety of diseases with varying clinical responses

(6,7,128–130). IFN-neutralizing Ab of IgG type were first demonstrated in a patient suffering from nasopharyngeal carcinoma treated with human fibroblast IFN (123). Later, IFN Ab were observed in three patients with metastatic cancers treated with human leukocyte IFN (15). Since then, many investigations have focused on the immunogenicity of different IFN species and the possible clinical significance of neutralizing Ab (131).

Therapeutic Importance. Response failure in patients with hairy cell leukemia and malignant carcinoid tumors treated with recombinant IFN-α2a has been associated with neutralizing IFN Ab (132,133). Clinical resistance to IFN of various degrees was present in 6 of 16 patients with neutralizing Ab. Furthermore the neutralizing Ab did not inhibit the antiviral effects of natural IFN-α1a in vitro. At the same time, response failure was observed in a patient with Philadelphia chromosome–positive chronic myeloid leukemia correlating with the development of neutralizing Ab (134). No cross reactivity was observed against human leukocyte IFN-α, and switching of treatment to this product resulted in hematological remission.

The immunogenicity of two recombinant IFN-α subtypes and lymphoblastoid IFN-α was compared in a multicenter study employing 296 patients with different forms of chronic hepatitis (128). IFN-α2a was significantly more immunogenic (20% seroconversion) compared with IFN-α2b (7%) or lymphoblastoid IFN-α (1%). Recently, neutralizing IFN-α Ab were reported in 39% of patients with breakthrough hepatitis during treatment with recombinant IFN-α2a and IFN-α2b compared with 3% in complete responders, suggesting that the development of neutralizing Ab may cause response failure at least in some patients (135).

The vast majority of therapy-induced IFN-α Ab are of IgG type, indicating a secondary immune response. Epitopes recognized by neutralizing IFN-α2 Ab have been localized within the N-terminal functional domain of the molecule (136).

Generally, recombinant IFN-α seems to be more immunogenic than pharmaceutic preparations of native IFN. For example, in patients with an initial response to recombinant IFN-α therapy and a subsequent response failure associated with development of neutralizing Ab, a clinical effect has been regained after change of therapy to natural leukocyte-derived IFN-α (137,138). This difference may be due to the lack of glycosylation in the *E. coli*–derived recombinant cytokines, the many subtypes in natural IFN preparations, or to different formulations of the cytokine preparations. That the latter is important is supported by the finding that protein aggregates are crucial for the antigenicity of IFN-α in normal and transgenic mice (139). This has recently lead to a change in the formulation of recombinant IFN-α2a to avoid immunogenicity.

Patients suffering from cutaneous T-cell lymphoma with preexisting IFN-α Ab have been shown to be particularly at risk of developing high levels of neutralizing Ab during therapy with recombinant IFN-α2 (140). It appears therefore that some patients develop specific Ab to the IFN species used for therapy and not cross reacting with natural IFN, whereas others appear to have a boost of production of preexisting Ab to native IFN.

Antibodies to Recombinant IFN-β

IFN-β–neutralizing Ab were first demonstrated in high-risk malignant melanoma patients receiving combination therapies of recombinant IFN-β and IFN-γ (141). Although no Ab to IFN-γ were found, 56% developed IFN-β–neutralizing Ab; notably in patients receiving IFN-β subcutaneously. In 1993, IFN-β was shown to reduce the exacerbation frequency and decrease lesion formation visualized by magnetic resonance (MR) scans in patients with multiple sclerosis (MS) (142). On this basis, prolonged therapy with two recombinant IFN-β products has been approved for the treatment of MS: a glycosylated form with the predicted natural amino acid sequence (IFN-β1a) and a nonglycosylated form with a Met-1 deletion and a Cys-17 to Ser mutation (IFNβ1b) (143).

Therapeutic Importance. Many studies have focused on the immunogenicity of IFN-β preparations in relapsing-remitting MS (144–146). The reported frequencies and titers of these Ab vary depending on the IFN-β preparations and assays for antibody detection. The clinical importance of the Ab has also been difficult to evaluate owing to the natural relapsing-remitting course of the disease. However, the relapse rates and MR scan results have been at placebo levels in patients with neutralizing Ab (144). An effect on the final clinical outcome has yet to be demonstrated.

We have found that a significant number of MS patients develop neutralizing IFN-β Ab within 6 months of IFN-β therapy (Ross et al., unpublished data). IFNβ1β appears to be more immunogenic than IFNβ1α, and high-dose IFN-β1α therapy induced a faster appearance of IFN-β Ab. The clinical significance of these neutralizing Ab is not known.

Antibodies to Recombinant IFN-γ

Because of its ability to activate MØ and NK cells, and its immunoregulatory effects, INF-γ has been used to treat patients with various diseases, including cancers, chronic infections, and systemic mastocytosis. Although INF-γ Ab seem to occur rarely in cancer patients, case reports mention the appearance of INF-γ Ab after 4–6 months of therapy in patients with systemic mastocytosis

(147,148). Secondary treatment failure could be associated with the appearance of Ab that neutralized the antiviral activity of recombinant human INF-γ in vitro. In one patient, the Ab were of polyclonal origin with a predominance of IgG1 and IgG2 and small amounts of IgA and IgM (148).

IL-2

IL-2 has been used primarily in cancer therapy, but it has also been used in congenital and acquired immune deficiencies, in infections, and as an adjuvant to vaccines (149,150). Success has been obtained in the treatment of metastatic renal cell carcinoma and malignant melanoma, although this is often hampered by dose-dependent severe toxicities. Both native and recombinant IL-2 have been used and both have turned out to be immunogenic, although to a varying extent; only a few reports have focused on the clinical consequences (9,150–154).

Antibodies to Recombinant and Native IL-2

Development of exclusively nonneutralizing IL-2 Ab was originally found in patients treated with an *E. coli*–expressed, unglycosylated, recombinant peptide which differs from native IL-2 by lacking the N-terminal arginine and by a serine for cysteine substitution at position 125 (38). These Ab did not cross react with the native cytokine, and the 125 position substitution did not appear to be the immunogenic epitope. The immunogenicity seems to be low when polyethylene Glycol–conjugated recombinant IL-2 is used (155,156), but it may increase if recombinant IFN-α is given simultaneously (9,152).

The de novo production of IL-2 Ab of IgG type was found in all of 14 patients with renal cell carcinoma, whereas natural IL-2 administration produced Ab in 1 of 5 patients (152). Although this suggests that recombinant IL-2 is more antigenic than endogenous and glycosylated IL-2, few patients developed Ab that were likely to affect the clinical efficacy of IL-2.

A comprehensive trial-summarizing study demonstrated that half the patients with metastatic cancer treated with the same recombinant IL-2 preparation developed IL-2 Ab detected by ELISA (9). In vitro neutralizing activity was demonstrated in 6% of patients receiving the drug by intravenous route and in 10% of those receiving it subcutaneously. This, however, did not appear to affect the frequency or duration of the clinical responses.

In general, the rate of treatment-induced IL-2 antibodies has varied from 0 to 100% and frequently exceeds 50% in patients exposed to recombinant IL-2. IL-2 Ab are polyclonal in nature and predominantly composed of IgM and IgG types.

Therapeutic Importance. Many studies have failed to show a significant correlation between IL-2 Ab development and clinical response in cancer patients (9,41). Short-term treatment with IL-2, often advocated because of significant side effects, usually leads to the development of nonneutralizing Ab with questionable in vivo significance (153,157). Also, IL-2 Ab titers in most cases drop quickly after cessation of therapy, suggesting a lack of induction of an autoreactive response (152).

A potential in vivo modulation of IL-2–sensitive immune functions has been shown in few reports. One of 10 patients with HIV-related malignancies developed IL-2 Ab that seemed to abrogate the usual IL-2–induced increase in blood NK cell counts (157). Recent work has also demonstrated that the appearance of IL-2 IgG Ab precedes a decrease in soluble IL-2 receptor levels in HIV patients, indicating an attenuation of IL-2–induced immune activation in vivo (C. Meyer, unpublished data). It has been suggested that patients with IL-2 IgG Ab may experience high and prolonged levels of recombinant IL-2 (153). Further analyses are needed to determine whether a carrier effect of IL-2 Ab may develop in some patients and whether this may have clinical implications.

Colony-stimulating Factors

Recombinant human erythropoietin (EPO), GM-CSF, and G-CSF have become extensively used in patients with various diseases. EPO selectively stimulates erythropoiesis and is used in the management of anemia in patients on dialysis and in other patients (158,159). GM-CSF and G-CSF stimulate the proliferation, differentiation, and maturation of myeloid progenitor cells and induce mobilization of stem cells and progenitor cells from bone marrow to peripheral blood (160). GM-CSF and G-CSF have most frequently been used to support granulopoesis in patients during or after myelosuppressive therapy, for harvesting blood stem cells for transplantation, and, in the case of GM-CSF, as immunostimulating agent in the treatment of cancer patients (161–163).

Antibodies to Recombinant Erythropoietin

Several studies have failed to demonstrate the development of EPO Ab in patients treated with recombinant human EPO. One study, however, reported the development of EPO Ab, and this was associated with rapidly progressive and EPO-resistant anemia (164).

Antibodies to Recombinant GM-CSF

Development of GM-CSF Ab detected by ELISA was first reported in a group of 16 patients with chemotherapy-resistant solid tumors or Hodgkin's disease (124). Whereas none of 3 patients treated with *E. coli*–derived GM-CSF developed Ab, 4 of 13 patients treated with a glycosylated form of human GM-CSF expressed in yeast cells developed IgG Ab capable of binding to recombinant GM-CSF. All four patients had accelerated clearance of the infused cytokine, as did one patient without GM-CSF Ab. The Ab had no in vitro neutralizing capacity; possibly because of the insensitivity of the biological assay.

The importance on Ab development of concomitant immunosuppressive treatment has been addressed in patients with multiple myeloma on combination therapy with repetitive cycles of cyclophosphamide, IFN-α, and glucocorticoid followed by *E. coli*–derived GM-CSF (116). Although GM-CSF Ab appeared in only 1 of 8 treated patients, 19 of 20 patients not on chemotherapy developed IgG Ab during GM-CSF therapy. The Ab levels progressively increased from the second treatment cycle and disappeared up to 30 weeks after completion of therapy. A similar transient appearance of induced GM-CSF Ab of IgG type has been observed in patients with lymphoid and plasma cell malignancies who after chemotherapy received *E. coli*–derived human GM-CSF (117).

Therapeutic Importance. In vitro neutralizing Ab have been detected only after repetitive administration of GM-CSF. In a study of patients with metastatic colorectal carcinoma, the neutralizing capacities of the sera correlated with the IgG anti–GM-CSF activity in direct ELISA (117). However, in a follow-up study, 40% of the Ab-positive patients had neutralizing Ab, which could be associated with a significant reduction in GM-CSF–induced expansion of leukocytes (165). Increased clearance of the infused GM-CSF and reduced frequencies of GM-CSF–induced side effects such as myalgia, fever, conjunctivitis, headache, and nausea have been reported following development of neutralizing GM-CSF Ab (116,166).

Thus, neutralizing Ab against GM-CSF may be induced by repetitive treatment in nonimmunocompromised patients with recombinant GM-CSF. It is uncertain whether the Ab induced by therapy cross bind to endogenous GM-CSF (116,124,165).

Antibodies to Recombinant G-CSF

There are very few reports on the induction of Ab in patients treated with recombinant G-CSF. Thus, 4 of 11 patients with severe lymphoid malignancies on chemotherapy developed Ab after repetitive therapy with *E. coli*–derived human G-CSF (115). The G-CSF Ab showed a transient appearance in ELISA with low levels detectable up to 8 months after treatment with the cytokine. The induced Ab did not influence the number of circulating neutrophils and hemopoietic recovery of the patients.

ASSAYS FOR CYTOKINE ANTIBODIES

Methodological Problems

Methods used for the detection of Ab to cytokines include antiviral neutralization assays (IFN), other cytokine bioassays, immunometric assays, and various blotting techniques. Serum Ab, however, do not always bind soluble polypeptides in a specific manner or with any appreciable affinity (26,53). Thus, heterophilic Ab may give false-positive results in ELISA consisting of two or more layers of Ab (167), and immunoblotting techniques and immunometric assays may show some degree of specificity even though the binding of Ab to ligand is weak and topographically unassociated with the specific binding sites of the Ab (see Fig. 1). For example, we and others have found that Western blotting and detection in direct ELISA are not suited for detection of IL-1α Ab because of strong, but nonspecific, immunoglobulin binding to IL-1α fixed to solid phases (45,51).

Without previous knowledge of the binding characteristics, competition in these assays by surplus cytokine in solution may also not discriminate between the native conformation, different degrees of denaturation of the cytokines, or other factors as a possible cause of the competitive activity. Although ELISA techniques can be used for screening purposes, verification of specific anticytokine Ab would require ligand binding to the Fab fragments of the Ab, using properly validated cytokine tracers, combined with saturation binding analyses and, at least in the case of natural Ab, demonstration of cross binding to the endogenous cytokines (168).

Another potential problem is that assays for anticytokine Ab often make use of unglycosylated, *E. coli*–derived cytokines. Since most native cytokines are glycosylated, the recombinant forms may possess other or concealed antigenic sites that could jeopardize the interpretation of the assays (see Fig. 1). Furthermore, even though glycosylated forms of some recombinant cytokines are available, their carbohydrate moieties may differ significantly from those of the wild-type cytokines. It is therefore possible that preexisting Ab to certain polysaccharide structures may mimic an anticytokine Ab by binding in vitro to a glycosylated recombinant cytokine, as has been shown in the case of IFN-β in mice and LIF in humans (89,169). These Ab may bind glycosylated cytokines both in direct binding assays, in solution, and in solid-phase assays (see Fig. 1).

When looking for therapy-induced anticytokine Ab, the relevant structural form to use in the assays is of course the drug given to the patient, usually a recombinant cytokine. However, whenever binding to Ab has been detected with a recombinant cytokine, demonstration of cross binding with the natural cytokine is always relevant (28,44,53,81).

THERAPEUTIC USE OF CYTOKINE ANTIBODIES

Passive Immunization

The mechanisms by which pooled IgG influence immunoinflammatory reactions are incompletely understood but may include antigen neutralization, Fc-receptor blockade, attenuation of complement activation, anti-idiotypic interactions, and binding to other molecules involved in immunoinflammatory processes (121,170).

Although it is still unclear whether anticytokine Ab contribute to the mode of action of IgG therapy, recent investigations support this notion. Thus, high-avidity ($K_d \approx 10^{-11}$ M) GM-CSF Ab in pharmaceutically prepared IgG block the binding of native GM-CSF to cellular receptors. Furthermore, the Ab are fully recovered after high-dose IgG therapy with a binding capacity of approximately 20 ng/mL of circulating GM-CSF (39). Similar observations have been obtained for IL-1α Ab in human IgG and, with lower binding capacities, Ab to IL-6 and IFN-α.

The relatively high levels of IFN-α–specific IgG in hyperimmune and normal IgG preparations may be relevant to treatment of a number of immunoinflammatory conditions with high-dose IgG (28). Whether IFN-α Ab in such preparations neutralize the effects of IFN-α in vivo and hence contribute to a positive therapeutic outcome is not yet clear, although highly likely. Hence, IgG preparations containing IFN-α Ab might have untoward effects in patients given IgG to combat infection, particularly if the infection is of viral origin.

Because IL-1α is associated with the cytoplasmic membrane of antigen-presenting cells appears to be particularly important in the triggering of T cells, the administration of IgG containing neutralizing Ab to membrane IL-1α might contribute to prompt immunosuppression in vivo. Also, IgG1 and IgG2 IL-1α Ab could trigger cytotoxic processes directed against both IL-1α–producing and IL-1α–responding cells, with a resultant rapid decrease in the number of circulating T and B cells.

If injection of neutralizing Ab to specific inflammatory cytokines contributes to the mode of action of high-dose IgG in various immunoinflammatory and hematological disorders, therapy with enriched or purified preparations of natural human IgG to selective cytokines might be worthwhile. On the other hand, IgG therapy of patients at risk of contracting infectious diseases, for example, patients with hypo- or agammaglobulinemia, might benefit from the use of pools selectively depleted of neutralizing Ab to, for example, IFN-α or IFN-β.

Active Immunization—Cytokine Vaccination

Several approaches to selective anticytokine therapy have been investigated in the past decade, including blockade of cytokine-receptor interactions by specific heterologous or humanized monoclonal Ab to the cytokines or their receptors, or by recombinant soluble receptors and naturally occurring or constructed receptor antagonists. A drawback with these approaches, and particularly troublesome if there is a need for repeated therapy, is the formation of Ab against the foreign proteins, including those generated by recombinant DNA techniques (6). In this regard, an interesting perspective is vaccination with specific human cytokines or genetically engineered human cytokines with or without biological activity.

Because of individual MHC-related restrictions in antigen presentation and different degrees of tolerance to a cytokine, covalent coupling of the mediator to a T-cell immunogen may be necessary for successful immunization. Thus, we have generated high levels of neutralizing IL-1α Ab in BCG (bacille Calmette-Guérin)–primed mice inoculated with a murine recombinant IL-1α–PPD (purified protein derivative) conjugate (M. Svenson et al. unpublished data). These Ab block receptor binding of both the precursor and mature forms of murine IL-1α and specifically neutralize the bioactivity of recombinant murine IL-1α in vivo. Another approach is vaccination with engineered cytokine analogues. This has recently been investigated in human IL-6 transgenic mice tolerant to and with high circulating levels of human IL-6 (171). The in vivo activity of the cytokine was completely neutralized by Ab induced by immunizing the animals with an engineered human IL-6 receptor antagonist.

CONCLUSIONS

Naturally occurring, specific and high-avidity Ab to IFN-α, IL-1α, IL-6, GM-CSF, and perhaps to other cytokines, interfere with biological and immunometric assays for these cytokines in vitro. These Ab have in some cases been shown to neutralize cytokines in vivo, but they may in some situations function as carriers and, paradoxically, prolong cytokine functions in vivo.

Inappropriate production/function of Ab to cytokines could be pathogenetically involved in infectious and other immunoinflammatory diseases, and the levels of certain anticytokine Ab may predict the outcome of certain diseases; for example, in chronic arthritis.

Anticytokine Ab may also contribute to the anti-inflammatory and immunosuppressive effects of human high-dose IgG therapy. However, although therapy with enriched or purified preparations of natural Ab to cytokines may prove to be useful, IgG therapy of patients with viral diseases or at risk of contracting infectious diseases may benefit from the use of IgG pools selectively depleted of neutralizing Ab; for example, to IFN.

Anticytokine Ab induced during therapy with human recombinant cytokines are of major concern, because their presence may result in loss of response to therapy and perhaps to chronic alterations in immune functions, including host responses to infectious agents and malignant tumors.

ACKNOWLEDGMENT

This work was supported by the Danish Biotechnology Program.

REFERENCES

1. K Bendtzen, M Svenson, V Jønsson, E Hippe. Autoantibodies to cytokines—friends or foes? Immunol Today 11:167–169, 1990.
2. MD Kazatchkine. Natural IgG autoantibodies in the sera of healthy individuals. J Interferon Res 14:165–168, 1994.
3. K Bendtzen, MB Hansen, C Ross, M Svenson. Cytokine autoantibodies. In: JB Peter, Y Shoenfeld, eds. Autoantibodies. Amsterdam: Elsevier, 1996, pp 209–216.
4. MB Hansen. Human cytokine autoantibodies. Characteristics, test procedures and possible physiological and clinical significance. APMIS 104, 59(suppl):1–33, 1996.
5. G Antonelli. In vivo development of antibody to interferons: an update to 1996. J Interferon Cytokine Res 17(suppl 1): S39–S46, 1997.
6. Multiauthors. A review of interferon immunogenicity. Based on a roundtable workshop held in London, United Kingdom, 9 February 1996. J Interferon Cytokine Res 17(suppl 1): S1–S55, 1997.

7. Special issue. Biotherapy 10, 1997.
8. K Bendtzen, MB Hansen, C Ross, M Svenson. High-avidity autoantibodies to cytokines. Immunol Today 19:209–211, 1998.
9. JGM Scharenberg, AGM Stam, BME von Blomberg, GJ Roest, PA Palmer, CR Franks, CJLM Meijer, RJ Scheper. The development of anti-interleukin-2 (IL-2) antibodies in cancer patients treated with recombinant IL-2. Eur J Cancer 30A:1804–1809, 1994.
10. B Klein, H Brailly. Cytokine-binding proteins: stimulating antagonists. Immunol Today 16:216–220, 1995.
11. FA Montero-Julian, B Klein, E Gautherot, H Brailly. Pharmacokinetic study of anti-interleukin-6 (IL-6) therapy with monoclonal antibodies: enhancement of IL-6 clearance by cocktails of anti-IL-6 antibodies. Blood 85:917–924, 1995.
12. MK Spriggs. One step ahead of the game: viral immunomodulatory molecules. Annu Rev Immunol 14:101–130, 1996.
13. KE Mogensen, PH Daubas, I Gresser, D Sereni, B Varet. Patient with circulating antibodies to α-interferon. Lancet 2:1227–1228, 1981.
14. S Panem, IJ Check, D Henriksen, J Vilcek. Antibodies to α-interferon in a patient with systemic lupus erythematosus. J Immunol 129:1–3, 1982.
15. PW Trown, MJ Kramer, RA Dennin, EV Connell, AV Palleroni, J Quesada, JU Gutterman. Antibodies to human leucocyte interferons in cancer patients. Lancet 1:81–87, 1983.
16. S Panem. Antibodies to interferon in man. Interferon 2:175–183, 1984.
17. B Pozzetto, KE Mogensen, MG Tovey, I Gresser. Characteristics of autoantibodies to human interferon in a patient with varicella-zoster disease. J Infect Dis 150:707–713, 1984.
18. O Prümmer, C Seyfarth, A Scherbaum, N Drees, F Porzsolt. Interferon-α antibodies in autoimmune diseases. J Interferon Res 9(suppl 1):S67–S74, 1989.
19. Y Ikeda, G Toda, N Hashimoto, N Umeda, K Miyake, M Yamanaka, K Kurokowa. Naturally occurring anti–interferon-α2a antibodies in patients with acute viral hepatitis. Clin Exp Immunol 85:80–84, 1991.
20. O Prümmer, N Frickhofen, W Digel, H Heimpel, F Porzsolt. Spontaneous interferon-α antibodies in a patient with pure red cell aplasia and recurrent cutaneous carcinomas. Ann Hematol 62:76–80, 1991.
21. O Prümmer, D Bunjes, M Wiesneth, R Arnold, F Porzsolt, H Heimpel. High-titre interferon-α antibodies in a patient with chronic graft-versus-host disease after allogeneic bone marrow transplantation. Bone Marrow Transplant 14: 483–486, 1994.
22. MB Hansen, V Andersen, K Rohde, A Florescu, C Ross, M Svenson, K Bendtzen. Cytokine autoantibodies in rheumatoid arthritis. Scand J Rheumatol 24: 197–203, 1995.
23. O Prümmer, D Zillikens, F Porzsolt. High-titer interferon-α antibodies in a patient with pemphigus foliaceus. Exp Dermatol 5:213–217, 1996.
24. A Meager, A Vincent, J Newsom-Davis, N Willcox. Spontaneous neutralising

antibodies to interferon-α and interleukin-12 in thymoma-associated autoimmune disease. Lancet 350:1596–1597, 1997.
25. C Ross, MB Hansen, T Schyberg, K Berg. Autoantibodies to crude human leucocyte interferon (IFN), native human IFN, recombinant human IFN-α2b and human IFN-γ in healthy blood donors. Clin Exp Immunol 82:57–62, 1990.
26. MB Hansen, M Svenson, K Bendtzen. Serum-induced suppression of interferon (IFN) activity. Lack of evidence for the presence of specific autoantibodies to IFN-α in normal human sera. Clin Exp Immunol 88:559–562, 1992.
27. LM Thurmond, MJ Reese. Immunochemical characterization of human antibodies to lymphoblastoid interferon. Clin Exp Immunol 86:514–519, 1991.
28. C Ross, M Svenson, MB Hansen, GL Vejlsgaard, K Bendtzen. High avidity IFN-neutralizing antibodies in pharmaceutically prepared human IgG. J Clin Invest 95:1974–1978, 1995.
29. A Caruso, C Bonfanti, D Colombrita, M de Francesco, C de Rango, I Foresti, F Gargiulo, R Gonzales, G Gribaudo, S Landolfo, N Manca, G Manni, F Pirali, P Pollara, G Ravizzola, G Scura, L Terlenghi, E Viani, A Turano. Natural antibodies to IFN-γ in man and their increase during viral infection. J Immunol 144:685–690, 1990.
30. A Caruso, I Foresti, G Gribaudo, C Bonfanti, P Pollara, A Dolei, oS Landolf, A Turano. Anti–interferon-γ antibodies in sera from HIV infected patients. J Biol Reg Homeost Agents 3:8–12, 1989.
31. A Caruso, A Turano. Natural antibodies to interferon-gamma. Biotherapy 10: 29–37, 1997.
32. RA Elkarim, M Mustafa, P Kivisakk, H Link, M Bakhiet. Cytokine autoantibodies in multiple sclerosis, aseptic meningitis and stroke. Eur J Clin Invest 28:295–299, 1998.
33. RA Elkarim, C Dahle, M Mustafa, R Press, LP Zou, C Ekerfelt, J Ernerudh, H Link, M Bakhiet. Recovery from Guillain-Barré syndrome is associated with increased levels of neutralizing autoantibodies to interferon-γ. Clin Immunol Immunopathol 88:241–248, 1998.
34. KL Bost, BH Hahn, MS Saag, GM Shaw, DA Weigent, JE Blalock. Individuals infected with HIV possess antibodies against IL-2. Immunology 65:611–615, 1988.
35. E Monti, A Pozzi, L Tiberio, D Morelli, A Caruso, ML Villa, A Balsari. Purification of interleukin-2 antibodies from healthy individuals. Immunol Lett 36: 261–266, 1993.
36. L Tiberio, A Caruso, A Pozzi, L Rivoltini, D Morelli, E Monti, A Balsari. The detection and biological activity of human antibodies to IL-2 in normal donors. Scand J Immunol 38:472–476, 1993.
37. JL Trotter, CA Damico, AL Trotter, KG Collins, AH Cross. Interleukin-2 binding proteins in sera from normal subjects and multiple sclerosis patients. Neurology 45:1971–1974, 1995.
38. M Allegretta, MB Atkins, RA Dempsey, EC Bradley, MW Konrad, A Childs, SN Wolfe, JW Mier. The development of anti-interleukin-2 antibodies in pa-

tients treated with recombinant human interleukin-2 (IL-2). J Clin Immunol 6: 481–490, 1986.
39. M Svenson, MB Hansen, C Ross, M Diamant, K Rieneck, H Nielsen, K Bendtzen. Antibody to granulocyte-macrophage colony-stimulating factor is a dominant anti-cytokine activity in human IgG preparations. Blood 91:2054–2061, 1998.
40. RP Whitehead, D Ward, L Hemingway, GPd Hemstreet, E Bradley, M Konrad. Subcutaneous recombinant interleukin 2 in a dose escalating regimen in patients with metastatic renal cell adenocarcinoma. Cancer Res 50:6708–6715, 1990.
41. J Atzpodien, A Korfer, CR Franks, H Poliwoda, H Kirchner. Home therapy with recombinant interleukin-2 and interferon-α2b in advanced human malignancies. Lancet 335:1509–1512, 1990.
42. M Svenson, LK Poulsen, A Fomsgaard, K Bendtzen. IgG autoantibodies against interleukin 1α in sera of normal individuals. Scand J Immunol 29:489–492, 1989.
43. H Suzuki, T Akama, M Okane, I Kono, Y Matsui, K Yamane, H Kashiwagi. Interleukin-1–inhibitory IgG in sera from some patients with rheumatoid arthritis. Arthritis Rheum 32:1528–1538, 1989.
44. M Svenson, MB Hansen, K Bendtzen. Distribution and characterization of autoantibodies to interleukin 1α in normal human sera. Scand J Immunol 32:695–701, 1990.
45. H Suzuki, J Kamimura, T Ayabe, H Kashiwagi. Demonstration of neutralizing autoantibodies against IL-1α in sera from patients with rheumatoid arthritis. J Immunol 145:2140–2146, 1990.
46. P Galley, J-P Mach, S Carrel. Characterization and detection of naturally occurring antibodies against IL-1α and IL-1β in normal human plasma. Eur Cytokine Netw 2:329–338, 1991.
47. MB Hansen, M Svenson, K Bendtzen. Human anti-interleukin 1α antibodies. Immunol Lett 30:133–140, 1991.
48. N Mae, DJ Liberato, R Chizzonite, H Satoh. Identification of high-affinity anti-IL-1α autoantibodies in normal human serum as an interfering substance in a sensitive enzyme-linked immunosorbent assay for IL-1α. Lymphokine Cytokine Res 10:61–68, 1991.
49. J-H Saurat, J Schifferli, G Steiger, J-M Dayer, L Didierjean. Anti–interleukin-1α autoantibodies in humans: characterization, isotype distribution, and receptor-binding inhibition—higher frequency in Schnitzler's syndrome (urticaria and macroglobulinemia). J Allergy Clin Immunol 88:244–256, 1991.
50. G Sunder-Plassmann, PL Sedlacek, R Sunder-Plassmann, K Derfler, K Swoboda, V Fabrizii, MM Hirschl, P Balcke. Anti–interleukin-1α autoantibodies in hemodialysis patients. Kidney Int 40:787–791, 1991.
51. H Suzuki, T Ayabe, J Kamimura, H Kashiwagi. Anti-IL-1α autoantibodies in patients with rheumatic diseases and in healthy subjects. Clin Exp Immunol 85:407–412, 1991.
52. M Svenson, MB Hansen, L Kayser, ÅK Rasmussen, CM Reimert, K Bendtzen.

Effects of human anti–IL-1α autoantibodies on receptor binding and biological activities of IL-1. Cytokine 4:125–133, 1992.
53. M Svenson, MB Hansen, K Bendtzen. Binding of cytokines to pharmaceutically prepared human immunoglobulin. J Clin Invest 92:2533–2539, 1993.
54. MB Hansen, M Svenson, K Abell, K Varming, HP Nielsen, A Bertelsen, K Bendtzen. Sex- and age-dependency of IgG auto-antibodies against IL-1α in healthy humans. Eur J Clin Invest 24:212–218, 1994.
55. H Satoh, R Chizzonite, C Ostrowski, G Ni-Wu, H Kim, B Fayer, N Mae, R Nadeau, DJ Liberato. Characterization of anti–IL-1α autoantibodies in the sera from healthy humans. Immunopharmacology 27: 107–118, 1994.
56. JS van der Zee, P van Swieten, RC Aalberse. Serologic aspects of IgG4 antibodies. II. IgG4 antibodies form small, nonprecipitating immune complexes due to functional monovalency. J Immunol 137:3566–3571, 1986.
57. JS van der Zee, P van Swieten, RC Aalberse. Inhibition of complement activation by IgG4 antibodies. Clin Exp Immunol 64:415–422, 1986.
58. PW Parren, PA Warmerdam, LC Boeije, J Arts, NA Westerdaal, A Vlug, PJ Capel, LA Aarden, JG van de Winkel. On the interaction of IgG subclasses with the low affinity FcγRIIa (CD32) on human monocytes, neutrophils, and platelets. Analysis of a functional polymorphism to human IgG_2. J Clin Invest 90:1537–1546, 1992.
59. PA Warmerdam, IE van den Herik-Oudijk, PW Parren, NA Westerdaal, JG van de Winkel, PJ Capel. Interaction of a human FcγRIIb1 (CD32) isoform with murine and human IgG subclasses. Int Immunol 5:239–247, 1993.
60. SK Dower, JE Sims, DP Cerretti, TA Bird. The interleukin-1 system: receptors, ligands and signals. Chem Immunol 51:33–64, 1992.
61. P Jouvenne, F Fossiez, P Garrone, O Djossou, J Banchereau, P Miossec. Increased incidence of neutralizing autoantibodies against interleukin-1α (IL-1α) in nondestructive chronic polyarthritis. J Clin Immunol 16:283–290, 1996.
62. P Garrone, O Djossou, F Fossiez, J Reyes, S Ait-Yahia, C Maat, S Ho, T Hauser, JM Dayer, J Greffe, P Miossec, S Lebecque, F Rousset, J Banchereau. Generation and characterization of a human monoclonal autoantibody that acts as a high affinity interleukin-1α specific inhibitor. Mol Immunol 33:649–658, 1996.
63. A Ferrante, LJ Beard, RG Feldman. IgG subclass distribution of antibodies to bacterial and viral antigens. Pediatr Infect Dis J 9:S16–24, 1990.
64. Y Itoh, T Okanoue, S Sakamoto, K Nishioji, K Kashima, Y Ohmoto. Serum autoantibody against interleukin-1α is unrelated to the etiology or activity of liver disease but can be raised by interferon treatment. Am J Gastroenterol 90: 777–782, 1995.
65. P Jouvenne, F Fossiez, J Banchereau, P Miossec. High levels of neutralizing autoantibodies against IL-1α are associated with a better prognosis in chronic polyarthritis: a follow-up study. Scand J Immunol 46:413–418, 1997.
66. C Homann, MB Hansen, N Graudal, P Hasselqvist, M Svenson, K Bendtzen, AC Thomsen, P Garred. Anti–interleukin-6 autoantibodies in plasma

are associated with an increased frequency of infections and increased mortality of patients with alcoholic cirrhosis. Scand J Immunol 44:623–629, 1996.

67. K Bendtzen, MB Hansen, C Ross, LK Poulsen, M Svenson. Cytokines and autoantibodies to cytokines. Stem Cells 13:206–222, 1995.
68. A Mantovani, M Muzio, P Ghezzi, C Colotta, M Introna. Regulation of inhibitory pathways of the interleukin-1 system. Ann NY Acad Sci 840:338–351, 1998.
69. M Svenson, MB Hansen, P Heegaard, K Abell, K Bendtzen. Specific binding of interleukin 1 (IL-1)β and IL-1 receptor antagonist (IL-1ra) to human serum. High-affinity binding of IL-1ra to soluble IL-1 receptor type I. Cytokine 5:427–435, 1993.
70. M Svenson, S Nedergaard, PMH Heegaard, TD Whisenand, WP Arend, K Bendtzen. Differential binding of human interleukin-1 (IL-1) receptor antagonist to natural and recombinant soluble and cellular IL-1 type I receptors. Eur J Immunol 25:2842–2850, 1995.
71. A Fomsgaard, M Svenson, K Bendtzen. Auto-antibodies to tumour necrosis factor α in healthy humans and patients with inflammatory diseases and Gram-negative bacterial infections. Scand J Immunol 30:219–223, 1989.
72. EWB Jeffes, EK Ininns, KL Schmitz, RS Yamamoto, CA Dett, GA Granger. The presence of antibodies to lymphotoxin and tumor necrosis factor in normal serum. Arthritis Rheum 32:1148–1152, 1989.
73. M Sioud, A Dybwad, L Jespersen, S Suleyman, JB Natvig, Ø Førre. Characterization of naturally occurring autoantibodies against tumour necrosis factor-α (TNF-α): in vitro function and precise epitope mapping by phage epitope library. Clin Exp Immunol 98:520–525, 1994.
74. L Tsuchiyama, T Wong, J Kieran, P Boyle, D Penza, GD Wetzel. Comparison of anti-TNFα autoantibodies in plasma and from EBV transformed lymphocytes of autoimmune and normal individuals. Hum Antibodies Hybridomas 6: 73–76, 1995.
75. D Caccavo, O Leri, GM Ferri, P Perinelli, D de Luca, A Afeltra. Anti–TNF-α antibodies are not associated with Helicobacter pylori induced gastritis. Eur Rev Med Pharmacol Sci 1:111–113, 1997.
76. H-G Leusch, G Sitzler, S Markos-Pusztai. Failure to demonstrate TNFα-specific autoantibodies in human sera by ELISA and Western blot. J Immunol Methods 139:145–147, 1991.
77. P Boyle, KJ Lembach, GD Wetzel. I. A novel monoclonal human IgM autoantibody which binds recombinant human and mouse tumor necrosis factor-α. Cell Immunol 152:556–568, 1993.
78. P Boyle, KJ Lembach, GD Wetzel. II. The B5 monoclonal human autoantibody binds to cell surface TNFα on human lymphoid cells and cell lines and appears to recognize a novel epitope. Cell Immunol 152:569–581, 1993.
79. MB Hansen, M Svenson, M Diamant, K Bendtzen. Anti–interleukin-6 antibodies in normal human serum. Scand J Immunol 33:777–781, 1991.

80. H Takemura, H Suzuki, K Yoshizaki, A Ogata, T Yuhara, T Akama, K Yamane, H Kashiwagi. Anti–interleukin-6 autoantibodies in rheumatic diseases. Increased frequency in the sera of patients with systemic sclerosis. Arthritis Rheum 35:940–943, 1992.
81. MB Hansen, M Svenson, M Diamant, K Bendtzen. High-affinity IgG autoantibodies to IL-6 in sera of normal individuals are competitive inhibitors of IL-6 in vitro. Cytokine 5:72–80, 1993.
82. M Diamant, MB Hansen, K Rieneck, M Svenson, K Yasukawa, K Bendtzen. Stimulation of the B9 hybridoma cell line by soluble interleukin-6 receptors. J Immunol Methods 173:229–235, 1994.
83. H Suzuki, H Takemura, K Yoshizaki, Y Koishihara, Y Ohsugi, A Okano, Y Akiyama, T Tojo, T Kishimoto, H Kashiwagi. IL-6-anti–IL-6 autoantibody complexes with IL-6 activity in sera from some patients with systemic sclerosis. J Immunol 152:935–942, 1994.
84. MB Hansen, M Svenson, K Abell, K Yasukawa, M Diamant, K Bendtzen. Influence of interleukin-6 (IL-6) autoantibodies on IL-6 binding to cellular receptors. Eur J Immunol 25:348–354, 1995.
85. MB Hansen, M Svenson, M Diamant, K Abell, K Bendtzen. Interleukin-6 autoantibodies: possible biological and clinical significance. Leukemia 9:1113–1115, 1995.
86. G Sunder-Plassmann, S Kapiotis, C Gasche, U Klaar. Functional characterization of cytokine autoantibodies in chronic renal failure patients. Kidney Int 45: 1484–1488, 1994.
87. J Bravo, D Staunton, JK Heath, EY Jones. Crystal structure of a cytokine-binding region of gp130. EMBO J 17:1665–1674, 1998.
88. JP Gaillard, R Bataille, H Brailly, C Zuber, K Yasukawa, M Attal, N Maruo, T Taga, T Kishimoto, B Klein. Increased and highly stable levels of functional soluble interleukin-6 receptor in sera of patients with monoclonal gammopathy. Eur J Immunol 23:820–824, 1993.
89. MB Hansen, M Svenson, M Diamant, C Ross, K Bendtzen. Interleukin-6 (IL-6) autoantibodies and blood IL-6 measurements. Blood 85:1145, 1995.
90. R Aston, WB Cowden, GL Ada. Antibody-mediated enhancement of hormone activity. Mol Immunol 26:435–446, 1989.
91. M Mihara, Y Koishihara, H Fukui, K Yasukawa, Y Oshugi. Murine anti-human IL-6 monoclonal antibody prolongs the half-life in circulating blood and thus prolongs the bioactivity of human IL-6 in mice. Immunology 74:55–59, 1991.
92. Z-Y Lu, J Brochier, J Wijdenes, H Brailly, R Bataille, B Klein. High amounts of circulating interleukin (IL)-6 in the form of monomeric immune complexes during anti–IL-6 therapy. Towards a new methodology for measuring overall cytokine production in human in vivo. Eur J Immunol 22:2819–2824, 1992.
93. E Martens, C Dillen, W Put, H Heremans, J van Damme, A Billiau. Increased circulating interleukin-6 (IL-6) activity in endotoxin-challenged mice pretreated with anti–IL-6 antibody is due to IL-6 accumulated in antigen-antibody complexes. Eur J Immunol 23:2026–2029, 1993.

94. LT May, R Neta, LL Moldawer, JS Kenney, K Patel, PB Sehgal. Antibodies chaperone circulating IL-6: paradoxical effects of anti-IL-6 ''neutralizing'' antibodies in vivo. J Immunol 151:3225–3236, 1993.
95. LT May, H Viguet, JS Kenney, N Ida, AC Allison, PB Sehgal. High levels of ''complexed'' interleukin-6 in human blood. J Biol Chem 267:19698–19704, 1992.
96. E Dicou, V Nerriere, V Labropoulou. Naturally occurring antibodies against nerve growth factor in human and rabbit sera: comparison between control and herpes simplex virus–infected patients. J Neuroimmunol 34:153–158, 1991.
97. E Dicou, D Hurez, V Nerriere. Natural autoantibodies against the nerve growth factor in autoimmune diseases. J Neuroimmunol 47:159–167, 1993.
98. MM Zanone, JP Banga, M Peakman, M Edmonds, PJ Watkins. An investigation of antibodies to nerve growth factor in diabetic autonomic neuropathy. Diabet Med 11:378–383, 1994.
99. E Dicou, P Vermersch, I Penisson-Besnier, F Dubas, V Nerrière. Anti-NGF autoantibodies and NGF in sera of Alzheimer patients and in normal subjects in relation to age. Autoimmunity 26:189–194, 1997.
100. E Dicou, V Nerriere. Evidence that natural autoantibodies against the nerve growth factor (NGF) may be potential carriers of NGF. J Neuroimmunol 75: 200–203, 1997.
101. P Peichl, M Ceska, H Broell, F Effenberger, IJD Lindley. Human neutrophil activating peptide/interleukin 8 acts as an autoantigen in rheumatoid arthritis. Ann Rheum Dis 51:19–22, 1992.
102. I Sylvester, T Yoshimura, M Sticherling, JM Schröder, M Ceska, P Peichl, EJ Leonard. Neutrophil attractant protein-1–immunoglobulin G immune complexes and free anti–NAP-1 antibody in normal human serum. J Clin Invest 90:471–481, 1992.
103. JE Crabtree, P Peichl, JI Wyatt, U Stachl, IJD Lindley. Gastric interleukin-8 and IgA IL-8 autoantibodies in Helicobacter pylori infection. Scand J Immunol 37:65–70, 1993.
104. S Reitamo, A Remitz, J Varga, M Ceska, F Effenberger, S Jimenez, J Uitto. Demonstration of interleukin 8 and autoantibodies to interleukin 8 in the serum of patients with systemic sclerosis and related disorders. Arch Dermatol 129: 189–193, 1993.
105. I Sylvester, AF Suffredini, AJ Boujoukos, GD Martich, RL Danner, T Yoshimura, EJ Leonard. Neutrophil attractant protein-1 and monocyte chemoattractant protein-1 in human serum. Effects of intravenous lipopolysaccharide on free attractants, specific IgG autoantibodies and immune complexes. J Immunol 151:3292–3298, 1993.
106. J Amiral, A Marfaing-Koka, M Wolf, MC Alessi, B Tardy, C Boyer-Neumann, AM Vissac, E Fressinaud, M Poncz, D Meyer. Presence of autoantibodies to interleukin-8 or neutrophil-activating peptide-2 in patients with heparin-associated thrombocytopenia. Blood 88:410–416, 1996.
107. A Kurdowska, EJ Miller, JM Noble, RP Baughman, MA Matthay, WG Brels-

ford, AB Cohen. Anti-IL-8 autoantibodies in alveolar fluid from patients with the adult respiratory distress syndrome. J Immunol 157:2699–2706, 1996.
108. EJ Leonard. Plasma chemokine and chemokine-autoantibody complexes in health and disease. Methods 10:150–157, 1996.
109. KN Lai, JK Shute, IJ Lindley, FM Lai, AW Yu, PK Li, CK Lai. Neutrophil attractant protein-1 interleukin 8 and its autoantibodies in IgA nephropathy. Clin Immunol Immunopathol 80:47–54, 1996.
110. S Kissler, C Susal, G Opelz. Anti-MIP-1α and anti-RANTES antibodies: new allies of HIV-1? Clin Immunol Immunopathol 84:338–341, 1997.
111. JK Shute, B Vrugt, IJ Lindley, ST Holgate, A Bron, R Aalbers, R Djukanovic. Free and complexed interleukin-8 in blood and bronchial mucosa in asthma. Am J Respir Crit Care Med 155:1877–1883, 1997.
112. K Bendtzen, MB Hansen, M Diamant, C Ross, M Svenson. Naturally occurring autoantibodies to interleukin-1α, interleukin-6, interleukin-10 and interferon-α. J Interferon Res 14:157–158, 1994.
113. C Menetrier-Caux, F Briere, P Jouvenne, E Peyron, F Peyron, J Banchereau. Identification of human IgG autoantibodies specific for IL-10. Clin Exp Immunol 104:173–179, 1996.
114. JE Tanner, F Diaz-Mitoma, CM Rooney, C Alfieri. Anti-interleukin-10 antibodies in patients with chronic active Epstein-Barr virus infection. J Infect Dis 176: 1454–1461, 1997.
115. L Laricchia-Robbio, S Moscato, A Genua, AM Liberati, RP Revoltella. Naturally occurring and therapy-induced antibodies to human granulocyte colony-stimulating factor (G-CSF) in human serum. J Cell Physiol 173:219–226, 1997.
116. P Ragnhammar, HJ Friesen, JE Frödin, AK Lefvert, M Hassan, A Osterborg, H Mellstedt. Induction of anti-recombinant human granulocyte-macrophage colony-stimulating factor (Escherichia coli–derived) antibodies and clinical effects in nonimmunocompromised patients. Blood 84:4078–4087, 1994.
117. RP Revoltella, L Laricchia-Robbio, S Moscato, A Genua, AM Liberati. Natural and therapy-induced anti–GM-CSF and anti–G-CSF antibodies in human serum. Leuk Lymphoma 26(suppl 1):29–34, 1997.
118. A Meager. Natural autoantibodies to interferons. J Interferon Cytokine Res 17(suppl 1):S51–S53, 1997.
119. M Pedersen, H Permin, C Bindslev-Jensen, K Bendtzen, S Norn. Cytokine-induced histamine release from basophils of AIDS patients. Interaction between cytokines and specific IgE antibodies. Allergy 46:129–134, 1991.
120. B Heilig, C Fiehn, M Brockhaus, H Gallati, A Pezzutto, W Hunstein. Evaluation of soluble tumor necrosis factor (TNF) receptors and TNF receptor antibodies in patients with systemic lupus erythematodes, progressive systemic sclerosis, and mixed connective tissue disease. J Clin Immunol 13:321–328, 1993.
121. JM Dwyer. Manipulating the immune system with immune globulin. N Engl J Med 326:107–116, 1992.
122. M Toyoda, X Zhang, A Petrosian, OA Galera, S-J Wang, SC Jordan. Modula-

tion of immunoglobulin production and cytokine mRNA expression in peripheral blood mononuclear cells by intravenous immunoglobulin. J Clin Immunol 14:178–189, 1994.

123. A Vallbracht, J Treuner, B Flehmig, KE Joester, D Niethammer. Interferon-neutralizing antibodies in a patient treated with human fibroblast interferon. Nature 289:496–497, 1981.
124. JG Gribben, S Devereux, NSB Thomas, M Keim, HM Jones, AH Goldstone, DC Linch. Development of antibodies to unprotected glycosylation sites on recombinant human GM-CSF. Lancet 335:434–437, 1990.
125. GG Steinmann, F Rosenkaimer, G Leitz. Clinical experiences with interferon-α and interferon-γ. Int Rev Exp Pathol 34 Pt B:193–207, 1993.
126. G Giannelli, G Antonelli, G Fera, S Del Vecchio, E Riva, C Broccia, O Schiraldi, F Dianzani. Biological and clinical significance of neutralizing and binding antibodies to interferon-α (IFN-α) during therapy for chronic hepatitis C. Clin Exp Immunol 97:4–9, 1994.
127. P von Wussow, R Hehlmann, T Hochhaus, D Jakschies, KU Nolte, O Prümmer, H Ansari, J Hasford, H Heimpel, H Deicher. Roferon (rIFN-α2a) is more immunogenic than intron A (rIFN-α2b) in patients with chronic mylogenous leukemia. J Interferon Res 14:217–219, 1994.
128. G Antonelli, M Currenti, O Turriziani, F Dianzani. Neutralizing antibodies to interferon-α: relative frequency in patients treated with different interferon preparations. J Infect Dis 163:882–885, 1991.
129. V Bocci. What roles have anti-interferon antibodies in physiology and pathology? Res Clin Lab 21:79–84, 1991.
130. S Baron, DH Coppenhaver, F Dianzani, WR Fleischmann Jr, TK Hughes Jr, GR Klimpel, DW Niesel, GJ Stanton, SK Tyring. Interferon: principles and medical applications. Galveston: University of Texas Medical Branch, 1992.
131. Special issue. J Interferon Res 9(suppl 1):S9–S74, 1989.
132. RG Steis, JW Smith, WJ Urba, JW Clark, LM Itri, LM Evans, C Schoenberger, DL Longo. Resistance to recombinant interferon alfa-2a in hairy-cell leukemia associated with neutralizing anti-interferon antibodies. N Engl J Med 318: 1409–1413, 1988.
133. K Öberg, GV Alm. Development of neutralizing interferon antibodies after treatment with recombinant interferon-α_{2b} in patients with malignant carcinoid tumors. J Interferon Res 9(suppl 1):S45–S49, 1989.
134. M Freund, P von Wussow, J Knüver-Hopf, H Mohr, U Pohl, G Exeriede, H Link, HJ Wilke, H Poliwoda. Treatment with natural human interferon α of a CML-patient with antibodies to recombinant interferon α-2b. Blut 57:311–315, 1988.
135. V Leroy, M Baud, C de Traversay, M Maynard-Muet, P Lebon, J-P Zarski. Role of anti-interferon antibodies in breakthrough occurrence during α2a and 2b therapy in patients with chronic hepatitis C. J Hepatol 28:375–381, 1998.
136. KU Nolte, G Gunther, P von Wussow. Epitopes recognized by neutralizing

therapy-induced human anti–interferon-α antibodies are localized within the N-terminal functional domain of recombinant interferon-α2. Eur J Immunol 26:2155–2159, 1996.

137. PV von Wussow, D Jakschies, M Freund, R Hehlmann, F Brockhaus, H Hochkeppel, M Horisberger, H Deicher. Treatment of anti–recombinant interferon-α2 antibody positive CML patients with natural interferon-α. Br J Haematol 78:210–216, 1991.

138. M Milella, G Antonelli, T Santantonio, M Currenti, L Monno, N Mariano, G Angarano, F Dianzani, G Pastore. Neutralizing antibodies to recombinant α-interferon and response to therapy in chronic hepatitis C virus infection. Liver 13:146–150, 1993.

139. A Braun, L Kwee, MA Labow, J Alsenz. Protein aggregates seem to play a key role among the parameters influencing the antigenicity of interferon α (IFN-α) in normal and transgenic mice. Pharm Res 14:1472–1478, 1997.

140. C Ross, M Svenson, B Windelborg, MB Hansen, K Thestrup-Pedersen, GL Vejlsgaard. Characterization of anti-IFN antibodies in patients with atopic dermatitis during rIFN-α2A therapy. Arch Dermatol Res 2000. (In press.)

141. R Dummer, W Müller, F Nestle, J Wiede, J Dues, I Haubitz, W Wolf, E Bill, G Burg. Formation of neutralizing antibodies against natural interferon-β, but not against recombinant interferon-γ during adjuvant therapy for high-risk malignant melanoma patients. Cancer 67:2300–2304, 1991.

142. The IFNB Multiple Sclerosis Study Group and The University of British Columbia MS/MRI Analysis Group. Interferon β-1b in the treatment of multiple sclerosis: final outcome of the randomized controlled trial. Neurology 45:1277–1285, 1995.

143. L Runkel, W Meier, RB Pepinsky, M Karpusas, A Whitty, K Kimball, M Brickelmaier, C Muldowney, W Jones, SE Goelz. Structural and functional differences between glycosylated and non-glycosylated forms of human interferon-β (IFN-β). Pharm Res 15:641–649, 1998.

144. The IFNB Multiple Sclerosis Study Group and the University of British Columbia MS/MRI Analysis Group. Neutralizing antibodies during treatment of multiple sclerosis with interferon β-1b: experience during the first three years. Neurology 47:889–894, 1996.

145. P Kivisäkk, GV Alm, WZ Tian, D Matusevicius, S Fredrikson, H Link. Neutralising and binding anti–interferon-β-1 (IFN-β-1b) antibodies during IFN-β-1b treatment of multiple sclerosis. Mult Scler 3:184–190, 1997.

146. G Antonelli, F Bagnato, C Pozzilli, E Simeoni, S Bastianelli, M Currenti, F de Pisa, C Fieschi, C Gasperini, M Salvetti, F Dianzani. Development of neutralizing antibodies in patients with relapsing-remitting multiple sclerosis treated with IFN-β1a: J Interferon Cytokine Res 18:345–350, 1998.

147. C Fiehn, O Prummer, H Gallati, B Heilig, W Hunstein. Treatment of systemic mastocytosis with interferon-γ: failure after appearance of anti–IFN-γ antibodies. Eur J Clin Invest 25:615–618, 1995.

148. O Prümmer, C Fiehn, H Gallati. Anti-interferon-γ antibodies in a patient under-

going interferon-γ treatment for systemic mastocytosis. J Interferon Cytokine Res 16:519–522, 1996.
149. R Whittington, D Faulds. Interleukin-2: a review of its pharmacological properties and therapeutic use in patients with cancer. Drugs 46:446–514, 1993.
150. O Prümmer. Treatment-induced antibodies to interleukin-2. Biotherapy 10:15–24, 1997.
151. EL Hänninen, A Korfer, M Hadam, C Schneekloth, I Dallmann, T Menzel, H Kirchner, H Poliwoda, J Atzpodien. Biological monitoring of low-dose interleukin 2 in humans: soluble interleukin 2 receptors, cytokines, and cell surface phenotypes. Cancer Res 51:6312–6316, 1991.
152. H Kirchner, A Korfer, P Evers, MM Szamel, J Knuver-Hopf, H Mohr, CR Franks, U Pohl, K Resch, M Hadam, H Poliwoda, J Atzpodien. The development of neutralizing antibodies in a patient receiving subcutaneous recombinant and natural interleukin-2. Cancer 67:1862–1864, 1991.
153. JE Kolitz, GY Wong, K Welte, VJ Merluzzi, A Engert, T Bialas, A Polivka, EC Bradley, M Konrad, C Gnecco, et al. Phase I trial of recombinant interleukin-2 and cyclophosphamide: augmentation of cellular immunity and T-cell mitogenic response with long-term administration of rIL-2. J Biol Response Mod 7:457–472, 1988.
154. GI Kirchner, A Franzke, J Buer, W Beil, M Probst-Kepper, F Wittke, K Overmann, S Lassmann, R Hoffmann, H Kirchner, A Ganser, J Atzpodien. Pharmacokinetics of recombinant human interleukin-2 in advanced renal cell carcinoma patients following subcutaneous application. Br J Clin Pharmacol 46:5–10, 1998.
155. NV Katre. Immunogenicity of recombinant IL-2 modified by covalent attachment of polyethylene glycol. J Immunol 144:209–213, 1990.
156. H Teppler, G Kaplan, KA Smith, AL Montana, P Meyn, ZA Cohn. Prolonged immunostimulatory effect of low-dose polyethylene glycol interleukin 2 in patients with human immunodeficiency virus type 1 infection. J Exp Med 177: 483–492, 1993.
157. ZP Bernstein, MM Porter, M Gould, B Lipman, EM Bluman, CC Stewart, RG Hewitt, G Fyfe, B Poiesz, MA Caligiuri. Prolonged administration of low-dose interleukin-2 in human immunodeficiency virus-associated malignancy results in selective expansion of innate immune effectors without significant clinical toxicity. Blood 86:3287–3294, 1995.
158. I Quirt, S Micucci, LA Moran, J Pater, G Browman. Erythropoietin in the management of patients with nonhematologic cancer receiving chemotherapy. Systemic Treatment Program Committee. Cancer Prev Control 1:241–248, 1997.
159. B Sowade, O Sowade, J Mocks, W Franke, H Warnke. The safety of treatment with recombinant human erythropoietin in clinical use: a review of controlled studies. Int J Mol Med 1:303–314, 1998.
160. PE Tarr. Granulocyte-macrophage colony-stimulating factor and the immune system. Med Oncol 13:133–140, 1996.
161. A Samanci, Q Yi, J Fagerberg, K Strigard, G Smith, U Ruden, B Wahren, H

Mellstedt. Pharmacological administration of granulocyte/macrophage-colony-stimulating factor is of significant importance for the induction of a strong humoral and cellular response in patients immunized with recombinant carcinoembryonic antigen. Cancer Immunol Immunother 47:131–142, 1998.
162. GC Avanzi, M Gallicchio, G Saglio. Hematopoietic growth factors in autologous transplantation. Biotherapy 10:299–308, 1998.
163. EM Johnston, J Crawford. Hematopoietic growth factors in the reduction of chemotherapeutic toxicity. Semin Oncol 25:552–561, 1998.
164. SS Prabhakar, T Muhlfelder. Antibodies to recombinant human erythropoietin causing pure red cell aplasia. Clin Nephrol 47:331–335, 1997.
165. M Wadhwa, C Bird, J Fagerberg, R Gaines-Das, P Ragnhammar, H Mellstedt, R Thorpe. Production of neutralizing granulocyte-macrophage colony-stimulating factor (GM-CSF) antibodies in carcinoma patients following GM-CSF combination therapy. Clin Exp Immunol 104:351–358, 1996.
166. P Ragnhammar, M Wadhwa. Neutralising antibodies to granulocyte-macrophage colony stimulating factor (GM-CSF) in carcinoma patients following GM-CSF combination therapy. Med Oncol 13:161–166, 1996.
167. J Grassi, CJ Roberge, Y Frobert, P Pradelles, PE Poubelle. Determination of IL 1α, IL1β and IL2 in biological media using specific enzyme immunometric assays. Immunol Rev 119:125–145, 1991.
168. M Svenson, P Herbrink. Measurement of cytokine antibodies. Test development. Biotherapy 10:87–92, 1997.
169. JJ Sedmak, SE Grossberg. High levels of circulating neutralizing antibody in normal animals to recombinant mouse interferon-β produced in yeast. J Interferon Res 9(suppl 1):S61–S65, 1989.
170. MD Kazatchkine, G Dietrich, V Hurez, N Ronda, B Bellon, F Rossi, SV Kaveri. V region–mediated selection of autoreactive repertoires by intravenous immunoglobulin (i.v.Ig). Immunol Rev 139:79–107, 1994.
171. L Ciapponi, D Maione, A Scoumanne, P Costa, MB Hansen, M Svenson, K Bendtzen, T Alonzi, G Paonessa, R Cortese, G Ciliberto, R Savino. Induction of interleukin-6 (IL-6) autoantibodies through vaccination with an engineered IL-6 receptor antagonist. Nature Biotechnol 15:997–1001, 1997.

4
Anti–Tumor Necrosis Factor Therapies in Crohn's Disease and Rheumatoid Arthritis

Elena G. Hitraya and Thomas F. Schaible
Centocor Inc., Malvern, Pennsylvania

INTRODUCTION

Cytokines are small intercellular signaling molecules that have emerged as important targets for disease management. Tumor necrosis factor-α (TNF-α) is a potent proinflammatory cytokine. This chapter will review the biology of TNF-α, focusing on its role in chronic inflammatory diseases such as Crohn's disease and rheumatoid arthritis, and also review anticytokine therapies targeted to neutralize TNF-α with emphasis on clinical studies conducted with a monoclonal antibody against TNF-α, infliximab. Published results of clinical trials of other TNF-α–neutralizing agents will also be summarized.

TNF-α

Cytokines are protein mediators, or signaling molecules, that are produced by the cells of the activated immune system. Among these mediators are proinflammatory cytokines including interferons, interleukins, and TNF-α. TNF-α is one of the most important and potent of the inflammatory mediators.

The biological activity of TNF-α was first observed over 100 years ago when physicians noted that, in some cases, cancer patients experienced shrinkage of certain tumors if they also had a serious bacterial infection. In the 1890s, William Coley reported a therapeutic benefit derived from treating patients with inoperable neoplastic disease with repeated inoculations of toxins prepared from gram-positive and gram-negative bacteria (1). These observations came to be attributed to the induction by the bacterial toxins of a host-soluble factor that caused the death, or necrosis, of tumor cells (2). This factor was a proinflammatory cytokine that was eventually named TNF-α. More recently, the wide range of biological activities ascribed to TNF-α suggest that this cytokine is a primary mediator of inflammation and modulates immune responses. Continued, overexpression of TNF-α can result in the chronic inflammation which is characteristic of immune disorders such as Crohn's disease and rheumatoid arthritis.

A wide variety of stimuli can induce TNF-α production by immune cells. Endotoxin, gram-negative bacteria, gram-positive bacteria, tumor cells, viruses such as human immunodeficiency virus (HIV), ionizing radiation, cytokines such as interleukin-1 (IL-1) and interferon-γ (IFN-γ), phorbol myristyl acetate, and various stress-related responses can all induce TNF-α production in at least some cell types (3). The TNF-α gene is one of the first genes expressed in T or B lymphocytes after these cells have been stimulated through their antigen receptors (4). Interestingly, TNF-α itself induces its own synthesis (5).

TNF-α is produced predominantly in activated macrophages and T cells as a 26-kD transmembrane protein that is released into the circulation as a 17-kD protein subunit after cleavage by proteolytic enzymes. These monomers self-associate into a homotrimer of 51 kD that is the active form of soluble TNF-α (Fig. 1). The 26-kD transmembrane form exists as a bioactive trimer at the cell surface, since transgenic mice that overexpress a modified, proteolysis-resistant form of transmembrane TNF-α develop inflammatory diseases (proliferative synovitis, chronic inflammatory arthritis) (6).

TNF-α and other cytokines act as mediators by binding to specific receptors on other cells. Two receptors for TNF-α have been identified and are designated the p55 receptor and the p75 receptor. The p55 receptor is expressed on almost all cell types, whereas the p75 receptor is expressed primarily on hematological, lymphoid, and endothelial cells. The p55 and p75 receptors appear to signal by different pathways, with the p55 receptor signaling most of the proinflammatory and cytotoxic effects of TNF-α, whereas p75 appears to play more of a role in proliferative responses. Soluble forms of the

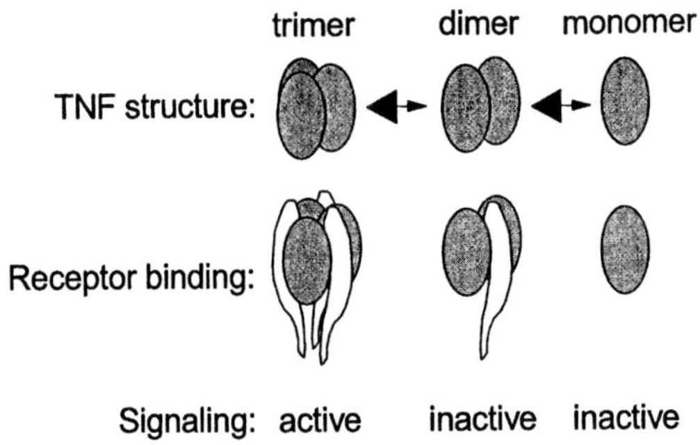

Figure 1 Schematic illustration of the relationship between TNF-α and receptor activation. TNF-α monomers undergo self-association to form dimers and trimers. Only the trimer form possesses biological activity. The double arrows are asymmetrical to indicate that trimer formation is favored under normal equilibrium. Receptor-ligand binding occurs at the interface between TNF-α monomer subunits.

p55 and p75 receptors are shed from the cell surface, and they may influence the circulating life span and the effects of TNF-α in vivo.

The nature of the interaction between TNF-α and its receptors was revealed by studies using mutant forms of the cytokine. These studies suggested that the receptors bind TNF-α in the cleft between two adjacent subunits of the TNF-α trimer. This was confirmed by Banner et al. (7), who determined that a single TNF-α trimer could simultaneously bind three receptor molecules (see Fig. 1). These results, combined with other studies (8), and the demonstration that the TNF-α trimer is the only active species, strongly suggested that TNF-α binding to cell surface receptors leads to intracellular signaling by cross linking two or three receptors. TNF-α–induced intracellular signaling can lead to the induction of genes and to the production of their gene products.

TNF-α can induce different genes including (a) transcription factors, such as NF-κB and c-jun/AP-1; (b) adhesion molecules, such as E-selectin and intercellular adhesion molecule (ICAM-1); (c) cytokines, such as IL-1, IL-6, and IL-8; (d) cytokine receptors, such as IL-1 receptor and IL-6 receptor; and (e) various inflammatory mediators, such as stromelysin, collagenase, C3 complement protein, and nitric oxide synthase (3,9). TNF-α may also reduce transcription of certain genes. For example, it inhibits collagen gene transcription, which is consistent with its proinflammatory role of inducing matrix metalloproteinase (MMP-1) that degrades collagen. Some typical cellular responses to TNF-α are listed in Table 1.

Table 1 Some Cellular Responses Attributed to TNF

Cell type	Induces
Macrophage	IL-1, IL-6, IL-8, IL-12
	M-CSF, GM-CSF
	Intracellular calcium
	Tumor cell killing
Neutrophil	Cell activation
	Generation of superoxide anion
	Release of platelet-activating factor
Endothelial	Nitric oxide synthase
	IL-1
	ICAM-1, VCAM-1
	E-selectin, P-selectin
	Release of platelet-activating factor
Fibroblast	Metalloproteinases
	Proliferation
	IL-6

IL, interleukin; M- or GM-CSF, macrophage or granulocyte-macrophage colony-stimulating factor; ICAM or VCAM, intercellular or vascular cell adhesion molecule; TIMP, tissue inhibitor of metalloproteinases.

ROLE OF TNF-α IN DISEASE

Considering the variety of stimuli that can induce TNF-α and the pleiotropic activities of TNF-α, it is not surprising that numerous types of disease have been associated with elevated concentrations of this cytokine. Many bacterial, viral, and parasitic infections are believed to trigger TNF-α production. In addition to conditions induced by exogenous agents, TNF-α may have an important role in autoimmune diseases.

Some of the symptoms of autoimmune diseases, tissue transplantations, asthma, and other types of stress-related conditions have been attributed to TNF-α (3). TNF-α has been implicated in several autoimmune diseases including insulin-dependent diabetes mellitus, multiple sclerosis, and rheumatoid arthritis (10–12). The inflammatory bowel diseases of Crohn's disease and ulcerative colitis are also caused by immune disregulation, and a role for TNF-α in both the acute and chronic phases of these inflammatory processes has been documented. In addition to the direct cytotoxic effects of TNF-α on cells, a significant portion of the proinflammatory effects of TNF-α may be due to its induction of chemokines and cell surface adhesion molecules that can lead to inflammatory cell recruitment.

Crohn's Disease

A strong association between TNF-α and local inflammation is suggested by the elevated concentrations of both TNF-α mRNA and TNF-α protein in the intestinal mucosa and stools of Crohn's disease patients (13,14). Moreover, anti–TNF-α antibodies reduce the severity of disease in mouse colitis models (15) and in naturally occurring colitis found in cotton-top tamarins (16). Increasing evidence indicates that Crohn's disease is characterized by an immune response controlled by $CD4^+$ T helper type 1 (Th1) cells (17,18). The Th1 cells produce IFN-γ, TNF-α, and IL-2 that contribute to inflammation and a T-cell cytotoxic response. In contrast, Th2 cells produce IL-4, IL-5, and IL-10 which induce the production of particular subclasses of immunoglobulin and are associated with allergic responses. Interestingly, Th1 and Th2 cells are antagonistic to each other, and thus an immune response to a particular antigen can usually be categorized as predominantly Th1 or Th2 in nature. Crohn's disease seems to be predominantly Th1 in nature. The predominance of Th1 was suggested by a reconstituted SCID mouse model in which colitis was ameliorated by treatment with antibodies to IFN-γ and TNF-α or by IL-10; all of which dampen the Th1 response and increase the Th2 response (15). The cellular mechanism of Crohn's disease may be caused by an exaggerated Th1 response to mucosal stimuli or persistent T cell activation due either to excessive proliferation or to a reduced level of T cell apoptosis. Regardless of the reason for a predominant and persistent Th1 response, TNF-α apparently plays an important role by perpetuating the chronic inflammatory state that is characteristic of Crohn's disease.

Rheumatoid Arthritis

Increased levels of TNF-α mRNA and protein have been detected in synovial fluid from joints of rheumatoid arthritis (RA) patients and synovial cell cultures (12,19,20). Feldmann et al. (20) proposed that in RA, TNF-α is near the beginning stages of a sequential induction cascade that involves upregulation of other proinflammatory cytokines such as IL-1 (21), IL-6, IL-8, and granulocyte-macrophage colony-stimulating factor (GM-CSF) (22,23). Additionally, other effector mechanisms such as activation of inflammatory cells, induction of chemokine production, and induction of adhesion molecules likely contribute to the inflammatory activities attributed to TNF-α. Successful amelioration of disease following specific inhibition of TNF-α activity in animal models of induced arthritis confirmed that TNF-α was an appropriate target for therapeutic intervention (24).

ANTI–TNF-α MONOCLONAL ANTIBODY (INFLIXIMAB, REMICADE)

TNF-α is a potential target for specific anticytokine therapy. Several strategies have been developed to neutralize the deleterious effects of TNF-α in disease states. One of these, based on monoclonal antibody (mAb) technology, led to the development of agents that bind to antigen with high specificity and affinity, thus neutralizing its biological activity. The advent of hybridoma technology (25) has permitted production of mAbs in commercial quantities. Initially, mAbs were purely murine in construct, but these had the potential of developing human antimouse antibodies when administered to the patients. Advances in biotechnology have made it possible to develop ''man-made'' and humanized mAbs which are much less immunogenic but still retain the high neutralizing activity for their targets.

Centocor Inc. (Malvern, PA) has developed a genetically engineered form of the mouse monoclonal antibody, A2, which is specific for human TNF-α. Human constant domains of both the heavy and light chains were substituted for the murine structural equivalents (26). This genetic replacement strategy yields a molecule, chimeric A2 (or cA2), that retains the full antigen (TNF-α) binding properties of the original A2 antibody. The human constant domain structures confer decreased immunogenicity and optimum antibody effector functioning within the human immune system. The final infliximab antibody possesses a normally glycosylated IgG1 structure with an approximate molecular weight of 150,000 D and the pharmacokinetic characteristics of a human immunoglobulin. The generic name for cA2 is infliximab and its trademark is REMICADE.

As described previously, TNF-α initiates intracellular signaling by cross linking TNF receptors present in the cell membrane. The therapeutic benefit of infliximab is likely due, in part, to a blockade of TNF-α–mediated cellular events by preventing the association of TNF-α with its receptors (26–28). Infliximab blocks the biological activity of both soluble and transmembrane TNF-α. Infliximab does not neutralize TNF-β (lymphotoxin-α), related cytokine that utilizes the same receptors as TNF-α (26). Cells expressing transmembrane TNF-α bound by infliximab can be lysed in vitro by complement or effector cells (27). Infliximab inhibits the functional activity of TNF-α in a wide variety of in vitro bioassays utilizing human fibroblasts, endothelial cells, neutrophils (28), B and T lymphocytes (29,30), and epithelial cells (31). Since infliximab does not cross react with TNF-α in species other than human and chimpanzees, animal reproduction studies have not been conducted with infliximab. In a developmental toxicity study conducted in mice using an analogous antibody that selectively inhibits the functional activity of mouse TNF-α, no evidence of maternal toxicity, embryotoxicity, or teratogenicity was observed (31–34).

Clinical Pharmacology Results

To detect and measure infliximab in patients' sera without interference from the large amounts of normal immunoglobulin, specific enzyme immunoassays were employed. The infliximab detection assays use immunological reagents that specifically detect the murine portions of infliximab and do not interact with the human immunoglobulin structural domains. The lower limit of detection of infliximab in these assays was 0.5 μg/mL or less (31).

The single-administration pharmacokinetics of infliximab were investigated in several clinical trials. The infliximab pharmacokinetic profile is similar to that of an intact IgG1 immunoglobulin and is likely metabolized by the same pathways as for human immunoglobulins. Weight-adjusted intravenous injections of 1, 5, 10, or 20 mg infliximab per kilogram yielded linear, dose-dependent increases in the pharmacokinetic parameters of maximum serum concentration (C_{max}) and area under the curve (AUC). In contrast, the volume of distribution (Vd_{ss}) and clearance were relatively independent of the administered dose and indicated that infliximab is predominantly distributed within the vascular compartment. Infliximab also exhibits a prolonged terminal half-life of 9.5 days that is comparable with that of a human immunoglobulin (35). No evidence of accumulation was observed after repeated dosing in patients with fistulizing disease given 5 mg/kg infliximab at weeks 0, 2, and 6 or in patients with moderate or severe Crohn's disease retreated with four infusions of 10 mg/kg infliximab at 8-week intervals (36).

Pharmacodynamic investigations indicate that infliximab binds to its intended target, TNF-α, in vivo and inhibits the deleterious functions of this inflammatory cytokine. Treatment with infliximab reduced infiltration of inflammatory cells and TNF-α production in inflamed areas of the intestine (37) and reduced the proportion of mononuclear cells from the lamina propria able to express TNF-α and INF-γ (38). After treatment with infliximab, patients with Crohn's disease had decreased levels of serum IL-6 and C-reactive protein compared to baseline (31). Peripheral blood lymphocytes from infliximab-treated patients, however, showed no decrease in proliferative responses to in vitro mitogenic stimulation when compared to cells from untreated patients (30).

Treatment of Crohn's Disease with Infliximab

The results of the nonclinical studies discussed above indicated that infliximab can efficiently neutralize human TNF-α. The data suggested that infliximab is a well-tolerated therapeutic agent for chronic inflammatory diseases such as Crohn's disease and RA.

Open-label Studies

The first patient to be treated with infliximab was a young girl with severe steroid refractory Crohn's colitis. Infliximab was administered twice at a dose of 10 mg/kg. Treatment was followed by a rapid decrease in the signs and symptoms of Crohn's disease (diarrhea, abdominal pain, fever, anorexia). In addition, dramatic healing of mucosal ulcers occurred and no apparent side effects were observed (39).

Encouraged by these results, an open-label 8-week study evaluating the effects of a single infusion of infliximab in 10 patients was conducted (40): Eight patients received 10 mg/kg and two patients received 20 mg/kg infliximab. These 10 patients with active, steroid-dependent Crohn's disease were assessed based on clinical evaluation using the Crohn's disease activity index (CDAI), Crohn's disease endoscopic index of severity (CDEIS), and evaluation of acute-phase reactants, such as C-reactive protein (CRP) and erythrocyte sedimentation rate (ESR). The CDAI is the most commonly used measurement tool for evaluating clinical disease activity in clinical trials in Crohn's disease patients. The CDAI (41) consists of variables measuring clinical signs and symptoms of Crohn's disease and has been repeatedly validated (42). The index incorporates eight items: the number of liquid or very soft stools, the patient's assessment of abdominal pain/cramping, the patient's assessment of general well-being, extraintestinal manifestations of Crohn's disease, the presence of an abdominal mass, use of antidiarrheal drugs, hematocrit, and body weight. These items yield a composite score ranging from 0 to approximately 600. Higher scores indicate more disease activity. Patients with a score <150 are considered to have inactive disease, whose patients with a score >450 are critically ill.

Nine patients were evaluable for assessment of efficacy; all nine patients showed a rapid reduction in the CDAI that was maintained through the last evaluation at week 8. The one patient who could not be evaluated had a perforated bowel (which may have been present before treatment) identified at 10 days postinfusion. Results for the CDAI in individual patients are shown in Figure 2. Eight of the nine patients (88.9%) were in clinical remission (CDAI <150) at the 2-week evaluation visit and at all follow-up visits through 8 weeks (40).

Using the CDEIS scoring system (43,44), endoscopic evaluations at baseline and 4 and 8 weeks showed mucosal healing in all infliximab-treated patients. The median CDEIS of 15.8 at baseline was improved to 5.6 at the 4-week evaluation and remained at 5.6 at the 8-week evaluation. An example of the endoscopic benefit provided by a single infusion of infliximab is shown in Figure 3. Reductions in CRP and ESR were seen as early as 24 and 72 h, respectively, following infliximab infusion. For both parameters, further reductions occurred and were sustained through 8 weeks (40). A decrease in circulating IL-6 concentration, in secreted phospholipase A_2 activities, and in mucosal expression of chemokines monocyte chemoattractant protein-1 (MCP-1), Macrophage inflammatory protein-1-alpha (MIP-1a), and regulated upon activation, normal T-cell expressed, and presumably excreted (RANTES) was also demonstrated, indicating potent anti-inflammatory effects of infliximab therapy (45).

A multicenter, open-label study was conducted at six centers in the United States and Europe to evaluate the dose-escalating (1, 5, 10, and 20 mg/kg) effects of infliximab therapy (46). Response parameters included the

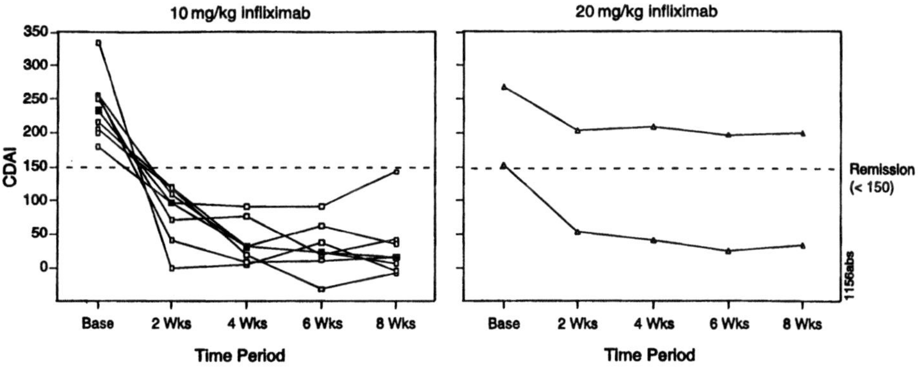

Figure 2 Plots of CDAI scores over time for each patient in the 10–mg/kg (left panel) and 20–mg/kg (right panel) treatment groups. In the left panel, open symbols represent the individual patient's scores and the closed symbols represent the median scores of all patients who received the 10–mg/kg dose. A CDAI score below 150 is associated with remission.

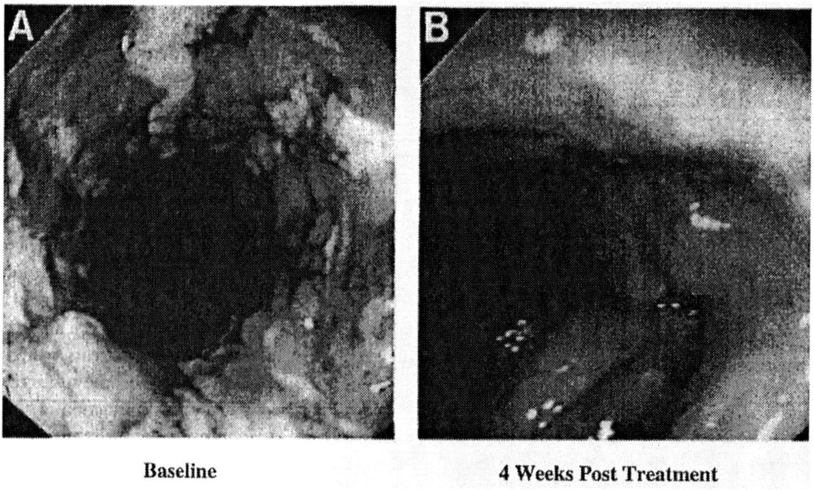

Figure 3 Healing of colonic ulcerations in a patient after treatment with infliximab at enrollment (A) and 4 weeks after infusion of infliximab (B). Photographs were obtained from videotapes, allowing comparison of the exact same location. (From ref. 40.)

CDAI, the inflammatory bowel disease questionnaire (IBDQ, a quality of life instrument), and the endoscopist's lesion severity assessment based on a 10-cm visual analog scale scoring system. A total of 20 patients (5 patients per group) were evaluable for the efficacy analysis. Among the 20 evaluable patients, 18 (90.0%) achieved a clinical response (decrease of ≥70 points in the baseline CDAI during the 4 weeks following infusion.) The response rate did not differ among the four dose groups, although the loss of response over time was greater in the 1–mg/kg infliximab dose group. In addition, endoscopic improvement was observed at 4 weeks in the 5–, 10–, and 20–mg/kg infliximab dose groups and was sustained through the last evaluation at 8 weeks. In contrast, improvement in the endoscopy score was not observed in the 1 mg/kg infliximab-treated patients. The lesser treatment benefit observed with the 1 mg/kg infusion led to the study of doses of ≥5 mg/kg in the subsequent placebo-controlled trials (46).

Controlled Trials

Infliximab in Active Crohn's Disease. This study was a multicenter, randomized, double-blind, placebo-controlled trial conducted in 18 centers in North America and Europe. Patients were randomly assigned to receive one of four single-infusion treatment regimens: placebo, 5, 10, or 20 mg/kg infliximab (47). One hundred and eight patients were enrolled into the study. To be eligible, patients were required to be receiving concurrent therapy with corticosteroids, aminosalicylates, or the immunomodulatory agents 6-mercaptopurine (6-MP) or azathioprine (AZA) or had failed therapy with aminosalicylates, 6-MP/AZA, methotrexate (MTX), or cyclosporine. The data collected at baseline to characterize the patient population confirmed that enrolled patients had moderate to severe Crohn's disease (median CDAI of 306 with a range of 215–437) not responding adequately to conventional therapies.

The primary endpoint was at 4 weeks and was defined as a reduction in the CDAI of 70 points or more. The key secondary endpoint was the number of patients achieving clinical remission; defined as a reduction in CDAI below 150. Patients not achieving the primary endpoint were offered an open-label infusion of 10 mg/kg after the 4-week evaluation visit. Each of the three infliximab treatment groups demonstrated a statistically significant larger proportion of patients achieving a clinical response (81.5, 50.0, and 64.3% for the 5-, 10-, and 20–mg/kg dose groups, respectively) than the placebo group (16.7%) (47). No evidence of a dose-relationship of the clinical response was observed; in fact, the highest clinical response rate was observed in the 5–mg/kg infliximab group (81.5%). When the three infliximab treatment groups were combined into a single group (n = 83), 65.1% of these patients had a clinical response compared to 16.7% of the 24 placebo-treated patients evaluated ($P < .001$). Approximately one-third (32.5%) of the infliximab-treated patients achieved a clinical remission compared to 4.2% of placebo-treated patients ($P = .005$), with the highest remission rate being observed in the 5–mg/kg infliximab group (48.1%). The clinical response and remission rates for each treatment group and the three infliximab treatment groups combined are shown in Figure 4 for the evaluation visits at week 2 and week 4. There data indicate that the onset of the clinical benefit is rapid; occurring by the first evaluation visit at 2 weeks.

In terms of signs and symptoms of Crohn's disease, infliximab was effective in reducing diarrhea, abdominal pain, and cramping and also improved the patient's general well-being. The highest level of benefit was observed in the group of patients receiving 5 mg/kg infliximab. To further assess the effectiveness of infliximab in Crohn's disease, the IBDQ quality of life instrument and CRP were measured throughout the study. For each of the CDAI, IBDQ, and CRP outcome measurements, a consistent benefit compared to placebo was present by the first posttreatment evaluation visit, as shown in Figure 5. With the 5–mg/kg dose, median values for the CDAI, IBDQ, and CRP were at levels associated with disease remission (47).

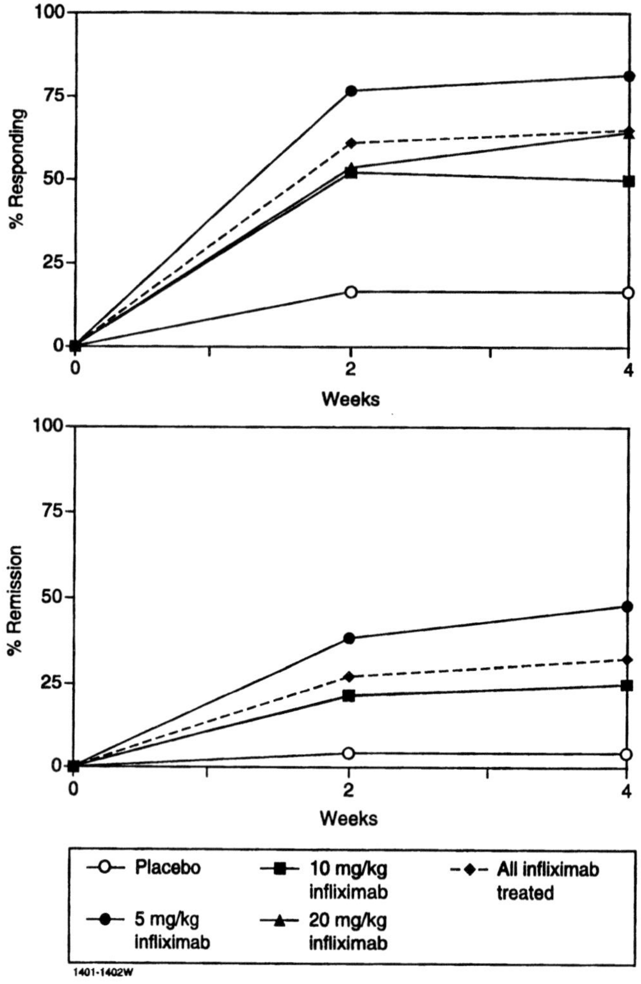

Figure 4 Clinical response rates (≥70-point decrease in CDAI from baseline, top panel) and clinical remission rates (CDAI <150, bottom panel) for each treatment group and the three infliximab group combined. Responses occurred by the first post-infusion evaluation (within 2 weeks following a single infusion of infliximab). (From ref. 47.)

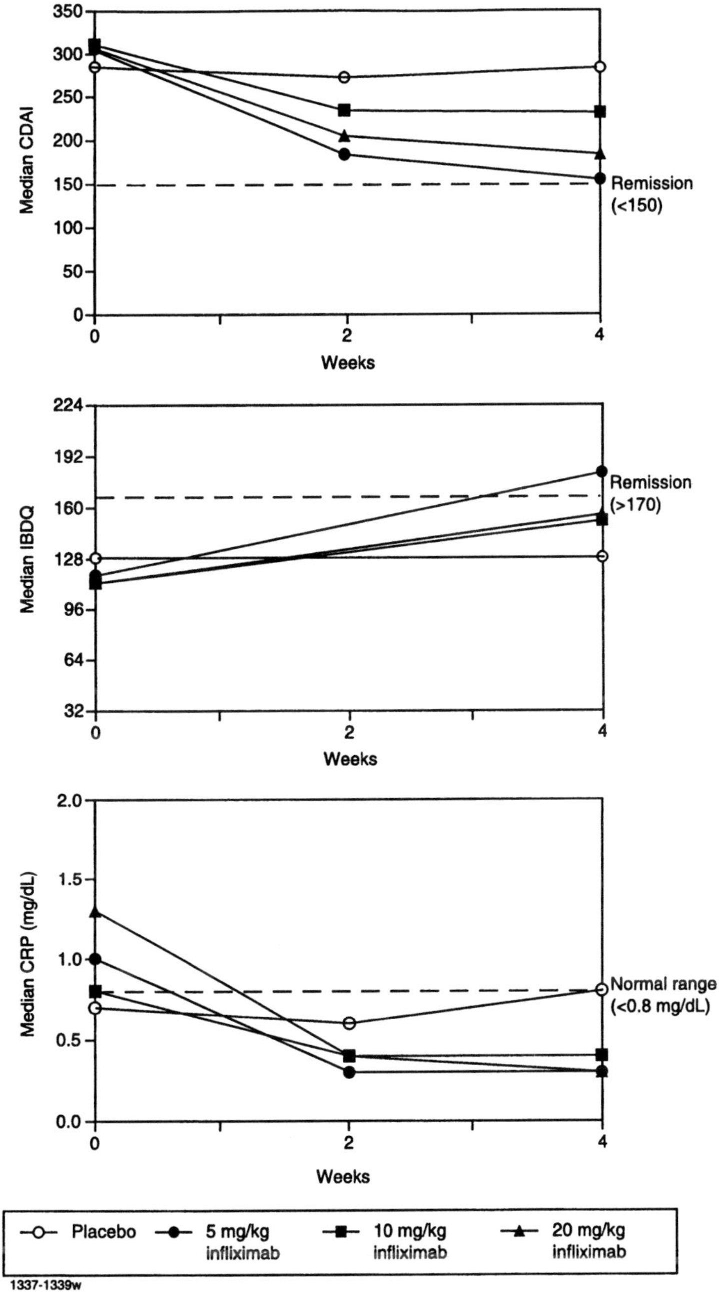

Figure 5 Median results for the CDAI (top panel), IBDQ (middle panel) and CRP (bottom panel) in all patients through the 4-week evaluation period for the active Crohn's disease study.

The effectiveness of infliximab in promoting mucosal healing was also evaluated in a subgroup of 30 patients enrolled at European centers in whom endoscopy was performed at baseline and week 4. CDEIS (42,43) was employed to evaluate the severity of mucosal inflammation. Similar to the results from the earlier open-label studies, a substantial reduction in mucosal inflammation was demonstrated in each of the infliximab treatment groups but not in the placebo group (48).

Repeated Treatment with Infliximab. Patients who showed a clinical response at 8 weeks following the initial treatment with infliximab were eligible for enrollment in a repeated treatment phase in which patients were rerandomized at week 12 to either four infusions of 10 mg/kg infliximab or placebo given at 8-week intervals. Patients were followed for efficacy evaluations through week 48. Seventy-three of the 108 patients enrolled in the initial treatment phase of the study were enrolled into this phase. Retreatment with infliximab every 8 weeks in patients with moderate to severe Crohn's disease maintained the initial treatment benefit over 48-week study period. Retreatment with placebo resulted in a gradual loss of the initial treatment benefit over 48-week study period. Importantly, measurements of serum infliximab concentration demonstrated stable pharmacokinetics throughout the retreatment period (49).

In summary, the benefit of treatment with infliximab was evident in all of the outcome measurements obtained, which included a clinical index of signs and symptoms (CDAI), a quality of life measurement (IBDQ), and a serum marker of inflammatory activity (CRP). The onset of benefit with infliximab treatment was rapid; reductions in serum concentration of an acute-phase reactant, CRP, occurred within 24 hs following infusion of infliximab and definitive clinical benefit was uniformly observed in all trials by the first evaluation visit at 2 weeks following infusion. The relevance of the reductions in disease activity produced by infliximab activity are further underscored by the low placebo response rates that were observed in the studies. Based on the clinical experience in these three trials, a dose of 5 mg/kg is considered the lowest effective dose in terms of a consistent clinical, laboratory, and endoscopic benefit in the treatment of Crohn's disease. The 5–mg/kg dose of infliximab was approved by the U.S. Food and Drug Administration (FDA) in August, 1998, for the indication of treatment of moderately to severely active Crohn's disease for the reduction of signs and symptoms in patients who have an inadequate response to convential therapy.

Infliximab in Enterocutaneous Fistulae as a Complication of Crohn's Disease. This was a multicenter, randomized, double-blind, placebo-controlled trial conducted at 12 centers throughout the United States and

Europe. A total of 94 patients with single or multiple draining enterocutaneous fistulas of at least 3 months' duration were enrolled into the study and randomized to each of the following three treatment groups: placebo, 5 mg/kg and 10 mg/kg infliximab. Therapy was administered at weeks 0, 2, and 6 (50). This more frequent dosing interval was implemented because it was believed that fistulas would be more refractory to treatment.

The three treatment groups were well balanced; no statistically significant differences among treatment groups were observed for demographical characteristics, baseline disease characteristics, or concomitant medications at baseline. Approximately 90% of patients had at least one perianal fistula, whereas the remaining patients had at least one abdominal fistula. On enrollment, approximately 45% of patients had 1 fistula and 55% had multiple fistulas ranging between 2 and 13 in number. The majority of patients had these draining fistulas despite receiving medical treatment for Crohn's disease.

The primary endpoint was at least a 50% reduction in the number of open fistulas for two consecutive visits (efficacy visits were at weeks 2, 6, 10, 14, and 18). A fistula was considered closed when it was no longer draining despite gentle compression. As shown in Table 2, approximately two-thirds (61.9%) of the 63 patients treated with infliximab achieved the primary endpoint (P = .002) versus the placebo response of 25.8%. In addition, approximately one-half (46.0%) of the 63 infliximab-treated patients achieved the secondary endpoint of a complete response (absence of any draining fistulas for at least two consecutive evaluation visits). This was also statistically sig-

Table 2 Patients Achieving Primary Endpoint for Fistula Closure and a Complete Response in Fistulizing Crohn's Disease Trial

		Dose of Infliximab		Infliximab-treated
	Placebo (n = 31)	5 mg/kg (n = 31)	10 mg/kg (n = 32)	patients (n = 63)
Patients achieving primary endpoint[a]	8 (25.8%)	21 (67.7%)	18 (56.3%)	39 (61.9%)
P-value vs placebo		0.002	0.021	0.002
Patients with complete response[b]	4 (12.9%)	17 (54.8%)	12 (37.5%)	29 (46.0%)
P-value vs placebo		0.001	0.041	0.001

[a] A ≥ 50% reduction from baseline in the number of open fistulas observed for two consecutive evaluation visits (i.e., at least 1 month).

[b] Absence of any draining fistulas for at least two consecutive evaluation visits.

nificantly greater than the placebo response rate (12.9%, $P = .001$). The best clinical benefit was observed with the 5–mg/kg dose (>50% reduction in a number of open fistulas in 68% of patients; complete response in 55% of patients) (50).

Treatment with infliximab was associated with a rapid onset of response. The majority of patients responding to infliximab treatment demonstrated the response by 2 weeks following infusion (median number of days to onset of response was 14 days for infliximab-treated patients compared to 42 days for the few responding placebo-treated patients). Through 18 weeks of follow-up, the median duration of benefit was approximately 3 months (50).

In summary, treatment of patients with enterocutaneous fistulas with infliximab produced a rapid and profound benefit in the closure and healing of these fistulas. These findings represent the first time that a therapeutic agent has demonstrated a statistically significant clinical benefit (closure of fistulas) in a controlled clinical trial setting. The onset of benefit was rapid, with the majority of responding patients demonstrating fistula closure by the first evaluation at 2 weeks. The benefit of treatment lasted at least 12 weeks in the majority of patients who responded to treatment. A regimen dose of 5 mg/kg given at 0, 2, and 6 weeks was approved by the FDA for the treatment of patients with fistulizing Crohn's disease for the reduction in the number of draining enterocutaneous fistulas.

Treatment of RA with Infliximab

The effectiveness of infliximab treatment in patients with RA has been studied in four clinical trials (three randomized, double-blind, placebo-controlled and one open label). A total of 203 patients have been enrolled in these trials.

Open-label Study

An open-label trial evaluated clinical response to infliximab in patients with advanced chronic erosive RA unresponsive to disease-modifying drugs (DMARD) such as methotrexate (MTX), gold, and salazopyrine (51). Twenty patients with a diagnosis of RA (52) were administered either an infusion of 10 mg/kg infliximab that was repeated 2 weeks later or four infusions of 5 mg/kg infliximab administered every 4 days. Patients were required to have discontinued DMARD therapy for at least 4 weeks prior to treatment and to be off DMARD therapy for the duration of the study. Concurrent treatment with nonsteroidal anti-inflammatory drugs (NSAID) or steroids (≤10 mg/day) was permitted.

Treatment with infliximab led to rapid improvement in every patient in all assessments of disease activity used for monitoring patients, such as the number of tender joints and number of swollen joints, duration of morning stiffness, combined grip strength of both hands, assessment of pain, ESR, and the CRP concentration. Improvement for these measurements ranged from 57 to 94%. The onset of benefit was rapid, being evident 1 week after the first infusion. Therapy was well tolerated by the patients. This trial provided the first positive efficacy experience in patients using an anti–TNF-α therapy.

Controlled Study

To further establish the efficacy and safety profile of infliximab a randomized, double-blind, placebo-controlled trial was performed in four European centers. Seventy three patients with active advanced RA were randomized to the following three treatment groups: single infusion of placebo or infliximab at a dose of 1 mg/kg or 10 mg/kg (53). Patients had to have failed prior treatment with at least one DMARD and discontinued DMARD therapy at least 4 weeks prior to treatment. Concurrent treatment with NSAID or steroids (≤12.5 mg/day) was permitted.

The primary endpoint of the study was the achievement at week 4 of a Paulus 20% response, a composite of six clinical, observational, and laboratory variables including the number of the tender and swollen joints, the duration of morning stiffness, the patient's and physician's assessment of disease severity, and ESR. Patients were considered to have responded if at least four of the six variables improved, defined as at least 20% improvement in the continuous variables, and at least two grades of improvement or improvement from grade 2 to 1 in the two disease-severity assessments. The more stringent Paulus 50% criteria also was used for the assessment.

All patients were to have active disease as defined by the presence of at least 6 swollen joints of a total of 58 joints and the presence of at least three of the following: duration of morning stiffness of at least 45 min, the presence of 6 or more tender/painful joints of a total of 60 joints, ESR of at least 28 mm/h, or CRP of at least 20 mg/L.

Substantial and highly significant differences were seen in response rates between infliximab and placebo. Only 2 of 24 placebo patients (8.3%) achieved a 20% Paulus response at week 4. By contrast, 19 of 24 patients (79.2%) treated with 10 mg/kg infliximab achieved a response by week 4 ($P < .0001$ compared with placebo). The response rates in the 1–mg/kg group were intermediate, with 11 of 25 patients (44.0%) responding at week 4 ($P = .0083$). Analysis of the Paulus 50% response showed similar differences between the groups, with 14 of 24 high-dose infliximab patients responding compared with 2 of 24 patients in the placebo group ($P = .0005$).

Improvement in both clinical and laboratory disease parameters was seen as early as 3 days after infusion, but maximal improvements were seen at week 2 for both the 1– and 10–mg/kg infliximab treatment groups. Maximal improvement ranged from 55 to 98%, with most measurements of disease activity improving 70% or more. Detailed time-response profiles for four disease-activity assessments are shown in Figure 6. There was no difference in magnitude of improvement between the 1– and 10–mg/kg infliximab treat-

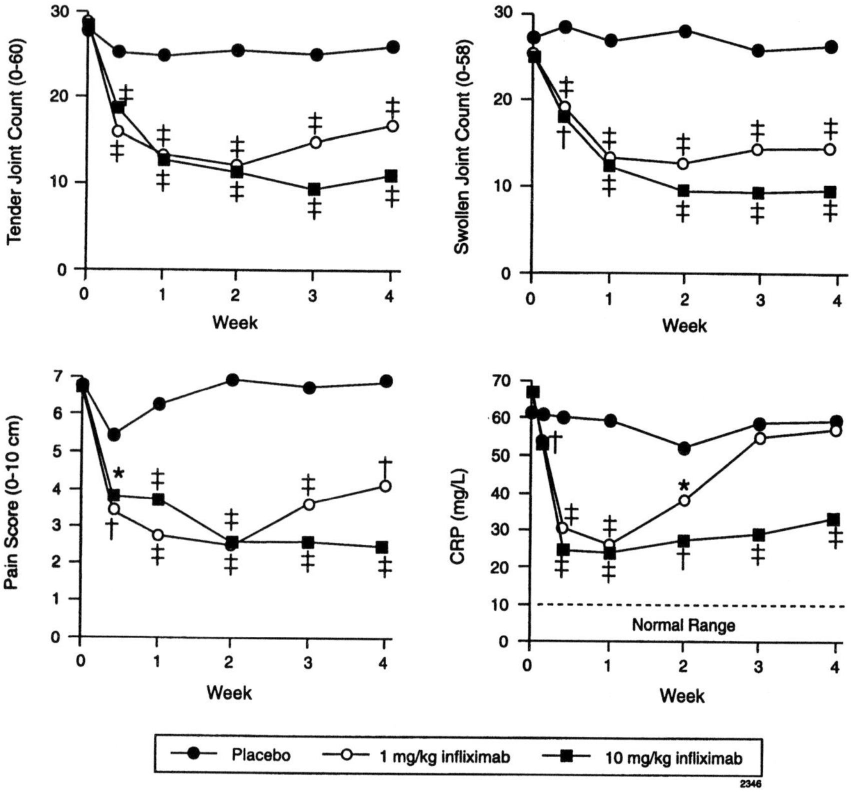

Figure 6 Rheumatoid arthritis disease activity assessments. Values are means of 24 patients at each point (25 for 1–mg/kg group). ● = placebo, ○ = 1 mg/kg infliximab, and ■ = 10 mg/kg infliximab. Significance versus placebo: $*P < .05$, $^{\dagger}P < .01$, $^{\ddagger}P < .001$. (Modified from ref. 53.)

ment groups, but the duration of response was longer in patients receiving the 10–mg/kg dose (53). The results also suggested that a multiple infusion regimen would be necessary to control chronic inflammatory processes in RA.

Controlled Study of Infliximab With or Without Concomitant Methotrexate Therapy

The next trial evaluated the efficacy and safety of multiple infusions of infliximab both with and without methotrexate (MTX), which is currently one of the most effective DMARD used for the treatment of RA (54). The rationale for studying infliximab and MTX in combination is that it is likely that new anticytokine therapies would be used in combination with existing drugs (55).

Patients with active RA who were being treated with MTX for a minimum of 6 months before study entry were enrolled into this trial. A total of 101 patients were randomized to one of seven treatment groups. The patients received either intravenous infliximab at 1, 3, or 10 mg/kg, with or without MTX 7.5 mg/day, or intravenous placebo plus MTX 7.5 mg/day at weeks 0, 2, 6, 10, and 14 and were followed through week 26 (56). All patients had to have active disease, defined as the presence of six or more swollen joints and at least one of the following: morning stiffness of at least 45 min, six or more tender/painful joints, ESR of at least 28 mm/h, or CRP of at least 15 mg/L.

Approximately 60–70% of patients receiving infliximab at 3 or 10 mg/kg with or without MTX achieved the 20% Paulus criteria for a median duration of 10.4 to >18.1 weeks (*P*-values vs placebo control <.001). An analysis of Paulus 50% responses, which gave an indication of the magnitude of the response, demonstrated initial response rates of about 50–60% and the benefit was generally sustained until exit from the study at 26 weeks (Fig. 7) (56). Longer periods of response were observed when infliximab and MTX were given together, although the differences between infliximab with and without MTX did not reach statistical significance. No dose-response relationship was noted between the 3– and 10–mg/kg doses.

At infliximab 1 mg/kg there was a significant difference among the treatment groups receiving MTX or not receiving MTX in duration of clinical response assessed by the Paulus 20% index (16.5 weeks with MTX vs 2.6 weeks without MTX, $P = .001$) (Fig. 8). These results show that the most pronounced synergistic effects between infliximab and MTX occur at low doses of infliximab.

A measure of the level of disease control during the trial was demonstrated by the reduction in the individual assessments of disease activity. Following therapy with infliximab, with or without MTX, there was improvement

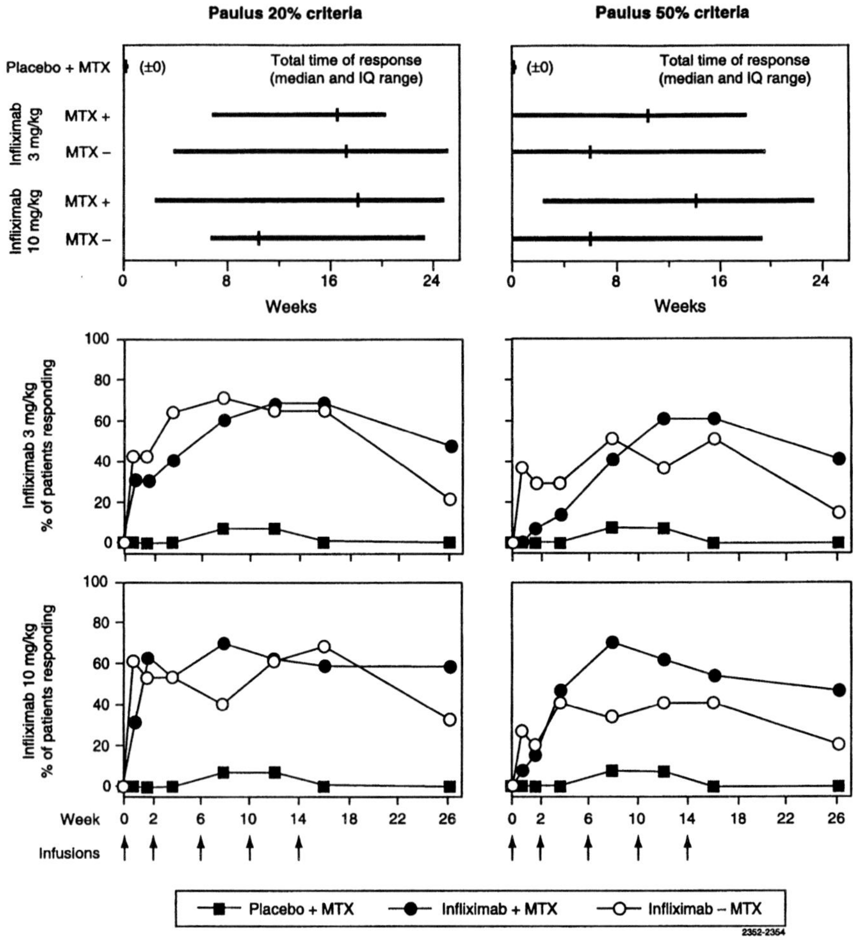

Figure 7 Efficacy of infliximab therapy based on Paulus 20 and 50% criteria. Results obtained with 3 and 10 mg/kg infliximab with (+) or without (−) methotrexate (MTX) and placebo + MTX. Results shown in the top panel are the total time of response, as median and interquartile (IQ) range, whereas the bottom panel represents the proportion (%) of patients responding at weeks 1, 2, 4, 8, 16, and 26, all patients being included at each time point. ● = infliximab + MTX; ○ = infliximab − MTX; ■ = placebo + MTX. Arrows indicate the times of infusion. (Modified from ref. 56.)

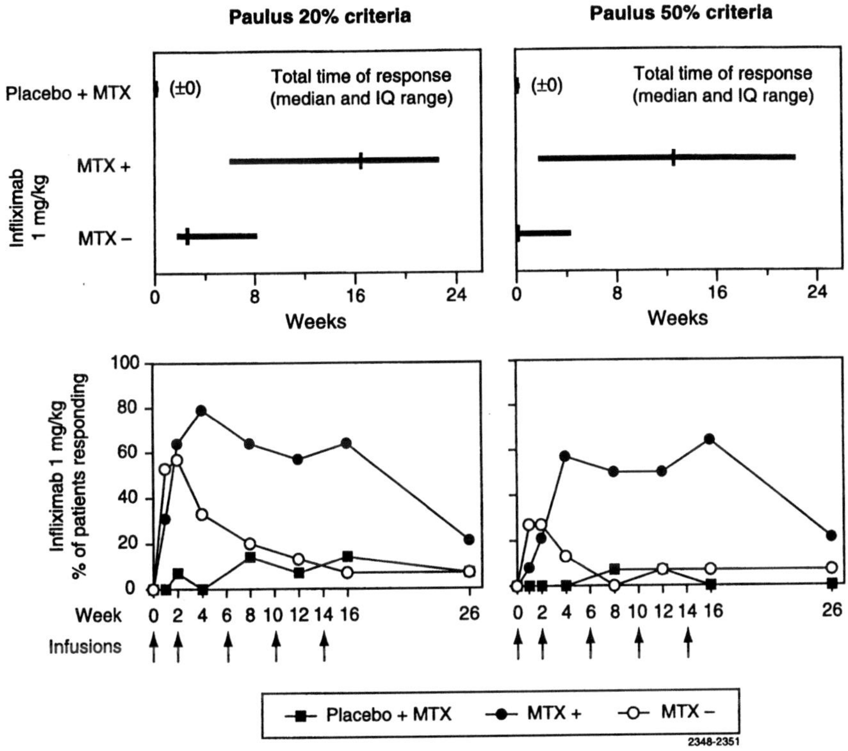

Figure 8 Efficacy of infliximab therapy based on Paulus 20 and 50% criteria. Results obtained with 1 mg/kg infliximab with (+) or without (−) methotrexate (MTX) and placebo + MTX. Results shown in the top panel are the total time of response, as median and interquartile (IQ) range, whereas the bottom panel represents the proportion (%) of patients responding at weeks 1, 2, 4, 8, 16, and 26, all patients being included at each time point. ● = infliximab + MTX; ○ = infliximab − MTX; ■ = placebo + MTX. Arrows indicate the times of infusion. The Paulus response is achieved by improvement in four of six of the following: 20 or 50% improvement in tender joint scores, swollen joint scores, duration of morning stiffness, erythrocyte sedimentation rate (ESR), or a two-grade improvement in the patient's and observer's assessment of disease severity. (Modified from ref. 52.)

in all of these measurements, which at 3 and 10 mg/kg generally exceeded 60–70% improvements from baseline. Changes in the swollen joint count, tender joint count, and CRP are shown in Figure 9 and demonstrate a marked and sustained improvement, achieving near remission levels, especially when infliximab and MTX were given in combination (56).

In summary, infliximab was effective and well tolerated when given as multiple infusions with the best results at 3 and 10 mg/kg alone or in combination at these doses with MTX. Synergy was observed with infliximab at 1 mg/kg when given with MTX. The results of the trial provided a strategy for

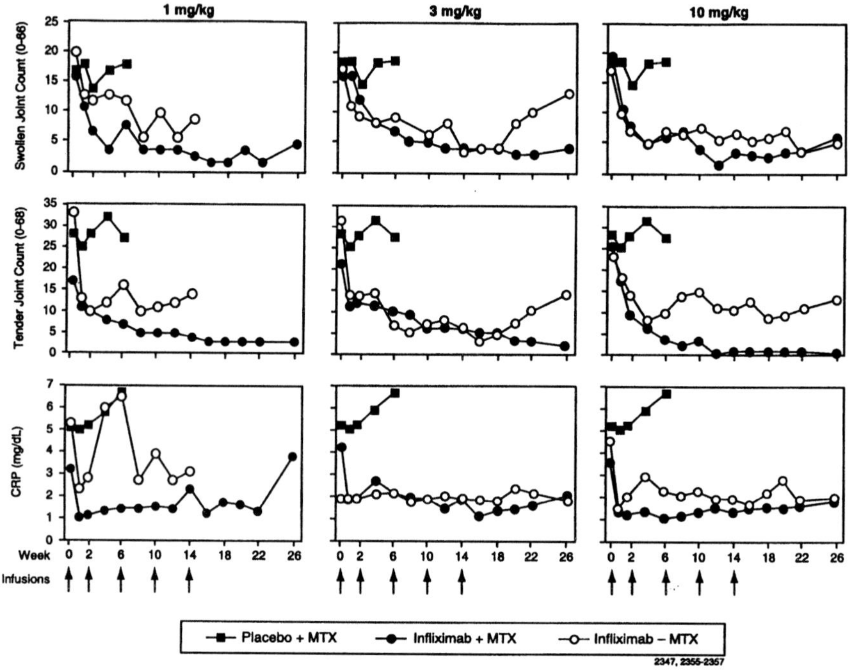

Figure 9 Serial measurements (median values) of the swollen joint count, tender joint count, and C-reactive protein (CRP) level, which are part of the ACR core set of outcome measures, before (day 0), during (weeks 1–14), and after (weeks 14–26) treatment. Arrows indicate the times of infusions. Median results are included only up to the point at which ≥50% of the original number of patients remained in the trial (up to week 6 for the placebo + methotrexate (MTX) group, and up to week 14 for the infliximab 1 mg/kg without MTX group). (Modified from ref. 56.)

further evaluation of the efficacy and safety of longer term treatment with infliximab.

Single-Dose Study of Infliximab in Patients with Active RA Receiving Methotrexate Therapy and Repeated-Dose Treatment Extension Study. This was another controlled study evaluating the safety and efficacy of a single dose of placebo, 5, 10, or 20 mg/kg of infliximab in patients with active RA receiving MTX, and also evaluated repeated treatments with three 10–mg/kg infusions of infliximab given 8 weeks apart.

Patients in the study had to have active disease despite having received treatment with MTX for at least 3 months before screening. Patients were stabilized at a dose of 10 mg/week MTX for at least 4 weeks before screening.

Twenty-eight patients received an initial infusion of placebo, 5, 10, or 20 mg/kg of infliximab (seven patients per treatment group). Twenty-three patients received retreatment with one or more open-label infusions of infliximab 10 mg/kg.

The American College of Rheumatology (ACR) definition of improvement was the basis for determination of clinical response (57). Other measures of efficacy consisted of the individual clinical and laboratory response parameters that comprise the ACR core set of outcome measurements (the number of swollen joints; number of tender joints; assessment of pain by a visual analog scale [VAS]; physician's and patient's global assessments [VAS]; health assessment questionnaire; ESR; and CRP).

The serum concentration of infliximab showed dose-proportional differences at any given follow-up visit following a single infusion of 5, 10, or 20 mg/kg. With infliximab retreatment, the serum concentration of infliximab demonstrated stable pharmacokinetics.

Single-dose Blinded Study. The proportion of patients who responded according to ACR 20% improvement at any time during the 12-week postinfusion period was significantly higher among all infliximab-treated patients (81.0%) compared to the placebo group (14.3%). There was no relationship to infliximab dose for the number of patients responding at any time (range 71.4–85.7% for the three infliximab treatment groups) (58).

Repeated Treatment Open-label Extension. The clinical benefit obtained during the single-dose blinded study following infliximab was maintained through the last evaluation (12 weeks after the last retreatment, a total of 40 weeks) of the retreatment extension (59).

This study provided additional dose-response information on beneficial effects of infliximab for treatment of RA, and suggested that repeated treatment with infliximab every 8 weeks was well tolerated, was associated with a stable pharmacokinetic profile, and was capable of sustaining the initial treatment benefit of infliximab.

SUMMARY OF SAFETY

Fourteen studies conducted in normal healthy volunteers, patients with RA, sepsis and ulcerative colitis, and Crohn's disease have been combined for the evaluation of infliximab safety. A total of 627 patients were enrolled in these studies. Patients were followed through approximately 12 weeks following their last infusion of infliximab for all adverse events and for up to 3 years (long-term follow-up) for major safety events defined as death, serious infections, malignancy, and new autoimmune disorders. Data are included in this summary for the follow-up of the 513 patients enrolled in the nonsepsis trials, of whom long-term follow-up was available on 445 patients.

MAJOR SAFETY EVENTS

Deaths

No patients died while on study during any of the nonsepsis trials. Nine deaths have been reported during the 3-year long-term follow-up period. Of the nine patients who died, two had Crohn's disease, and seven had RA. Causes of death included cardiovascular disease, malignancy or its treatment, or infection. The observed mortality over the course of long-term follow-up following infliximab treatment is within the expected incidence based on available data (60).

Malignancies

Nine patients developed cancer in the clinical trials of infliximab. Three patients developed solid tumors during a 3-year follow-up period: one was a metastatic breast cancer diagnosed 1 week after a single 10–mg/kg dose of infliximab, a Clark's II malignant melanoma developing 4.5 months after two infusions (5 and 10 mg/kg) of infliximab, and a papillary thyroid cancer in a patient with a history of Graves' disease, and a nodular thyroid at 20 months after a single 5–mg/kg infusion of infliximab. The medical histories in each of the patients with solid tumors suggested that malignancy was present at the time of infliximab infusion or arose from a documented premalignant condition.

One patient with Crohn's disease and one patient with rheumatoid arthritis developed non-Hodgkin's B-cell lymphomas. A single case of Hodgkin's disease and a single case of multiple myeloma (both occurred in rheumatoid arthritis patients) also occurred in infliximab-treated patients. None of the cases of B-cell cancer involved the central nervous system. The majority were nodal presentations. These B-cell malignancies occurred in patients with RA and Crohn's disease of long duration who had chronic exposure to immunosuppressants such as AZA and MTX, factors that are known to increase the risk of lymphoproliferative disorders (61–64). No relationship to the dose of infliximab or duration of exposure was observed.

Infections

Because TNF is a mediator of inflammation and modulates cellular immune response, it may be important in the defense against some intracellular pathogens. Therefore, it is possible that anti-TNF therapeutic agents, such as infliximab, may influence the ability to mount an inflammatory response against intracellular pathogens.

Infections were reported more often in infliximab-treated patients than in placebo-treated patients (21 vs 11%). In infliximab-treated patients, urinary tract infections, bronchitis, pharyngitis, moniliasis, upper respiratory tract infection, sinusitis, and herpes zoster were each reported by more than 1% of patients. In placebo-treated patients, urinary tract infections and upper respiratory tract infections were the most frequently reported infections; also reported in more than 1% of patients. No differences in the occurrence of serious infections were observed between placebo- and infliximab-treated patients. It is important to note that infliximab-treated patients were followed on study for nearly twice as long as placebo-treated patients (22.3 weeks vs 12.2 weeks, respectively). When a common follow-up period was used in the placebo-controlled Crohn's disease and RA trials, there were no differences between infection rates between infliximab-treated and placebo-treated patients. In addition, no relationship between the number of infliximab infusions and the rate of infections was seen.

Finally, it is worth nothing that in a trial of infliximab for the treatment of sepsis, the 28-day mortality rate was 41.2% for placebo patients and 26.6% for infliximab-treated patients.

Infusion Reactions

A reaction to an infusion was defined as any adverse experience reported during an infusion or within 2 hs following the end of the infusion. Overall, 15.9%

of infliximab-treated patients and 6.5% of placebo-treated patients had an infusion reaction. The most commonly reported infusion reactions among infliximab-treated patients were headache, nausea, dizziness, flushing, pruritus, urticaria, and chest pain. Among 1207 infliximab infusions given (an average of 2.7 infusions per patient), 9 infusion reactions resulted in immediate discontinuation of the study agent. Four of these reactions were assessed as serious. In all four cases, minimal to no treatment was required and all patients recovered completely.

Autoantibodies: Antinuclear Antibodies and Antibody Against Double-Stranded DNA (Anti-dsDNA)

Because of the role of autoimmunity in RA, autoantibody data were obtained from all of the RA studies and from the two controlled trials in patients with Crohn's disease. Approximately one-third of infliximab-treated patients changed antinuclear antibody (ANA) status during the trials, resulting in a net increase in the percentage of infliximab-treated patients positive for ANA from 23.8% at screening to 35.9% at last evaluation. Similar trends were seen in both Crohn's disease and RA patients.

When samples positive for ANA were analyzed for anti–double-stranded DNA (anti-dsDNA) antibodies, it was found that 8.7% of infliximab-treated patients became positive for anti-dsDNA at some point during follow-up. The rates of conversion to positivity for anti-dsDNA antibodies were similar in Crohn's disease and RA. Anti-dsDNA antibody titers normalized with discontinuation of infliximab treatment.

Although anti-TNF treatment with infliximab is associated with the development of anti-dsDNA antibodies, clinical manifestations in patients who developed these antibodies have been rare. Two patients in 453 nonsepsis patients treated with infliximab developed clinical signs of lupus. One was a RA patient who developed dyspnea and pericarditis with resolution of her symptoms within 6–8 weeks of initiation of treatment with oral steroids. The other was a Crohn's disease patient who developed lupus arthritis, which also responded to corticosteroids and symptoms resolved within 6 months after the last infliximab infusion. Follow-up of patients through 3-years following the last infusion of infliximab revealed no additional cases of lupus-like conditions. A number of drugs have been shown to induce the development of anti-dsDNA antibodies and lupus-like conditions similar in nature to the infrequent rates observed with infliximab (65–70).

Human Antichimeric Antibody Responses

Human antichimeric antibody (HACA) responses defined as an increase in specific anti-infliximab antibodies in posttreatment samples relative to baseline employing an enzyme immune assay. In Crohn's disease protocols, 13.4% of patients were HACA positive (most were low titer, <1:40). In the RA trial in which patients received multiple treatments of 1, 3, or 10 mg/kg infliximab with or without MTX, 17.4% of patients were HACA positive. In the later trial, the development of HACA was related to both the dose of infliximab and concurrent treatment with MTX. The rate of HACA responses was inversely proportional to the dosage; thus, HACA formation occurred in 53, 21, and 7% of patients treated with 1, 3, and 10 mg/kg infliximab without MTX. Concurrent treatment with MTX further diminished the appearance of HACA with rates of 15, 7, and 0% at the three dosage levels of infliximab. Patients who developed HACA were more likely to experience a reaction to infliximab infusion (39 vs 14% in patients who did not develop HACA, respectively). At doses of infliximab developed for commercial use (≥3 mg/kg), there was insufficient evidence to evaluate whether HACA diminished the efficacy of treatment.

OTHER ANTI-TNF THERAPIES FOR THE TREATMENT OF CROHN'S DISEASE AND RA

Humanized Anti-TNF Monoclonal Antibody

Humanized anti-TNF monoclonal antibody (CDP571, Celltech, Slough, England) was generated by grafting the antigen-binding site from a murine antibody into a human IgG4 framework, which does not have the capability of mediating cell lysis through complement activation and antibody-dependent cellular cytotoxicity. CDP571 antibody consists of 95% human and 5% murine sequences. This strategy of building the antibody should theoretically minimize the development of antiantibody responses that may facilitate the clearance of molecules and affect the duration of the response.

The effect of CDP571 antibody was evaluated in a multicenter, double-blind trial in 36 patients with severe RA. The patients treated with a high dose of CDP571 (10 mg/kg) demonstrated a significant improvement after 1 or 2 weeks in several clinical and laboratory parameters including tender joints score, pain score, ESR, and a composite index of disease activity which lasted 4–8 weeks (71); however, no significant difference was noted in the swollen joint score. Treatment with a low dose of CDP571 (1 mg/kg) showed loss of response by week 4. Comparative dose-response relationships suggested that the potency of CDP571 was less than that of infliximab. Whether this reflects differences in epitope specificity, affinity, or isotype (IgG4 vs IgG1) is unknown, but further investigation could illuminate the mode of action of the two anti–TNF-α antibodies (19). Despite the fact that CDP571 is a humanized monoclonal antibody, anti-CDP571 antibodies were detected in a large number of patients studied (71). Like other anti–TNF-α agents, CDP571 also induced ANA, anticardiolipin, and anti-dsDNA antibody production (72). Larger trials with CDP571 for the treatment of RA have not been reported.

In another double-blind, placebo-controlled study, 31 patients with active Crohn's disease on concomitant steroid therapy were randomly assigned to receive a single infusion of either placebo or CDP571 at a dose of 5 mg/kg (73). In 30 evaluable patients, there was a fall in the median CDAI from 263 to 167 at the week 2 measurement for those patients randomized to CDP571 (n = 20). At week 8 posttreatment, the median CDAI of both active group and the placebo group was approximately 220. CRP concentrations fell in the CDP571-treated patients, but not significantly, and the reduction was not sustained through 8 weeks. Low levels of anti-CDP571 antibodies were reported in 7 of 21 patients (73). The results of this study further confirmed that TNF-α is an important target for the treatment of Crohn's disease.

Recombinant Human Soluble p75 TNF Receptor Fusion Protein (Etanercept)

Recombinant constructs of the TNF receptor have also been developed as an anti–TNF-α therapeutic modality. Etanercept is a dimeric fusion protein consisting of the extracellular ligand-binding portion of the human 75 kD (p75) TNF receptor linked to the Fc portion of human IgG1 (TNFR:Fc, Immunex Corp., Seattle, Washington). This agent received FDA approval in November, 1998, for the treatment of moderate to severe RA in patients who had failed DMARD therapy. In a multicenter, double-blind, randomized trial of 180 patients with RA who had failed at least one DMARD, patients were treated with subcutaneous injections of placebo or etanercept at doses of 0.25, 2.0, or 16 mg/m^2 twice weekly for 3 months (74). Treatment with etanercept led to significant reductions in disease activity, and the therapeutic effects of etanercept were dose related. In addition, laboratory markers of disease activity, such as CRP and ESR, were improved by treatment with etanercept. Treatment with etanercept was well-tolerated; patients receiving etanercept reported more mild upper respiratory tract infections and injection site reactions than patients treated with placebo. In the FDA-approved package insert of drug-prescribing information (75), antibodies to etanercept were reported to be present at least once in sera of 16% of RA patients. Similar to the results from other clinical trials with anti-TNF agents, the development of anti-dsDNA antibodies was reported. Anti-dsDNA antibodies were detected in 15% of patients treated with etanercept compared to 4% of placebo-treated patients. The observed rates and incidences of malignancies in patients treated with Enbrel were similar to those expected for the population studied, and they were similar to those reported for other anti-TNF-α treatments, including infliximab.

Recombinant Human Soluble p55 TNF Receptor Fusion Protein (Lenercept)

Lenercept (TNF-R55-IgG1, Ro 45-2081, Roche) is a glycosylated fused protein consisting of two human p55 receptors linked to a human IgG1 Fc region. Several small clinical trials have been performed treating RA patients with lenercept. Clinical measures of disease activity (swollen and tender joint counts, physician and patient's global assessment of disease activity, patient's assessment of pain) showed decreases after lenercept treatment in all groups receiving more than 2.5 mg lenercept given intravenously as a single infusion. Improvement was observed from 24 h postdose which lasted up to 28 days (76). Antibodies to lenercept were detected in 28% of the patients, although it was reported to have no influence on TNF-binding capacity, on clinical efficacy, or on safety (77). Although initial studies with lenercept were encouraging in the treatment of RA patients, further clinical development of this agent by Roche has not been pursued.

CONCLUSIONS

The results of clinical trials with several anti–TNF-α therapeutic modalities have demonstrated that these products are effective agents for the treatment of chronic inflammatory conditions such as Cronh's disease and RA. Infliximab, a chimeric monoclonal antibody against TNF-α, has been shown to be effective in reducing the signs and symptoms of both Crohn's disease and RA and has received FDA approval for the treatment of Crohn's disease and is under regulatory review for the treatment of RA. The results also showed that infliximab therapy is safe as a single or as a multiple infusion regimen. A number of other biological products directed against TNF-α have been evaluated in human clinical trials for treatment of Crohn's disease and RA. The results from these clinical trials also demonstrated the effectiveness of an anti–TNF-α therapeutic approach. A TNF receptor fusion protein construct (etanercept) has received FDA approval for the treatment of RA. Thus, we are entering a new era of treatment of chronic inflammatory conditions, where targeting a specific molecule responsible for initiating or promulgating inflammatory processes reduces the clinical symptoms and signs of the disease. Long-term clinical trials are on the way to establish the long-term safety and efficacy of new biological agents and to define the precise place of these therapies in the clinical management of these chronic, incurable disorders.

REFERENCES

1. WB Coley. The therapeutic value of the mixed toxins of the streptococcus of erysipelas in the treatment of inoperable malignant tumors, with a report of 100 cases. Am J Med Sci 112:251–281, 1896.
2. EA Carswell, LJ Old, RL Kassel, S Green, N Fiore, B Williamson. An endotoxin induced serum factor which causes necrosis of tumors. Proc Natl Acad Sci USA 72(9):3666–3670, 1975.
3. BB Aggarwal, K Natarajan: Tumor necrosis factors: developments during the last decade. Eur Cytokine Netw 7:93–124, 1996.
4. AE Goldfeld, EK Flemington, VA Boussiotis, CM Theodos, RG Titus, JL Strominger, SH Speck. Trascription of the tumor necrosis factor α gene is rapidly induced by anti-immunoglobulin and blocked by cyclosporin A and FK506 in human B cells. Proc Natl Acad Sci USA 89:12198–12201, 1992.
5. G Hensel, D-N Mannel, K Pfizenmaier, M Kronke. Autocrine stimulation of TNF-alpha mRNA expression in HL-60 cells. Lymphokine Res 6:119–125, 1987.
6. S Georgopoulos, D Plows, G Kollias. Transmembrane TNF is sufficient to induce localized tissue toxicity and chronic inflammatory arthritis in transgenic mice. J Inflamm 46:86–97, 1996.
7. DW Banner, A D'Arcy, W James, R Gentz, H-J Schoenfeld, C Broger, H Loetscher, W Lesslauer. Crystal structure of the soluble human 55 kd TNF receptor-human TNFβ complex: implications for TNF receptor activation. Cell 73: 431–445, 1993.
8. P Vandenabeele, W Declercq, R Beyaert, W Fiers. Two tumor necrosis factor receptors: structure and function. Trends Cell Biol 5:392–399, 1995.
9. B Beutler. Tumor Necrosis Factor. The Molecules and Their Emerging Roles in Medicine. New York: Raven Press, 1992.
10. BB Aggarwal, J Vilcek, eds. Tumor Necrosis Factors: Structure, Function, and Mechanisms of Action. Immunology Series. New York: Marcel Dekker, 1991.

11. FM Brennan, RN Maini, M Feldmann. TNFα—a pivotal role in rheumatoid arthritis. Br J Rheumatol 31:293–298, 1992.
12. M Feldman, FM Brennan, RN Maini. Role of cytokines in rheumatoid arthritis. Annu Rev Immunol 14:397–440, 1996.
13. CP Braegger, S Nicholls, SH Murch, S Stephens, TT MacDonald. Tumour necrosis factor alpha in stool as a marker of intestinal inflammation. Lancet 339:89–91, 1992.
14. EJ Breese, CA Michie, SW Nicholls, SH Murch, CB Williams, P Domizio, JA Walker-Smith, TT MacDonald. Tumor necrosis factor alpha-producing cells in the intestinal mucosa of children with inflammatory bowel disease. Gastoenterology 106:1455–1466, 1994.
15. F Powrie, MW Leach, S Mauze, S Menon, LB Caddle, RL Coffman. Inhibition of Th1 responses prevents inflammatory bowel disease in scid mice reconstituted with $CD45RB^{hi}$ $CD4^{+}T$ cells. Immunity 1:553–562, 1994.
16. PE Watkins, BF Warren, S Stephens, P Ward, R Foulkes. Treatment of ulcerative colitis in the cottontop tamarin using antibody to tumour necrosis factor alpha. Gut 40:628–633, 1997.
17. SJH Van Deventer. Tumour necrosis factor and Crohn's disease. Gut 40:443–448, 1997.
18. SJ Simpson, GA Hollander, E Mizoguchi, D Allen, AK Bhan, B Wang, C Terhorst. Expression of pro-inflammatory cytokines by TCR alpha beta+ and TCR gamma delta+ T cell sin an experimental model of colitis. Eur J Immunol 27: 17–25, 1997.
19. M Feldmann, MJ Elliott, JN Woody, RN Maini. Anti TNFα therapy of rheumatoid arthritis. In Adv Immunol 64:283–350, 1997.
20. M Feldmann. Cell cooperation in the antibody response. In: Roitt I, Brostoff J, Male D, eds. Immunology. 4th ed. London: Mosby, 1996, pp 8.1–8.16.
21. FM Brennan, D Chantry, A Jackson, RN Maini, M Feldmann. Inhibitory effect of TNFα antibodies on synovial cell interleukin-1 production in rheumatoid arthritis. Lancet 2:244–247, 1989.
22. DM Butler, RN Maini, M Feldman, FM Brennan. Modulation of proinflammatory cytokine release in rheumatoid synovial membrane cell cultures. Comparison of monoclonal anti TNF-alpha antibody with the interleukin-1 receptor antagonist. Eur Cytokine Netw 6:225–230, 1995.
23. C Haworth, FM Brennan, D Chantry, M Turner, RN Maini, M Feldmann. Expression of granulocyte-macrophage colony stimulating factor in rheumatoid arthritis: Regulation by tumor necrosis factor α. Eur J Immunol 21:2575–2579, 1991.
24. RO Williams, M Feldmann, RN Maini. Anti-tumor necrosis factor ameliorates joint diseases in murine collagen-induced arthritis. Proc Natl Acad Sci USA 89: 9784–9788, 1992.
25. G Kohler, C Milstein. Continuous cultures of fused cells secreting antibody of predefined specificity. Nature 256:495–497, 1975.
26. DM Knight, H Trinh, J Le, S Siegel, D Shealy, M McDonough, B Scallon, MA

Moore, J Vilcek, P Daddona, J Ghrayeb. Construction and initial characterization of a mouse-human chimeric anti-TNF antibody. Mol Immunol 30:1443–1453, 1993.
27. BJ Scallon, MA Moore, H Trinh, DM Knight, J Ghrayeb. Chimeric anti-TNFα monoclonal antibody cA2 binds recombinant transmembrane TNFα and activates immune effector functions. Cytokine 7:251–259, 1995.
28. SA Siegel, DJ Shealy, MT Nakada, J Le, DS Woulfe, L Probert, G Kollias, J Ghrayeb, J Vilcek, PE Daddona. The mouse/human chimeric monoclonal antibody cA2 neutralizes TNF in vitro and protects transgenic mice from cachexia and TNF lethality in vivo. Cytokine 7:15–25, 1995.
29. VA Boussiotis, LM Nadler, JL Strominger, AE Goldfeld. Tumor necrosis factor α is an autocrine growth factor for normal human B cells. Proc Natl Acad Sci USA 91:7007–7011, 1994.
30. AP Cope, M Londei, NR Chu, SBA Cohen, MJ Elliott, FM Brennan, RN Maini, M Feldmann. Chronic exposure to tumor necrosis factor (TNF) in vitro impairs the activation of T cells through the T cell receptor/CD3 complex; reversal in vivo by anti-TNF antibodies in patients with rheumatoid arthritis. J Clin Invest 94:749–760, 1994.
31. Data on file, Centocor.
32. MW Marino, A Dunn, D Grail, M Inglese, Y Noguchi, E Richards, A Jungbluth, H Wada, M Moore, B Williamson, S Basu, LJ Old. Characterization of tumor necrosis factor-deficient mice. Proc Natl Acad Sci USA 94:8093–8098, 1997.
33. M Pasparakis, L Alexopoulou, V Episkopou, G Kollias. Immune and inflammation responses in TNFα deficient mice: a critical requirement for TNFα in germinal center formation and in the maturation of the humoral immune responses. Presentation at the 6th International TNF Congress. Eur Cytokine Netw 7:239, 1996.
34. P De Togni, J Goellner, NH Ruddle, PR Streeter, A Fick, S Mariathasan, SC Smith, R Carlson, LP Shornick, J Strauss-Schoenberger, JH Russell, R Karr, DD Chaplin. Abnormal development of peripheral lymphoid organs in mice deficient in lymphotoxin. Science 264:703–707, 1994.
35. TA Waldmann, W Strober, RM Blaese. Variations in the metabolism of immunoglobulins measured by turnover rates. In: Immunoglobulins. Biological Aspects and Clinical Uses. Washington, DC: National Academy of Sciences, pp 33–51, 1970.
36. REMICADE (infliximab) prescribing information.
37. FJ Baert, GR D'Haens, M Peeters, MI Hiele, TF Schaible, D Shealy, K Geboes, PJ Rutgeerts. Tumor necrosis factor α antibody (infliximab) therapy profoundly down-regulates the inflammation in Crohn's ileocolitis. Gastroenterology 116: 22–28, 1999.
38. SE Plevy, CS Landers, J Prehn, NM Carramanzana, RL Deem, D Shealy, SR Targan. A role for TNFα and mucosal T helper-1 cytokines in the pathogenesis of Crohn's disease. J Immunol 159:6276–6282, 1997.
39. HHF Derkx, J Taminiau, SA Radema, A Stronkhorst, C Wortel, GNJ Tytgat, et

al. Tumor necrosis factor antibody treatment in Crohn's Disease (letter). Lancet 342:173–174, 1993.
40. HM Van Dullemen, SJH Van Deventer, DW Hommes, HA Bijl, J Jansen, GNJ Tytgat, J Woody. Treatment of Crohn's disease with anti-tumor necrosis factor chimeric monoclonal antibody (cA2). Gastroenterology 109:129–135, 1995.
41. WR Best, JM Becktel, JW Singleton, F Kern. Development of a Crohn's disease activity index. National Cooperative Crohn's Disease Study. Gastroenterology 70:439–444, 1995.
42. SB Hanauer. Inflammatory bowel disease. N Engl J Med 334:841–848, 1996.
43. JY Mary, R Modigliani. Development and validation of an endoscopic index of the severity for Crohn's disease: a prospective multicentre study. Gut 30:983–989, 1989.
44. R Modigliani, JY Mary. Reproducibility of colonoscopic findings in Crohn's disease. A prospective multicenter study of interobserver variation. Dig Dis Sci 32:1370–1379, 1987.
45. SA Radema, H van Dulleman, M Mevissen, J Jansen, GNJ Tytgat, SJH van Deventer. Anti-TNFa therapy decreases production of chemokines in patients with Crohn's Disease. In: Cytokine production, immune activation, and neutrophil migration in inflammatory bowel disease. (PhD dissertation), Amsterdam: University of Amsterdam, The Netherlands, 1996.
46. RP McCabe, J Woody, SJH van Deventer, SR Targan, L Mayer, R van Hogezand, P Rutgeers, SB Hanauer, D Podolsky, CO Elson. A multicentral trial of cA2 anti-TNF chimeric monoclonal antibody in patients with active Crohn's disease. Gastroenterology 110(suppl):A962, 1996.
47. SR Targan, SR Hanauer, SJH van Deventer, L Mayer, DH Present, T Braakman, KL Dewoody, TF Schaible, PJ Rutgeerts, for the Crohn's Disease cA2 Study Group. A short-term study of chimeric monoclonal antibody cA2 to tumor necrosis factor α for Crohn's disease. N Engl J Med 337:1029–1035, 1997.
48. GR D'Haens, SJH van Deventer, R van Hogezand, DM Chalmers, T Braakman, T Schaible, P Rutgeerts, and the European cA2 Study Group. Anti-TNFα monoclonal antibody (cA2) produces endoscopic healing in patients with treatment-resistant, active Crohn's. Gastroenterology 114:A964, 1998.
49. P Rutgeerts, GD D'Haens, S van Deventer, D Present, L Mayer, S Hanauer, K DeWoody, T Schaible, S Targan, and the Crohn's Disease cA2 Study Group. Retreatment with anti-TNFa chimeric monoclonal antibody (cA2) effectively maintains cA2-induced remission in Crohn's disease. Gastroenterology 112: A1078, 1997.
50. DH Present, L Mayer, SJH van Deventer, P Rutgeerts, S Hanauer, SR Targan, K DeWoody, T Braakman, T Schaible and the Crohn's Disease cA2 Study Group. Retreatment with anti-TNFa chimeric antibody (cA2) effectively maintains cA2-induced remission in Crohn's disease. Am J Gastroenterol 92(suppl)A648, 1997.
51. MJ Elliott, RN Maini, M Feldmann, A Long-Fox, P Charles, P Katsikis, FM Brennan, J Walker, H Bijl, J Ghrayeb, JN Woody. Treatment of rheumatoid

arthritis with chimeric monoclonal antibodies to tumor necrosis factor α. Arthritis Rheum 36:1681–1690, 1993.
52. FC Arnett, SM Edworthy, DA Bloch, et al. American Rheumatism Association 1987 revised criteria for the classification of rheumatoid arthritis. Arthritis Rheum 31:315–324, 1988.
53. MJ Elliott, RN Maini, M Feldmann, JR Kalden, C Antoni, JS Smolen, B Leeb, FC Breedveld, JD MacFarlane, H Bijl, JN Woody. Randomised double-blind comparison of chimeric monoclonal antibody to tumour necrosis factor α (cA2) versus placebo in rheumatoid arthritis. Lancet 344:1105–1110, 1994.
54. ME Weinblatt, R Polisson, SD Blotner, JL Sosman, P Aliabadi, N Baker, BN Weissman. The effects of drug therapy on radiographic progression of rheumatoid arthritis: results of a 36-week randomized trial comparing methotrexate and auranofin. Arthritis Rheum 36:613–619, 1993.
55. GS Firestein, NJ Zvaifler. Anticytokine therapy in rheumatoid arthritis. N Engl J Med, 337:195–197, 1997.
56. RN Maini, FC Breedveld, JR Kalden, JS Smolen, D Davis, JG MacFarlane, C Antoni, B Leeb, MJ Elliott, JN Woody, TF Schaible, M Feldmann. Therapeutic efficacy of multiple intravenous infusions of anti–tumor necrosis factor α monoclonal antibody combined with low-dose weekly methotrexate in rheumatoid arthritis. Arthritis Rheum 41:1552–1563, 1998.
57. DT-Felson, JJ-Anderson, M-Boers, C-Bombardier, D-Furst, C-Goldsmith, LM-Katz, R-Lightfoot Jr; H-Paulus, V-Strand, et al. American College of Rheumatology. Preliminary definition of improvement in rheumatoid arthritis (see Comments). Arthritis Rheum 38:727–735, 1995.
58. AF Kavanaugh, JJ Cush, EW St.Clair, WJ McCune, TAJ Braakman, LA Nichols, PE Lipsky. Anti-TNFα monoclonal antibody (mAb) treatment of rheumatoid arthritis (RA) patients with active disease on methotrexate (MTX): results of a double-blind, placebo controlled multicentral trial. Arthritis Rheum 39(suppl): A575, 1996.
59. AF Kavanaugh, JJ Cush, EW St.Clair, WJ McCune, TAJ Braakman, LA Nichols, PE Lipsky. Anti–TNFα monoclonal antibody (mAb) treatment of rheumatoid arthritis (RA) patients with active disease on methotrexate (MTX): results of open label, repeated dose administration following a single dose double-blind, placebo controlled trial. Arthritis Rheum 39(suppl):A1296, 1996.
60. F Wolfe, DM Mitchell, JT Sibley, JF Fries, DA Bloch, CA Williams, PW Spitz, M Haga, SM Kleinheksel, MA Cathey. The mortality of rheumatoid arthritis. Arthritis Rheum 37:481–494, 1994.
61. P Prior. Cancer and rheumatoid arthritis: epidemiologic considerations. Am J Med 78(suppl 1A):15–21, 1985.
62. CA Williams, DA Bloch, J Sibley, et al. Lymphoma and leukemia in rheumatoid arthritis: are they associated with azathioprine, cyclophosphamide, or methotrexate? J Clin Rheumatol 2:64–72, 1996.
63. M Jones, D Symmons, J Finn, F Wolfe. Does exposure to immunosuppressive

therapy increase the 10 year malignancy and mortality risks in rheumatoid arthritis? A matched cohort study. Br J Rheumatol 35:738–745, 1996.
64. AJ Greenstein, GE Mullin, JA Strauchen, T Heimann, HD Janowitz, AH Aufses Jr, DB Sachar. Lymphoma in inflammatory bowel disease. Cancer 69:1119–1123, 1992.
65. MJ Fritzler. Drugs recently associated with lupus syndromes. Lupus 3:455–459, 1994.
66. A Flores, A Olivé, E Feliu, X Tena. Systemic lupus erythematosus following interferon therapy. Br J Rheumatol 33:787–792, 1994.
67. WB Graninger, W Hassfeld, BB Pesau, KP Machold, CC Zielinski, JS Smolen. Induction of systemic lupus erythematosus by interferon-γ in a patient with rheumatoid arthritis. J Rheumatol 18:1621–162, 1991.
68. EV Hess. Drug-related lupus. Curr Opin Rheumat 3:809–814, 1991.
69. EJ Price, PJW Venables. Drug-induced lupus. Drug Safety 12:283–290, 1995.
70. M Seitz, M Franke, H Kirchner. Induction of antinuclear antibodies in patients with rheumatoid arthritis receiving treatment with human recombinant interferon gamma. Ann Rheum Dis 47:642–644, 1988.
71. ECC Rankin, EHS Choy, D Kassimos, GH Kingsley, SM Sopwith, DA Isenberg, GS Panay. The therapeutic effects of an engineered human anti–tumour necrosis factor α antibody (CD571) in rheumatoid arthritis. Br J Rhematol 34:334–342, 1995.
72. ECC Rankin, CT Ravirajan, MR Ehrenstein, et al. Serological effects following treatment with an engineered human anti–TNF α antibody, CDP571 in patients with rheumatoid arthritis (RA). Br J Rhematol 34(suppl 1):101, 1995.
73. WA Stack, SD Mann, AJ Roy, P Heath, M Sopwith, J Freeman, G Holmes, R Long, A Forbes, MA Kamm, Hawkey. Randomised controlled trial of CDP571 antibody to tumour necrosis factor-α in Crohn's disease. Lancet 349:521–524, 1997.
74. LW Moreland, SW Baumgartner, MH Schiff, EA Tindall, RM Fleischmann, AL Weaver, RE Ettlinger, S Cohen, WJ Koopman, K Mohler, MB Widmer, CM Blosch. Treatment of rheumatoid arthritis with a recombinant human tumor necrosis factor receptor (p75)-Fc fusion protein N Engl J Med 337:141–147, 1997.
75. Enbrel, prescribing information.
76. F Hasler, L van de Putte, E Dumont, J Kneer, J Bock, S Dickinson, W Leeslauer, P van der Auwera. Safety and efficacy of TNF neutralization by lenercept (TNFR55-IgG1, Ro 45-2081) in patients with rheumatoid arthritis exposed to a single dose Arthritis Rheum 39:S243, 1996.
77. O Sander, R Rau, P van Riel, L van de Putte, F Hasler, M Baudin, E Ludin, T McAuliffe, S Dickinson, MR Kahny, W Lesslauer, P van der Auwera. Neutralization of TNF by lenercept (TNFR55-IgG1, Ro 45-2081) in patients with rheumatoid arthritis treated for 3 months: results of a European phase II trial Arthritis Rheum 39:S242, 1996.

5

Design and Development of Small-Molecule Inhibitor of Tumor Necrosis Factor-α

Ramachandran Murali, Akihiro Hasegawa, and Alan Berezov
University of Pennsylvania, Philadelphia, Pennsylvania

Kiichi Kajino*
Shiga University of Medical Sciences, Otsu, Japan

Wataru Takasaki
Drug Metabolism and Pharmacokinetics Research Laboratories, Sankyo Co., Ltd., Tokyo, Japan

INTRODUCTION

The most important function of the innate immune response is to eradicate foreign infectious agents such as microbes and viruses. In such cases, phagocytes and a variety of effector molecules accumulate at the site of infection and induce the secretion of several cytokines and other inflammatory mediators. This class of molecules has effects on subsequent immunological events such as inflammation and lymphoid apoptosis. Cytokines, secreted by macrophages in response to the foreign antigens such as microbes, are often called monokines, since they are produced by monocytes. Monokines include interleukin-1 (IL-1), interleukin-6 (IL-6), interleukin-8 (IL-8), interleukin-12 (IL-12), and tumor necrosis factor-α (TNF-α). Most of these molecules are pleiotropic (i.e., affect different biological functions). These effectors contribute to many elements and the inflammatory response at the site of infection.

As early as in 1893, Coley (1) observed beneficial inflammatory effects in terminally ill cancer patients. Subsequent investigations revealed that the curative effects were caused by induced release of TNF-α to the bacterial endotoxin (2). The name *tumor necrosis factor* was coined by Lloyd Old (2)

---

* Formerly at University of Pennsylvania, Philadelphia, Pennsylvania.

for its ability to trigger necrosis and involution of transplantable tumor (3). For some time, it was believed that the factor could be a chemotherapeutic agent against cancer. TNF-α was shown to be highly toxic to both humans and animals (3). In an unrelated experiment, cachectin isolated from waste body fluids of animals and humans with chronic diseases was determined to be the same as the necrosis factor. The effect of TNF-α on lipopolysaccharide (LPS)–induced biological functions led to the conclusion that TNF-α is also a strong mediator of shock, disseminated coagulator, metabolic acidosis, and end-organ damage brought about by LPS.

TNF has been the focus of research for several years and remains one of the most active areas of research. Early in 1980, TNF-α was cloned, sequenced, and purified. Subsequently, several biochemical and biological properties of TNF-α have been elucidated. Production of TNF-α has been found to promote chronic inflammatory disease. Overproduction or prolonged production of TNF-α has been observed to modify the anticoagulant properties of endothelial cells and neutrophils and promotes release of other proinflammatory cytokines such as IL-1 leading to cardiovascular collapse (4). On the other hand, low levels of TNF-α promote bone resorption, fever, and anemia (5). Normal production of TNF-α is thought to be critical for immune regulation against infection (6).

In addition to lysis of tumor and induction of sepsis, TNF-α has been shown to play a role in the development of other diseases. High concentrations of TNF-α in plasma correlate with the development of autoimmune processes (7–13) such as rheumatoid arthritis (RA) and Crohn's disease. Excessive secretion of TNF-α due to infection has also been shown to influence the development of pathophysiological conditions (6), including osteoporosis (14–16), cancer (17,18), and acquired immunodeficiency syndrome (AIDS) (19).

Multiple roles mediated by TNF-α in the development of autoimmune diseases prompted efforts to refocus on inhibition of TNF-α as a viable therapeutic agent for diseases other than cancer and sepsis (20). The molecule has been found to play a critical role in modulating programmed cell death (apoptosis) (7,21,22). Inhibition of TNF-α has proven to be of therapeutic value in some preliminary general studies (23,24). Rheumatoid arthritic symptoms have been improved through the use of anti-TNF-α antibody (13). Also, anti-TNF-α agents may be valuable in the treatment of bone resorption (25,26), obesity due to insulin resistance (27–29), and eye injury (30–32).

We have recently developed a novel anti-TNF-α peptidomimetic (33) based on a detailed structural knowledge of the TNF receptor (55 kD) and its complex with TNF-β (34–36). In this chapter, we will focus on the strategies used in the development of small molecular forms of anti-TNF-α and their status.

CURRENT APPROACHES IN ANTI-TNF-α DESIGN

An understanding of how the receptors work is essential for effective therapeutic design. The receptors for cytokines differ from growth factor receptors such as epidermal growth factor (EGF), prostaglandin F (PDGF), and ErbB2 receptors, because they lack endodomain with intracellular kinase domains. By association and aggregation with other members of receptors, cytokine receptors mediate a comparable complex pattern in modulating signal transduction and biological functions. Active cytokine receptors increase their phosphorylation by recruiting other components to assemble with the receptors (37). Functional redundancy observed in the cytokines' functions may be attributed to cross reactivity among the members of the superfamily of cytokines receptors (38). Cytokine receptors have been shown to share structural motifs such as, for example, interferon-γ receptor, immunoglobulin (Ig) superfamily fold, and chemokines, based on their structure, genetic organization, and cellular source (39). Ligand cytokines are found either in soluble forms or as membrane-associated molecules. Membrane-associated cytokines are often cleaved to perform the same functions as soluble ones and enable the two forms to perform complex biological functions by physical contact between cells and paracrine pathways.

TNF receptors (TNFR) exist in two forms: a 55-kD (TNFR-I) and 75-kD (TNFR-II) receptors. These receptors have been observed both as membrane-bound (TNFR) and soluble forms (sTNFR) and both show an ability to bind TNF-α. However, their biological functions differ. The functional role of TNFR-II is not quite clear, but studies indicate that TNFR-I is more critical for immune protection than TNFR-II (40,41).

Attempts to disable TNF-α functions are targeted at several levels and use different approaches. The methods extend from gene therapy to small molecule antagonists. The processes include (1) blocking the production of TNF-α synthesis through protease inhibitors, (2) interference with molecules involved in the signaling such as kinase inhibitors, (3) gene therapy to modulate nuclear factor-κB (NF-κB) expression, and (4) blocking excessive TNF-α in the plasma binding to its receptor either by antibody, soluble receptor, or small molecule antagonists. No one method has been proven to be uniformly effective. Some approaches in reducing excessive TNF-α have shown promise for therapeutic values in certain type of disease. For example, anti-TNF-α antibody, in general, has been shown to be effective against some arthritic problems (42,43) but not others (20,24). Nevertheless, the curative potential of anti-TNF-α agents may be improved by (1) understanding the role of TNF-α in the development of diseases and by (2) balancing the TNF-α concentration by precise intervention. To this purpose, it is important to address the current approaches and evaluate them for their advantages or disadvantages.

Modulation of Signal Transduction

The elucidation of the assembly of the TNF receptor–ligand complex and the associated signal transduction and biological function is an active area of research. Some progress has been made in identifying various molecules involved in the TNF-α-induced apoptosis pathways (22,44–46) and mechanisms of diseases (6,10,24,27,47–49).

TNF-α binding to its receptor leads to the activation of transcription factors such as transcription-activator proteins (AP-1) and NF-κB (50,51). These transcription factors play a role in the programmed cell death pathways (apoptosis). Thus, there has been considerable effort to control the expression of NF-κB for therapeutic use against cancers as well as other inflammatory processes as an alternate strategy to blocking TNF-α binding to its receptor per se. In this regard, molecules involved in signal transduction due to TNF-α are targeted. Protein tyrosine kinase (PTK) inhibitors such as genistein and erbstatin have been shown to block TNF-α–induced activation of NF-κB (52). Proteasome inhibitors may also be able to block the activation of some transcription factors (53–57). Other approaches including gene therapy (58–61) and antisense methods (41,62–64) have been noted to limit activation of transcription factors. These approaches are being actively pursued, and it may be too early to assess their merits.

Blocking Biosynthesis of TNF-α

Membrane-bound TNF-α is proteolytically processed and released as a soluble mature form. TNF-α is processed by proteases, known as matrix metalloproteinases (MMP) (65–68), members of a family of enzymes mainly involved in degradation of the extracellular matrix. Their biological role and relevance as therapeutic targets have been reviewed by others (68–70).

A relevant role for TNF-converting enzyme (TACE), which is a metalloproteinase disintegrin, has been identified in mice lacking the TACE gene. These mutant mice exhibited reduced production of TNF-α, suggesting that TACE is critical for releasing membrane-bound TNF-α (71,72). Subsequently, the gene was cloned (72). Natural and synthetic metalloproteinase inhibitors have been identified and some are in early clinical evaluation (70,71).

Small molecular inhibitors such as MMP inhibitors are nonselective agents. It has been shown that certain MMP inhibitors can block TNF-α not only synthesis but also TNFR cleavage (73). Recent crystal structure determination of MMP (74) may help to design more selective inhibitors. Some of the inhibitors are in early clinical trials mostly for cancer (69). Blocking p38 mitogen-activated protein kinase has been shown to reduce the production and accumulation of TNF-α (75,76).

Macromolecular Inhibitors of TNF-α

The classic approaches in the development of antagonists are either ligand mimics or substrates analogs. Antagonists may be discovered using high-throughput screening. Other approaches such as, for example, monoclonal antibody and minibodies are also being developed (13,42,77,78). These approaches require little knowledge of structure and function.

Monoclonal Antibody

Monoclonal antibodies have been proven to be successful in the treatment of several diseases (79). Monoclonal antibodies that bind TNF-α have shown promise in clinical trials (9,13,80,81). Although the monoclonal antibody approach has been promising in treating some cancers, there is no report on treating TNF-α–related diseases. Despite the limited success of monoclonal antibodies as therapeutic agents, it has some drawbacks: cost of production, humanization, and other related disadvantages associated with macromolecules (82).

Soluble Receptor

Soluble TNF receptor species have been detected in the plasma. It is thought that soluble receptors play a role of controlling the TNF-α activity and are necessary for normal immune regulation (6,48,83). Soluble receptors as therapeutic agents have been used in treating inflammatory diseases (40,47,81). In some cytokine systems, soluble receptors or antibodies do not neutralize the circulating cytokines but merely extend their half-life, thereby potentiating the effect (49).

One of the problems with soluble receptors as therapeutic agents is related to their long half-life; that is, the molecules will not be cleared from the body through the normal secretion process but may be active for a long time before the large molecules are degraded by proteases. This may have unfavorable effects on systemic immune system reactions in fighting infections. Since TNF-α has pleotropic effects in vitro and may even have more unexpected side effects in vivo. Further, these macromolecules may also induce neutralizing antibodies. Fusion proteins containing the constant domain of Ig (with a soluble receptor) may be particularly immunogenic, because they bind to the Fc receptors of antigen-presenting cells, thereby facilitating uptake and antigen presentation (49).

Disadvantages of Macromolecular Therapeutics

Although the advantage of macromolecules as drugs has been reported and it has been shown that they (1) are highly specific, (2) are selective, (3) often do not require extensive analysis of biodistribution and toxicity, as in the case of small molecules, and (4) have a long half-life, macromolecules also have some drawbacks: (1) commercial-scale production may be either difficult or costly, (2) purity may be difficult to achieve and microheterogeneity may be inevitable, (3) conformational stability may vary with environment of body fluids, and (4) they may be excluded from certain compartments such as the blood-brain barrier (82).

Most of the above disadvantages of macromolecules can be overcome by creating small molecular inhibitors. Small molecules have their own limitations, including their biodistribution and half-life. Often peptides are created first to assess biological effects, but when used as a template lead to further development of viable therapeutic agents (84–86).

Several macromolecular anti-TNF-α agents such as monoclonal antibodies and soluble receptors have already been used in clinical trials. Some successes have been reported in the treatment of certain diseases. In the treatment of sepsis, anti-TNF-α treatments did not reduce the mortality (20), but in Crohn's disease where chronic TNF-α synthesis is thought to occur, treatment by monoclonal antibodies has improved the disease (87). As mentioned, one potential side effect with macromolecular anti-TNF-α agents treatments such as antibodies is that they have a longer half-life, and the prolonged inhibition of a critical inflammatory agent such as TNF-α might compromise the natural immune response. A small molecule inhibitor may therefore be more suitable, not only for the development of viable drugs, but also for more controlled intervention of TNF-α in the plasma.

RATIONAL DESIGN OF SMALL MOLECULE ANTI-TNF-α

A knowledge-based approach to drug design has been shown to be effective. The design of a human immunodeficiency virus (HIV) protease inhibitors from the structure and function of the enzyme highlights the progress made in recent years in structural biology. Unlike conventional approaches, rational design offers not only a viable lead, but may also provide the opportunity to decipher the biological role of target molecules. Our approach to the design and development of the antireceptor antagonists stems from a combined structural analysis of relevant molecules, antibodies, ligands, and receptors. An understanding of the properties of molecular recognition at the atomic level has allowed us to engineer molecules that either can mimic the ligands or modulate the receptors' signaling function.

Recent progresses in crystallographic, nuclear magnetic resonance (NMR), and molecular modeling techniques have created a large three-dimensional structure database. These efforts have shown that related macromolecular structures are often highly conserved. Macromolecules, in general, have distinct components: (1) Scaffolds: required for stability of the folding (e.g., the framework regions in an antibody); (2) functional regions: mainly small regions in a molecule which are involved in molecular recognition (e.g., complementary determining region [CDR] loops in an antibody, β-turns, or long flexible loops); and (3) functional surfaces/cavities: folded proteins contain solvent accessible or occupied clefts or cavities which are required for their functions (e.g., active site in an enzyme is a large cavity where substrate binds). When the structure of either target protein or its associating members is known, it may be possible rationally to design agonists or antagonists (88,89).

Structure of Immunoglobulins

The immunoglobulin fold is one of the most common folds in proteins and receptors involved in immunological functions. Macromolecules which contain this fold range from antibody fragments to T-cell receptor complexes. Immunoglobulin domains were first identified in antibodies. The fold contains six to seven β-strands arranged like a barrel with extended CDR loops. The CDR loops are the functional secondary structures that are responsible for molecular recognition.

Molecular and crystallographic analyses of immunoglobulins have revealed that critical ligand-binding surfaces are predominantly the CDR loop projections. Canonical conformations of the CDR of the V_k light chain CDR, and two of three of the heavy chain CDR have been noted. The third CDR of the heavy chain, as a consequence of the complex genetic mechanism which influences its structure, has medium or long loops which have diverse patterns of interactions. In general, the canonical CDR, aside from the CDR3 of the heavy chain, have reverse turn conformations which sometimes have the regular features of β turns. In addition, the two constant domains, C1 and C2, although similarly fashioned, have different roles; the C1 domains are involved in antigen interactions, whereas C2 domains subserve Fc receptor and adhesive structures such as leukocyte function-associated antigen-3 (LFA-3), myclin-associated glycoprotein (MAG), CD2, neural cell adhesion molecule (NCAM), and intercellular adhesion molecule (ICAM) (90,91).

Conformational properties of CDR loops or reverse turns are considered to be important mediators in the biological activity of polypeptides. Turns provide suitable orientations of binding groups essential for bioactivity by stabilizing a folded conformation, and thus may be involved in both binding and recognition (92,93). Moreover, crystallographic studies of antigen-antibody complexes reveal that molecular recognition often requires an interface area of about 600–1000 $Å^2$. Yet, most of the intermolecular interactions are mediated by few residues; predominantly from the heavy chain CDR3 region and in some cases from light chains (94). The studies of small naturally secreted peptides, such as somatostatin and enkephalins, have shown that the structural aspects of β turns, namely, optimal disposition of side chains, leads to effective binding to receptors.

Structure of Receptors

Analysis of the evolution of biological molecules has revealed that the structural topology is more conserved across different species than the primary sequence. This is evident from the fact that many proteins of immunological interest and the large number of cell surface receptors and growth factors share the immunoglobulin fold. A common feature of receptor types shows that they often have predominantly immunoglobulin folds or cystine knot repeats. The immunoglobulin fold is characterized by six to eight β strands which are sandwiched against each other. Some of the β strands are stabilized by at least one disulfide bond. On the other hand, the cystine knot is characterized by three or four β strands which are stabilized by three disulfide bonds. Unlike the immunoglobulin fold, which has a unique globular topology, the cystine knot has been shown to adopt different topologies (95,96). In the receptors, cystine knots are arranged in a head-to-tail or elongated fashion in the receptors studied to date. In multidomain macromolecules, the role of individual subdomains can be considered either as framework or scaffolds or as functional units. In immunoglobulin and the cystine knot, β strands stabilized by disulfide bonds can be considered to be scaffolds and loops interleaved between them as functional units.

Further, irrespective of structural topology, the subdomain of all macromolecules is built by four major secondary structural elements: the alpha helix, β sheet, β turn, and loops. All the secondary structures are well defined and classified except loops. The loop structures are highly variable and often mobile. In many protein-protein and receptor-ligand complexes, the flexible loop structures are often involved in binding to their counterparts, and thus loops have been attributed to their role in molecular recognition and binding (97–99). An attempt to classify loops in the protein is still elusive, but in certain domain structures, it is possible to predict and classify small loops owing to the small length of the amino acid sequence involved (99–101). Thus, the functional units in biological molecules appear predominantly as loops or reverse turns between the structural framework regions.

In designing antireceptor small molecules, we have employed a variety of features such as structures of antibodies, receptors, ligands, and biochemical and biological data to design antireceptor antagonists. We have designed small molecule antagonists of CD4 (102,103) and TNF receptors (33). The structure of the CD4 receptor contains an immunoglobulin domain similar to an antibody, and TNF receptors contain ''cystine knot'' repeating domains. In the following sections, the design is illustrated for the TNF receptor. Although these approaches are described in the context of TNF receptors, they can be also used for other related receptors (86).

General Strategy in the Design of Peptidomimetics

Linear peptides possess too much inherent flexibility. Studies have shown that constraining the peptides enhanced their stability and, in some case, their affinity. Achieving structural stability, solubility in physiological relevant solutions, and bioviability are necessary properties for a therapeutic use. We have modified cyclic peptides to increase their stability and bioviability by addition of aromatic residues at the termini.

Placement of Constraining Cystine Residues

Placement of constraining cystine residues in the loop or β turn structure is critical. Constraining cystine residues in a cyclic peptide alters both ring size and the conformation of other residues. For this paired cystine residues are placed systematically at residues (Cα atoms separated at least by 6.2 Å) away from the critical residues using the program MODIP (R. Murali, unpublished data). The effect of disulfide closure on the loop structure of a peptide and its loop ring size are evaluated by a conformational search (100) followed by molecular energy minimization and dynamics (INSIGHT, Molecular Simulations, Inc., San Diego, California).

Aromatic Modification of Peptides

Cyclization of peptides confers structural and conformational rigidity which are critical for optimal interaction with macromolecules, but it does little to improve solubility. To increase stability, solubility, and other properties, a variety of strategies (85) have been adopted. For example, mixed anhydride coupling has been used (104), and DeGrado and colleagues have used a semi-rigid linker m-aminomethyl benzoic acid which links the two ends of the peptide in a simple reaction (105,106). These simple chemistries have allowed the development of different forms of constrained readily synthesized small molecules of known structure.

Distribution of hydrophobic residues and to a smaller extent hydrophilic residues has been observed at the protein-protein interfaces and within the antibody-combining site (107–109). A general explanation for such a disproportionate distribution at the interface is attributed to their role in stabilizing interactions. Owing to their large hydrophobicity, especially aromatic residues with planar π charges, they can exclude solvents at the interface and decrease the entropy for a higher binding property (109). Based on these observations, we proposed that bulky aromatics such as phenylalanine and tyrosine, known to decorate by enhancing the binding surface area, would protrude from the antigen-binding surface and promote forming either ordered water binding to main chain residues or exclusion of solvent in the binding site. These bound waters then lead to a network of water chains that become involved in bridging antigen and antibody. This concept provides a basis for more rational structural modeling and has been used to aid our creation of small forms of molecules.

Based on our studies of the dominant features of aromatics in the antigen-binding surface and their ability to lead to propagated water networks that facilitate binding and increase complementarity, we have used aromatic residues in our macrocyclic constructs. In our early studies, we found that addition of aromatic residues Phe or Tyr to the termini of peptides greatly increased the 50% efficiency of cyclization to almost 100% (Berezov and Murali, unpublished results). Furthermore, the aromatically modified cyclic peptides behaved far more efficiently than any small compound tested to date, with some possessing nanomolar (nM) affinity. The thermodynamic consequence of the bulky hydrophobic residues on ordering water molecules may be responsible for the improved relative affinity of the compounds. Thus, in our design of peptidomimetics, the peptides are generally flanked by aromatic residues at both N and C termini, but clearly other thermodynamically equivalent modifications could be employed.

Molecular Modeling

There are two major approaches: (1) a de novo folding design using energy minimization and molecular dynamics and (2) comparative modeling followed by energy minimization and molecular dynamics. These two approaches differ only in developing the trial or initial structures. The folding patterns are studied using energy minimization and molecular dynamics. Parameters used in the modeling of peptide mimics have been described earlier by our group (110). A detailed description of various methods used to design peptidomimetics is beyond the scope of this chapter. A brief summary of the strategy used by us is discussed.

Initial trial structures were developed using a database consisting of loops from proteins in the Brookhaven protein database (111). Based on the sequence similarity and the loop size, trial structures were selected. Each of the structures was evaluated for the loop size, relative orientation of the side chains, and solvent effects using a combination of energy minimization and molecular dynamics. In the simulation studies, both room temperature (300°K) and high temperature (900°K) are employed. Low-energy conformers are then subjected to further minimization and compared with the native conformation of the template. Each assigned score is based on the similarity (as measured by Cα atoms), relative disposition of critical amino acids with respect to their neighboring residues, predicted solubility, and ability to form oligomers. When required, original amino acid residues in the template are replaced in an iterative manner to conform to the above criteria.

Design of a TNF-α Inhibitor

Structure of the TNF Receptor Complex

Small molecule TNF antagonist molecules are designed from the combined structural knowledge of the TNF receptor complex and anti-TNF antibodies. Several good reviews describing the structural complex of the TNF receptor are reported in the literature (112,113). For readers who are not familiar with the structure of the TNF receptor complex, a brief description is provided below.

Active TNF is a trimeric molecule. Binding of TNF-α to its receptor promotes formation of a trimeric receptor complex. Recent crystallographic work reveals that TNF-α, TNF-β, and TNFR-I exist as trimers. The crystalline structure of the TNF receptor both in complexed and uncomplexed forms provides a general understanding by which these receptors bind to their ligands (34,35) (Fig. 1). TNF receptors are characterized by cysteine residue repeat. This type of repeat has been found in other protein species such as toxins and has come to be known as the cystine knot (95,114). The cystine knot in the TNF receptor family consists of 42 amino acid residues with six cysteine resi-

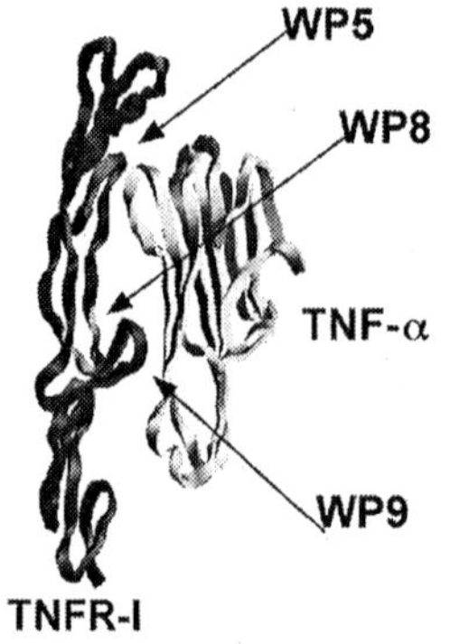

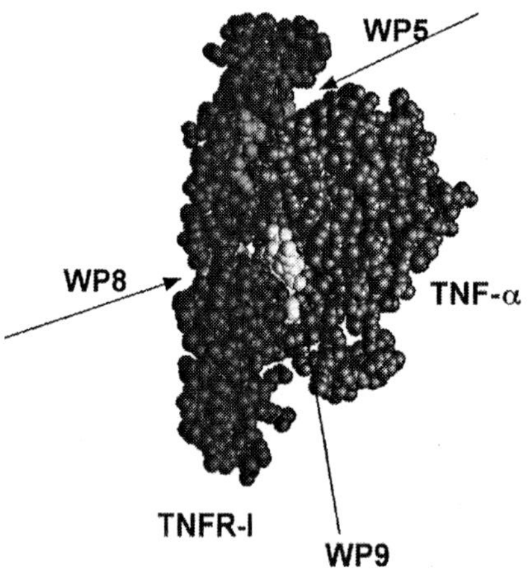

Figure 1 Complex of TNFR and TNF-α is shown as (a) ribbon and (b) space-filling models. The model is based on the crystal structure of TNFR and TNF-β (36). Crystal structure of TNF-α (34) was superimposed on TNF-β to create the complex. The three major sites considered for peptidomimetic design are shown: domain 1, WP5; domain 2, WP8; and domain 3, WP9.

dues forming three interchain disulfide bonds to create the structural motif. The three-dimensional structure of the TNF receptor reveals four cystine knot repeats with each repeat about 30 Å in length and arranged in a head-to-tail fashion exposing the loops on one side of the receptor. These loops appear to be involved either in oligomerization or in ligand binding (34,35). Although there is no direct evidence that the TNF receptors form dimers in solutions, the three-dimensional structure of uncomplexed TNF receptors shows receptors associated as either parallel or anti-parallel dimers. It has been argued that this may represent the oligomerization pattern of TNF receptors (115). Nevertheless, in the parallel dimeric form, the first and last cystine knot domains are involved in dimeric contact (116). In both crystalline structures, the membrane proximal domain is disordered; perhaps due to the lack of the transmembrane that normally holds this domain's structure in a stable form.

In the following sections, the basis and design strategy of an anti-TNF inhibitor is discussed. The design includes consideration of all the three components: (1) antibodies, (2) ligand, TNF-α, and (3) the TNF receptor.

Antibody as Template

Anti-TNF-α antibodies have been shown to block TNF-α and are being evaluated for clinical usefulness against rheumatoid arthritis (117). Often antibodies mirror competing antigens either structurally or sequencewise or both (118). Several antibody epitope-mapping studies suggested that the antibodies might bind to the interface of trimeric TNF-α. It was suggested that one of the mechanisms by which antibodies block TNF-α is by conformational restriction (34). At least four anti-TNF antibodies, Di62 (119), 007, 189, and 004 (120), have been reported and have been sequenced. Similarities between the primary structure of all relevant antibodies and receptor molecules have been compared. Peptides derived from the CDR3 region of the light chain of Di62 have been shown to possess antagonistic activity (119).

The anti-TNF-α monoclonal antibody Di62 analysis led to CDR-based TNF-α antagonists (119). The peptides inhibited TNF-α binding to both TNFR-I and TNFR-II receptors. Inhibition of apoptosis in L929 cells at micromolar concentrations (IC_{50} = 6-μM) were noted. Interestingly, these unconstrained peptides show much higher activity than constrained ones (119), suggesting that peptide oligomers may be the active form.

We have modeled the three-dimensional structure of all the four antibodies (101) using AbM (Oxford Molecule, Inc., Hunt Valley, Maryland). The structures of all the three CDR loops were analyzed. Comparison of the sequences of Di62 revealed about 36% homology to the TNF receptor in the region of residues 75–85 whereas the CDR3H of 004 antibodies share 50% homology to a region in TNF-α (128–139) which is located at the interface of the trimer in the ligand complex. Since Di63 and 004 antibodies have some similarities to receptor and TNF-α, respectively, the CDR3L of Di63 and CDR3H of 004 loops were considered as a template. Similarities in the loop structures were analyzed normally on a graphics workstation. The following peptides were designed from the CDR of the two monoclonal antibodies.

Mimic	Sequence	Antibody	Template	Original sequence
WP1	YCSQSVSNDCF	Di62	CDR3L	TASQSVSNDVV
WP2	YCSGDSLRRLCF	004	CDR3H	SGDSLRRLVYFDY

Ligand as Template

Mimics of ligands that bind to receptors with high affinity could act as competitive inhibitors to the ligand. In the crystalline structure of TNF-β complexed to its receptor, there is one water molecule found at the interface and this suggests that there is either a large conformational change in the molecules on forming a complex or the interface is solvent accessible. No significant conformational change was observed from the analysis of both complex and uncomplexed receptor-ligand structures. In order to open up the interface to water molecules, at least some part of the molecule must be flexible. A loop region (105–113) exposed at the bottom of the ligand may facilitate such an opening for solvents. Moreover, this loop is part of the binding region of the receptor. It is common that, in antibodies, one of the three loops is sufficient for binding. Thus, it is possible to use this loop as a template to mimic ligand binding to its receptor.

TNF-α and TNF-β are structurally very similar, although, in primary sequences, they are only 33% related. We have compared the structure of TNF-α and TNF-β by superimposing them on each other. Key residues are mapped onto the structure, and their interactions with the receptor were studied. The CDR3H of the 004 antibody has some similarities to regions in TNF-α (128–133). These regions are proximal to the receptor-binding site but are not exposed. Nevertheless, we designed peptide mimetics from this region (WP3) to see whether they possess any activity. Also, the loop region 105–113 was used as a template, and several peptidomimetics have been designed (WP4). The following peptidomimetics were designed.

Mimic	Sequence	Structural Template
WP3	YC KGDRLS CY	Antibody 004, CDR3H
WP4	YC AVSYQTKVN CF	TNF-α (105–113)
	YC RI AVSYQTKVN CF	
	YC SYQTKVN CF	
	YC VSYQTKVN CF	

Receptor as a Template

The functional form of the receptor is oligomeric. This means that the ligand and receptor have multiple contact sites. In the crystallographic analysis, whereas uncomplexed TNF receptors associate as dimers, complexed form associates as a trimer. There is no evidence to support that the dimeric form of TNF receptors exist in solution (116). It is possible that crystallization artifacts may create oligomeric complexes in the crystal lattice that have no relevance to the biologically active form. Thus, it is important to correlate the three-dimensional structure of the complex TNF receptor to the biological function (115).

The trimeric complex of the TNFR-I receptor and TNF-β has been determined at atomic resolution (36). TNF-α and TNF-β molecules share a similar three-dimensional structure and 33% sequence homology. A trimeric complex of TNFR-I and TNF-α was built by superimposing TNF-α over the TNF-β three-dimensional structure. Binding sites were identified in both TNFR–TNF-β and TNFR–TNF-α receptor complexes. Based on the analysis, we designated three sites for the design of peptidomimetics (see Fig. 1). These sites are referred to as WP5, WP8, and WP9. Each of these contact sites is located in different parts of the receptor: WP5 is located in the first domain; WP8 is located in the second domain; and WP9 is located in the third domain. At the three interaction sites, specific loops were identified: 60–67 (WP5), 76–83 (WP8), and 107–111 (WP9) as a template. Based on the technique described earlier, the following peptides were designed.

Mimic	Sequence	Template
WP5	YC FTASENH CY	Loop (60–67) in domain 1
WP5N	YC FTNSENH CY	
WP5R	YC FTRSENH CY	
WP5J	FC ASENH CY	
WP5JN	FC NSENH CY	
WP5JY	YC ASENH CY	
WP8L	YC RKELGQV CY	Loop (76–83) in domain 2
WP8J	YC RKEMG CY	
WP8JF	FC RKEMG CY	
WP98JP	YC KEPGQ CY	
WP9Q	YC WSQNL CY	Loop (107–111) in domain 3
WP9ELY	YC ELSQYL CY	

Evaluation of Peptidomimetics

Cyclic peptides derived from all the three surface loops of the TNF receptor, namely, the loop (56–73) in domain 1, (76–83) loops of domain 2, and the first loop (107–114) of domain 3 on the TNF receptor (33) have been tested for biological and biochemical functions.

The activity of the peptides were evaluated for (1) binding to the receptor, (2) ability to compete with TNF-α for binding to the receptor, and (3) measuring the ability to inhibit apoptosis (MTT = 3,(4-5 dimethyl thiazol-2-yl)-2-5-diphenyl tetrazolium bromide assay). Peptides derived from antibody (WP1, WP2) and ligand (WP3, WP4) showed a little or no effect, but the peptides derived from receptors consistently showed significantly higher activity. The results for the first generation of active peptides from the receptor are shown in Figure 2.

The cyclic peptide, WP9Q, was the most promising, as deduced from the initial screen. We have reengineered the peptide by careful analysis of the interaction site. A close look at the interaction site revealed that the binding site (WP9) is accessible to solvents (Fig. 3). The crystalline structure of TNFR–TNF-β shows that one of the charged residues, 109E, in the loop is disordered. The undetected side chain of 109E is within the contact distance

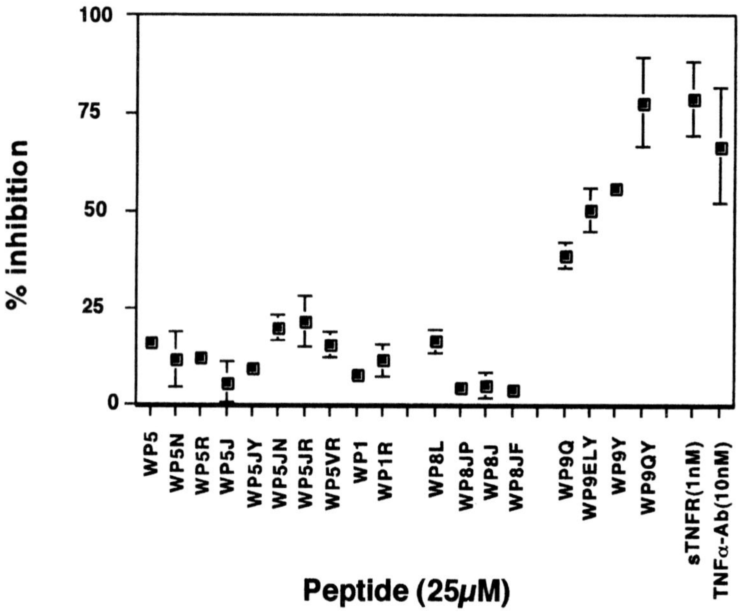

Figure 2 Peptides designed from antibody (WP1), the ligand TNF-α (WP4) and receptor (WP5, WP8, WP9) have been evaluated for their ability to compete with TNF-α for binding to the receptor. Only the peptides that show measurable activity are shown. Inhibition of ^{125}I–TNF-α and the receptor interaction by exocyclic peptidomimetics in competitive radioreceptor assay. Inhibitory activities were compared at 25 μM of each peptide, 1 nM of soluble TNF receptor and 10 nM of anti–TNF-α antibody.

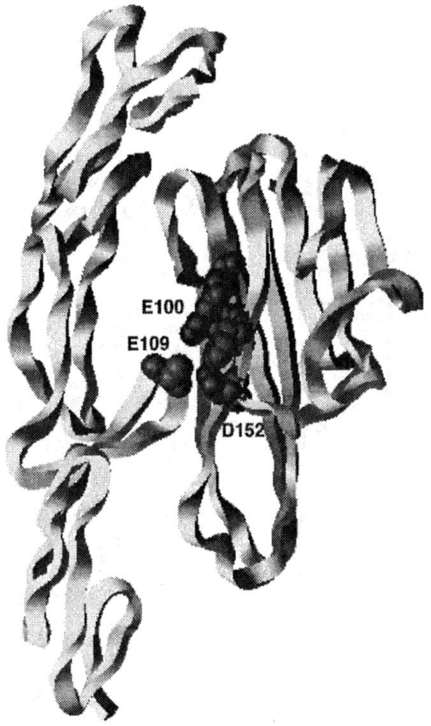

Figure 3 Analysis of WP9 site. The interaction site, WP9, of TNFR and TNF-β is shown using SETOR (132). This interaction site, WP9 is solvent accessible. At this site, E109 in the loop is disordered and the side chain atoms are not visible in the electron density (36). The interface at this site is negatively charged owing to contribution residues from TNF-α are shown. See text for further discussion.

of 150H, 152D, E100, and Q102. Thus, the interface is negatively charged at the WP9 site. Since the loop is exposed to solvents, water molecules may involve in the interaction between TNF-β and TNFR through glutamic acid in the loop (109E). Further, charged residues have been implicated in recruiting solvent molecules (121). Water molecules at the site of interaction provide stability (122–124). Macromolecules have a large surface area and one or two water molecules can provide stability. However, small molecules cannot mimic the large surface area of a protein. In such cases, water molecules tend to weaken the interactions owing to unstable secondary structure formation (125,126). So we modified the peptide with bulky aromatic residues and replaced a charged residue with a polar residue. Based on our modeling, the following peptides were designed.

Mimic	Sequence	Template
WP9Q	YC WSQNL CY	Loop (107–111) in domain 3
WP9QY	YC WSQYL CY	
WP9Y	YC WSQNY CY	

The peptidomimetic engineered from the third domain (WP9QY) was the most potent and inhibited TNF-α binding (IC_{50} = 7.5 μM) to its receptor (Fig. 4). Also, the peptidomimetic protected cells against TNF-α–induced cell

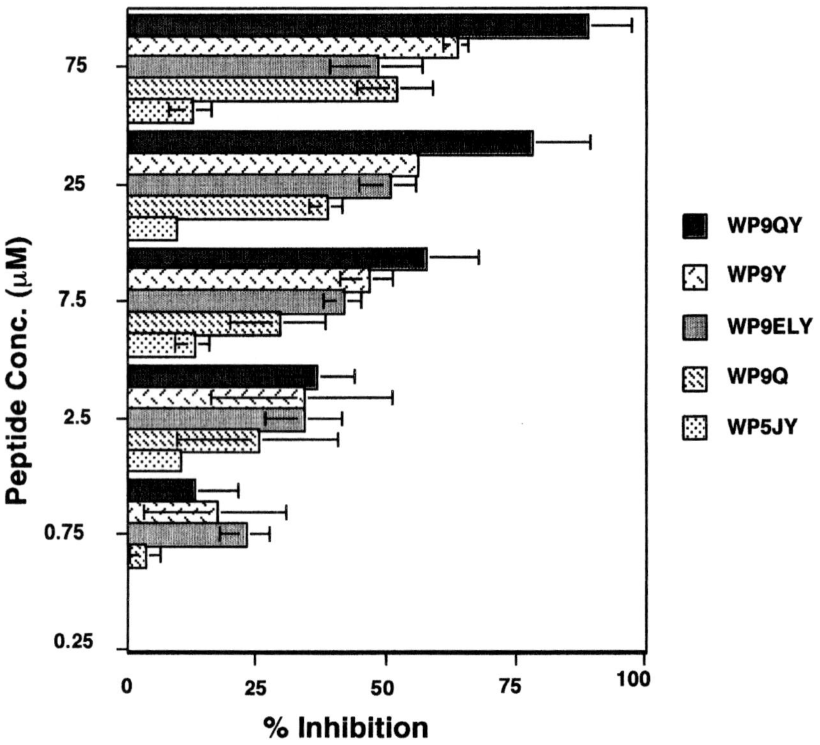

Figure 4 Evaluation of second generation peptide mimic from WP9 site. Inhibition (%) by several doses of peptides were calculated and plotted. The experimental conditions are same as explained in Figure 2. The results indicate the means and standard deviations derived from three independent experiments.

death when apoptosis was induced with 7 pg of TNF-α, suggesting that the peptide specifically binds to TNF-α (Fig. 5).

The peptidomimetic shows therapeutic values. These peptides are soluble and have been tested in vivo. In a relapsing experimental autoimmune encephalitis (EAE) model, mice were immunized with mouse spinal cord homogenate. The peptides have been administered at 150 μg/mL per day for 3 days after the onset of disease. The peptides have diminished the progression of disease but did not cure the disease (Murali and Greene, unpublished result).

Peptidomimetic (WP9QY) is one of the first peptides to demonstrate anti–TNF-α activity and can be further improved as a substitute for antibody. The method developed by us to design an antireceptor antagonist by combining the knowledge from antibodies, ligands, and receptors can be extended to other receptor systems which share structural homology with TNF.

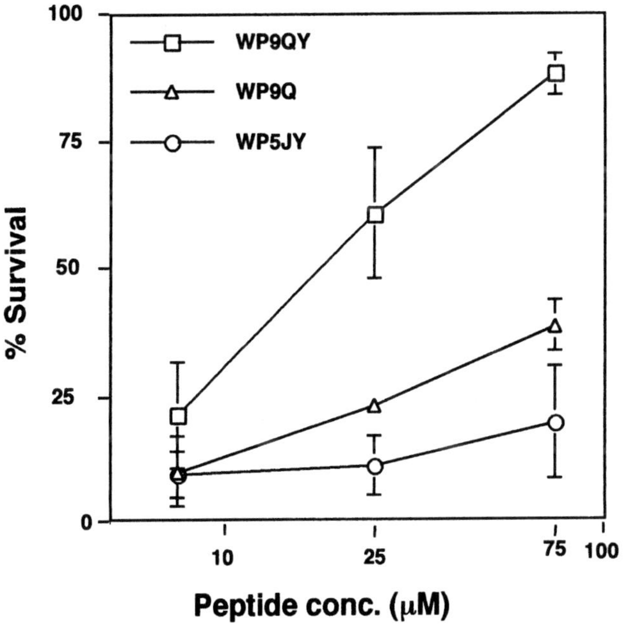

Figure 5 Inhibition of TNF-α–induced cytolysis of L929 cells by the antagonistic peptides. Absorbance obtained with 1 μg/mL of ACT-D alone and with ACT-D and 50 pg/mL of TNF-α were referred as 100% survival and 100% cytotoxicity, respectively. The results indicate the means and standard deviations derived from three independent experiments.

CONCLUSIONS AND PERSPECTIVES

Structure-based drug design has proved to be more cost effective and can be selective (88). Recent success in the development of HIV protease inhibitors (127,128) and anticoagulants are some examples where the drugs were designed and developed on a three-dimensional structure. The three-dimensional structure of TNF-β (35) and the TNF receptor (116) and its complex with TNF-β enabled us first to design anti-TNF-α small molecules that are specific and selective. Further, the structural study of the TNF receptor not only enhanced our understanding of their function but also led to the realization that the TNF receptor's topology is not unique but rather is shared by many other receptors (129). Now the TNF receptor and its ligand complex becomes the template for understanding other receptors and their function. This aspect led to the molecular modeling of other receptors (130,131) and thus accelerated the development of therapeutic agonists and antagonist for other receptors (86).

Several growth factors and tyrosine kinase receptors share the TNF structural organization (39,129), and the peptidomimetic design described here can be considered as a paradigm for development of peptidomimetics to other members such as CD40, Fas, and p185/neu/HER-2 (48,129).

ACKNOWLEDGMENT

We would like to thank Prof. Mark I. Greene for his suggestions and support. R.M. would like to thank Xycte Therapies, Inc. for partial support of this work.

REFERENCES

1. WB Coley. The treatment of malignant tumor by repeated innoculations of erysipelas: with a report original cases. Am J Med 105:487–511, 1893.
2. EA Carswell, LJ Old, RL Kassel, S Green, N Fiore, B Williamson. An endotoxin induced serum factor that causes necrosis of tumor. Proc Natl Acad Sci 72:3666–3670, 1975.
3. B Beutler. TNF, immunity and inflammatory disease: lessons of the past decade. J Invest Med 43:227–235, 1995.
4. F Bazzoni, B Beutler. Seminars in Medicine of the Beth Israel Hospital, Boston: the tumor necrosis factor ligand and receptor families. N Engl J Med 334:1717–1725, 1996.

5. B Beutler, A Cerami. Tumor necrosis, cachexia, shock, and inflammation: a common mediator. Annu Rev Biochem 57:505–518, 1988.
6. B Beutler, GE Grau. Tumor necrosis factor in the pathogenesis of infectious diseases. Crit Care Med 21:S423–S435, 1993.
7. B Beutler, F Bazzoni. TNF, apoptosis and autoimmunity: a common thread? Blood Cells Mol Dis 24:216–230, 1998.
8. JM Clements, JA Cossins, GM Wells, DJ Corkill, K Helfrich, LM Wood, R Pigott, G Stabler, GA Ward, AJ Gearing, KM Miller. Matrix metalloproteinase expression during experimental autoimmune encephalomyelitis and effects of a combined matrix metalloproteinase and tumour necrosis factor-alpha inhibitor. J Neuroimmunol 74:85–94, 1997.
9. DA Fox. Biological therapies: a novel approach to the treatment of autoimmune disease. Am J Med 99:82–88, 1995.
10. CM Hill, J Lunec. The TNF-ligand and receptor superfamilies: controllers of immunity and the trojan horses of autoimmune disease? Mol Asp Med 17: 455–&, 1996.
11. R Hohlfeld. Inhibitors of tumor necrosis factor-alpha: promising agents for the treatment of multiple sclerosis? Mult Scler 1:376–378, 1996.
12. MG Shire, GW Muller. TNF-alpha inhibitors and rheumatoid arthritis. Expert Opin Ther Pat 8:531–544, 1998.
13. M Takeno, T Sakane. Monoclonal antibodies for treating autoimmune diseases. Nippon Rinsho 55:1543–1548, 1997.
14. GA Rodan. Introduction to bone biology. Bone 13:S3–S6, 1992.
15. T Fujita, T Matsui, Y Nakao, S Shiozawa, Y Imai. Cytokines and osteoporosis. Ann NY Acad Sci 587:371–375, 1990.
16. K Kurokouchi, F Kambe, K Yasukawa, R Izumi, N Ishiguro, H Iwata, H Seo. TNF-alpha increases expression of IL-6 and ICAM-1 genes through activation of NF-kappaB in osteoblast-like ROS17/2.8 cells. J Bone Miner Res 13:1290–1299, 1998.
17. JY Blay, S Chouaib. Tumor necrosis factor alpha (cachectin). Biological properties and role in physiopathology. Presse Med 18:975–979, 1989.
18. V Chopra, TV Dinh, EV Hannigan. Circulating serum levels of cytokines and angiogenic factors in patients with cervical cancer. Cancer Invest 16:152–159, 1998.
19. G Herbein, LJ Montaner, S Gordon. Tumor necrosis factor alpha inhibits entry of human immunodeficiency virus type 1 into primary human macrophages: a selective role for the 75-kilodalton receptor. J Virol 70:7388–7397, 1996.
20. A Edward. Cytokine modifiers: Pipe dream or reality. Chest 113:224S–227S, 1998.
21. JP Revillard, L Adorini, M Goldman, D Kabelitz, H Waldmann. Apoptosis: potential for disease therapies. Immunol Today 19:291–293, 1998.
22. J Yuan. Transducing signals of life and death (Review). Curr Opin Cell Biol 9:247–251, 1997.

23. AG Porter. The prospects for therapy with tumour necrosis factors and their antagonists. Trends Biotechnol 9:158–162, 1991.
24. L Probert, K Selmaj. TNF and related molecules: trends in neuroscience and clinical applications (Review). J Neuroimmunol 72:113–117, 1997.
25. AA Deodhar, AD Woolf. Bone mass measurement and bone metabolism in rheumatoid arthritis: a review. Br J Rheumatol 35:309–322, 1996.
26. SE Ross, RO Williams, LJ Mason, C Mauri, L Marinova-Mutafchieva, AM Malfait, RN Maini, M Feldmann. Suppression of TNF-alpha expression, inhibition of Th1 activity, and amelioration of collagen-induced arthritis by rolipram. J Immunol 159:6253–6259, 1997.
27. M Halle, A Berg, H Northoff, J Keul. Importance of TNF-alpha and leptin in obesity and insulin resistance: a hypothesis on the impact of physical exercise. Exerc Immunol Rev 4:77–94, 1998.
28. Y Morimoto, K Nishikawa, M Ohashi. KB-R7785, a novel matrix metalloproteinase inhibitor, exerts its antidiabetic effect by inhibiting tumor necrosis factor-alpha production. Life Sci 61:795–803, 1997.
29. L Ragolia, N Begum. Protein phosphatase-1 and insulin action. Mol Cell Biochem 182:49–58, 1998.
30. AF De Vos, MA Van Haren, C Verhagen, R Hoekzema, A Kijlstra. Systemic anti-tumor necrosis factor antibody treatment exacerbates endotoxin-induced uveitis in the rat. Exp Eye Res 61:667–675, 1995.
31. G Sartani, PB Silver, LV Rizzo, CC Chan, B Wiggert, G Mastorakos, RR Caspi. Anti–tumor necrosis factor alpha therapy suppresses the induction of experimental autoimmune uveoretinitis in mice by inhibiting antigen priming. Invest Ophthalmol Vis Sci 37:2211–2218, 1996.
32. AD Dick, PG McMenamin, H Korner, BJ Scallon, J Ghrayeb, JV Forrester, JD Sedgwick. Inhibition of tumor necrosis factor activity minimizes target organ damage in experimental autoimmune uveoretinitis despite quantitatively normal activated T cell traffic to the retina. Eur J Immunol 26:1018–1025, 1996.
33. W Takasaki, Y Kajino, K Kajino, R Murali, MI Greene. Structure-based design and characterization of exocyclic peptidomimetics that inhibit TNF alpha binding to its receptor. Nat Biotechnol 15:1266–1270, 1997.
34. MJ Eck, SR Sprang. The structure of tumor necrosis factor-alpha at 2.6 A resolution. Implications for receptor binding. J Biol Chem 264:17595–17605, 1989.
35. MJ Eck, M Ultsch, E Rinderknecht, AM de Vos, SR Sprang. The structure of human lymphotoxin (tumor necrosis factor-beta) at 1.9-A resolution. J Biol Chem 267:2119–2122, 1992.
36. DW Banner, A D'Arcy, W Janes, R Gentz, HJ Schoenfeld, C Broger, H Loetscher, W Lesslauer. Crystal structure of the soluble human 55 kd TNF receptor-human TNF beta complex: implications for TNF receptor activation. Cell 73:431–445, 1993.
37. T Taniguchi. Cytokine signaling through nonreceptor protein tyrosine kinases. Science 268:251–255, 1995.

38. M Onishi, T Nosaka, T Kitamura. Cytokine receptors: structures and signal transduction. Int Rev Immunol 16:617–634, 1998.
39. JF Bazan. Emerging families of cytokines and receptors. Curr Biol 3:603–606, 1993.
40. TJ Evans, D Moyes, A Carpenter, R Martin, H Loetscher, W Lesslauer, J Cohen. Protective effect of 55- but not 75-kD soluble tumor necrosis factor receptor-immunoglobulin G fusion proteins in an animal model of gram-negative sepsis. J Exp Med 180:2173–2179, 1994.
41. K Pfeffer, T Matsuyama, TM Kundig, A Wakeham, K Kishihara, A Shahinian, K Wiegmann, PS Ohashi, M Kronke, TW Mak. Mice deficient for the 55 kd tumor necrosis factor receptor are resistant to endotoxic shock, yet succumb to L. monocytogenes infection. Cell 73:457–467, 1993.
42. EH Choy, GS Panayi, GH Kingsley. Therapeutic monoclonal antibodies. Br J Rheumatol 34:707–715, 1995.
43. JVMS Quinn, HT Ngo, GJMDF Slotman. Anti–tumor necrosis factor murine monoclonal antibody (tnfmab) blocks eicosanoid release, improves pulmonary and systemic pressures and hypermetabolism in septic shock, but may decrease tissue perfusion and exacerbate fluid flux. Crit Care Med 26:126A, 1998.
44. D Wallach, M Boldin, E Varfolomeev, R Beyaert, P Vandenabeele, W Fiers. Cell death induction by receptors of the TNF family: towards a molecular understanding (Review). FEBS Lett 410:96–106, 1997.
45. D Wallach. Cell death induction by TNF: a matter of self control (Review). Trends Biochem Sci 22:107–109, 1997.
46. MR Smith, H Kung, SK Durum, NH Colburn, Y Sun. TIMP-3 induces cell death by stabilizing TNF-alpha receptors on the surface of human colon carcinoma cells. Cytokine 9:770–780, 1997.
47. R FernandezBotran, CP M., MY H. Soluble cytokine receptors: Their roles in immunoregulation, disease, and therapy. Adv Immunol 63:269–336, 1996.
48. HJ Gruss, SK Dower. The Tnf ligand superfamily and its relevance for human diseases. Cytol Mol Ther 1:75–105, 1995.
49. R Hohlfeld. Biotechnological agents for the immunotherapy of multiple sclerosis. Principles, problems and perspectives. Brain 120:865–916, 1997.
50. HY Song, CH Regnier, CJ Kirschning, DV Goeddel, M Rothe. Tumor necrosis factor (TNF)–mediated kinase cascades: bifurcation of nuclear factor-kappaB and c-jun N-terminal kinase (JNK/SAPK) pathways at TNF receptor–associated factor 2. Proc Natl Acad Sci USA 94:9792–9796, 1997.
51. MA Kelliher, S Grimm, Y Ishida, F Kuo, BZ Stanger, P Leder. The death domain kinase RIP mediates the TNF-induced NF-kappaB signal. Immunity 8: 297–303, 1998.
52. K Natarajan, SK Manna, MM Chaturvedi, BB Aggarwal. Protein tyrosine kinase inhibitors block tumor necrosis factor-induced activation of nuclear factor-kappaB, degradation of IkappaBalpha, nuclear translocation of p65, and subsequent gene expression. Arch Biochem Biophys 352:59–70, 1998.
53. M Haas, S Page, M Page, FJ Neumann, N Marx, M Adam, HW Ziegler-Heit-

brock, D Neumeier, K Brand. Effect of proteasome inhibitors on monocytic IkappaB-alpha and -beta depletion, NF-kappaB activation, and cytokine production. J Leukoc Biol 63:395–404, 1998.

54. EB Traenckner, S Wilk, PA Baeuerle. A proteasome inhibitor prevents activation of NF-kappa B and stabilizes a newly phosphorylated form of I kappa B-alpha that is still bound to NF-kappa B. EMBO J 13:5433–5441, 1994.
55. U Ponnappan. Regulation of transcription factor NFkappa B in immune senescence. Front Biosci 3:D152–D168, 1998.
56. J Delic, P Masdehors, S Omura, JM Cosset, J Dumont, JL Binet, H Magdelenat. The proteasome inhibitor lactacystin induces apoptosis and sensitizes chemo- and radioresistant human chronic lymphocytic leukaemia lymphocytes to TNF-alpha–initiated apoptosis. Br J Cancer 77:1103–1107, 1998.
57. MA Fiedler, K Warnke-Dollries, JM Stark. Inhibition of TNF-alpha–induced NF-kappaB activation and IL-8 release in A549 cells with the proteasome inhibitor MG-132. Am J Respir Cell Mol Biol 19:259–268, 1998.
58. C Jobin, A Panja, C Hellerbrand, Y Iimuro, J Didonato, DA Brenner, RB Sartor. Inhibition of proinflammatory molecule production by adenovirus-mediated expression of a nuclear factor kappaB super-repressor in human intestinal epithelial cells. J Immunol 160:410–418, 1998.
59. J Kolls, K Peppel, M Silva, B Beutler. Prolonged and effective blockade of tumor necrosis factor activity through adenovirus-mediated gene transfer. Proc Natl Acad Sci USA 91:215–219, 1994.
60. JK Kolls, D Lei, S Nelson, WR Summer, S Greenberg, B Beutler. Adenovirus-mediated blockade of tumor necrosis factor in mice protects against endotoxic shock yet impairs pulmonary host defense. J Infect Dis 171:570–575, 1995.
61. R Trueb, G Brown, C Van Huffel, A Poltorak, M Valdez-Silva, B Beutler. Expression of an adenovirally encoded lymphotoxin-beta inhibitor prevents clearance of Listeria monocytogenes in mice. J Inflamm 45:239–247, 1995.
62. BB Aggarwal, L Schwarz, ME Hogan, RF Rando. Triple helix-forming oligodeoxyribonucleotides targeted to the human tumor necrosis factor (TNF) gene inhibit TNF production and block the TNF-dependent growth of human glioblastoma tumor cells. Cancer Res 56:5156–5164, 1996.
63. EM Bulger, I Garcia, RV Maier. Dithiocarbamates enhance tumor necrosis factor-alpha production by rabbit alveolar macrophages, despite inhibition of NF-kappaB. Shock 9:397–405, 1998.
64. AR Khaled, EJ Butfiloski, ES Sobel, J Schiffenbauer. Use of phosphorothioate-modified oligodeoxynucleotides to inhibit NF-kappaB expression and lymphocyte function. Clin Immunol Immunopathol 86:170–179, 1998.
65. N Watanabe, K Nakada, Y Kobayashi. Processing and release of tumor necrosis factor alpha. Eur J Biochem 253:576–582, 1998.
66. T So, A Ito, T Sato, Y Mori, S Hirakawa. Tumor necrosis factor-alpha stimulates the biosynthesis of matrix metalloproteinases and plasminogen activator in cultured human chorionic cells. Biol Reprod 46:772–778, 1992.
67. N Di Girolamo, MJ Verma, PJ McCluskey, A Lloyd, D Wakefield. Increased

matrix metalloproteinases in the aqueous humor of patients and experimental animals with uveitis. Curr Eye Res 15:1060–1068, 1996.
68. S Chandler, KM Miller, JM Clements, J Lury, D Corkill, DC Anthony, SE Adams, AJ Gearing. Matrix metalloproteinases, tumor necrosis factor and multiple sclerosis: an overview (Review). J Neuroimmunol 72:155–161, 1997.
69. LJ Denis, J Verweij. Matrix metalloproteinase inhibitors: present achievements and future prospects. Invest New Drugs 15:175–185, 1997.
70. SL Parsons, SA Watson, PD Brown, HM Collins, RJ Steele. Matrix metalloproteinases (see Comments). Br J Surg 84:160–166, 1997.
71. RA Black, CT Rauch, CJ Kozlosky, JJ Peschon, JL Slack, MF Wolfson, BJ Castner, KL Stocking, P Reddy, S Srinivasan, N Nelson, N Boiani, KA Schooley, M Gerhart, R Davis, JN Fitzner, RS Johnson, RJ Paxton, CJ March, DP Cerretti. A metalloproteinase disintegrin that releases tumour-necrosis factor-alpha from cells. Nature 385:729–733, 1997.
72. ML Moss, SL Jin, ME Milla, DM Bickett, W Burkhart, HL Carter, WJ Chen, WC Clay, JR Didsbury, D Hassler, CR Hoffman, TA Kost, MH Lambert, MA Leesnitzer, P McCauley, G McGeehan, J Mitchell, M Moyer, G Pahel, W Rocque, LK Overton, F Schoenen, T Seaton, JL Su, JD Becherer, et al. Cloning of a disintegrin metalloproteinase that processes precursor tumour-necrosis factor-alpha. Nature 385:733–736, 1997.
73. LM Williams, DL Gibbons, A Gearing, RN Maini, M Feldmann, FM Brennan. Paradoxical effects of a synthetic metalloproteinase inhibitor that blocks both p55 and p75 TNF receptor shedding and TNF alpha processing in RA synovial membrane cell cultures. J Clin Invest 97:2833–2841, 1996.
74. K Maskos, C Fernandez-Catalan, R Huber, GP Bourenkov, H Bartunik, GA Ellestad, P Reddy, MF Wolfson, CT Rauch, BJ Castner, R Davis, HRG Clarke, M Petersen, JN Fitzner, DP Cerretti, CJ March, RJ Paxton, RA Black, W Bode. Crystal structure of the catalytic domain of human tumor necrosis factor-alpha–converting enzyme. Proc Natl Acad Sci USA 95:3408–3412, 1998.
75. T Ishizuka, N Terada, P Gerwins, E Hamelmann, A Oshiba, GR Fanger, GL Johnson, EW Gelfand. Mast cell tumor necrosis factor alpha production is regulated by MEK kinases. Proc Natl Acad Sci USA 94:6358–6363, 1997.
76. SJ Ajizian, BK English, EA Meals. Specific inhibitors of p38 and extracellular signal-regulated kinase mitogen-activated protein kinase pathways block inducible nitric oxide synthase and tumor necrosis factor accumulation in murine macrophages stimulated with lipopolysaccharide and interferon-gamma. J Infect Dis 179:939–944, 1999.
77. P Pack, A Plückthun. Miniantibodies: use of amphipathic helices to produce functional, flexibly linked dimeric Fv fragments with high avidity in E. coli. Biochemistry 31:1579–1584, 1992.
78. G Winter, C Milstein. Man made antibodies. Nature 349:293–299, 1991.
79. TA Waldmann. Monoclonal antibodies in diagnosis and therapy. Science 252: 1657–1662, 1991.
80. RN Maini, M Elliott, FM Brennan, RO Williams, M Feldmann. TNF blockade

in rheumatoid arthritis: implications for therapy and pathogenesis. APMIS 105: 257–263, 1997.
81. K Peppel, D Crawford, B Beutler. A tumor necrosis factor (TNF) receptor–IgG heavy chain chimeric protein as a bivalent antagonist of TNF activity. J Exp Med 174:1483–1489, 1991.
82. MJ Cho, R Juliano. Macromolecular versus small-molecule therapeutics: drug discovery, development and clinical considerations. Trends Biotechnol 14:153–158, 1996.
83. CL Selinsky, KL Boroughs, WA Halsey, MD Howell. Multifaceted inhibition of anti-tumour immune mechanisms by soluble tumour necrosis factor receptor type I. Immunology 94:88–93, 1998.
84. GJ Moore. Designing peptide mimetics. Trends Pharmacol Sci 15:124–129, 1994.
85. T Kieber-Emmons, R Murali, MI Greene. Therapeutic peptides and peptidomimetics. Curr Opin Biotechnol 8:435–441, 1997.
86. R Murali, MI Greene. Structure-based design of immunologically active therapeutic peptides. Immunol Res 17:163–169, 1998.
87. A Eigler, B Sinha, G Hartmann, S Endres. Taming TNF: strategies to restrain this proinflammatory cytokine (Review). Immunol Today 18:487–492, 1997.
88. TL Blundell. Structure-based drug design. Nature 384:23–26, 1996.
89. TL Blundell. The Leon Goldberg Memorial Lecture. Structural molecular biology and drug discovery. Food Chem Toxicol 33:979–985, 1995.
90. WV Williams, HR Guy, DH Rubin, F Robey, JN Myers, T Kieber-Emmons, DB Weiner, MI Greene. Sequences of the cell-attachment sites of reovirus type 3 and its anti-idiotypic/antireceptor antibody: modeling of their three-dimensional structures. Proc Natl Acad Sci USA 85:6488–6492, 1988.
91. AF Williams. A year in the life of the immunoglobulin superfamily. Immunol Today 8:298–303, 1987.
92. BL Sibanda, TL Blundell, JM Thornton. Conformation of beta-hairpins in protein structures. A systematic classification with applications to modelling by homology, electron density fitting and protein engineering. J Mol Biol 206: 759–777, 1989.
93. HU Saragovi, D Fitzpatrick, A Raktabutr, H Nakanishi, M Kahn, MI Greene. Design and synthesis of a mimetic from an antibody complementarity-determining region. Science 253:792–795, 1991.
94. RM MacCallum, AC Martin, JM Thornton. Antibody-antigen interactions: contact analysis and binding site topography. J Mol Biol 262:732–745, 1996.
95. PD Sun, DR Davies. The cystine-knot growth-factor superfamily. Annu Rev Biophys Biomol Struct 24:269–291, 1995.
96. NQ McDonald, WA Hendrickson. A structural superfamily of growth factors containing a cystine knot motif. Cell 73:421–424, 1993.
97. JF Leszczynski, GD Rose. Loops in globular proteins: a novel category of secondary structure. Science 234:849–855, 1986.

98. J Janin, SJ Wodak. Structural domains in proteins and their role in the dynamics of protein function. Prog Biophys Mol Biol 42:21–78, 1983.
99. CS Ring, DG Kneller, R Langridge, FE Cohen. Taxonomy and conformational analysis of loops in proteins (published erratum appears in J Mol Biol 1992 Oct 5;227[3]:977). J Mol Biol 224:685–699, 1992.
100. RE Bruccoleri, E Haber, J Novotny. Structure of antibody hypervariable loops reproduced by a conformational search algorithm (published erratum appears in Nature 1988 Nov 17; 336[6196]:266). Nature 335:564–568, 1988.
101. C Madal, DB Kingery, JM Anchin, S Subramaniam, SD Linthicum. ABGEN: A knowledge-based automated approach for antibody structure modeling. Nat Bio Technol 14:323–328, 1996.
102. X Zhang, D Piatiertonneau, C Auffray, R Murali, A Mahapatra, FQ Zhang, CC Maier, H Saragovi, MI Greene. Synthetic Cd4 exocyclic peptides antagonize Cd4 holoreceptor binding and T-cell activation. Nat Biotechnol 14:472–475, 1996.
103. X Zhang, M Gaubin, L Briant, V Srikantan, R Murali, U Saragovi, D Weiner, C Devaux, M Autiero, D Piatier-Tonneau, MI Greene. Synthetic CD4 exocyclics inhibit binding of human immunodeficiency virus type 1 envelope to CD4 and virus replication in T lymphocytes. Nat Biotechnol 15:150–154, 1997.
104. HU Saragovi, MI Greene, RA Chrusciel, M Kahn. Loops and secondary structure mimetics: development and applications in basic science and rational drug design. Biotechnology 10:773–778, 1992.
105. AC Bach, CJ Eyermann, JD Gross, rMJ Bowe, RL Harlow, PC Weber, WF DeGrado. Structural studies of a family of high affinity ligands for GPIIb/IIIa. J Am Chem Soc 116:3207–3219, 1994.
106. S Jackson, R Harlow, A Dwivedi, A Parthasarathy, A Higley, J Krywko, A Rockwell, J Markwalder, G Wells, R Wexler, S Mousa, WF DeGrado. Template-constrained cyclic peptides: design of high-affinity ligands for GPIIb/IIIa. J Am Chem Soc 116:3220–3230, 1994.
107. S Jones, JM Thornton. Protein-protein interactions: a review of protein dimer structures (Review). Prog Biophys Mol Biol 63:31–65, 1995.
108. J Janin, S Miller, C Chothia. Surface, subunit interfaces and interior of oligomeric proteins. J Mol Biol 204:155–164, 1988.
109. TN Bhat, GA Bentley, G Boulot, MI Greene, D Tello, W Dall'Acqua, H Souchon, FP Schwarz, RA Mariuzza, RJ Poljak. Bound water molecules and conformational stabilization help mediate an antigen-antibody association. Proc Natl Acad Sci USA 91:1089–1093, 1994.
110. WV Williams, DA Moss, T Kieber-Emmons, JA Cohen, JN Myers, DB Weiner, MI Greene. Development of biologically active peptides based on antibody structure (published erratum appears in Proc Natl Acad Sci U S A 1989 Oct; 86[20]:8044). Proc Natl Acad Sci 86:5537–5541, 1989.
111. FC Bernstein, TF Koetzle, GJ Williams, EE Meyer, Jr., MD Brice, JR Rodgers, O Kennard, T Shimanouchi, M Tasumi. The Protein Data Bank: a computer-

based archival file for macromolecular structures. J Mol Biol 112:535–542, 1977.
112. HJ Gruss. Molecular, structural and biological characteristics of the tumor necrosis factor ligand superfamily (Review). Int J Clin Lab Res 26:143–159, 1996.
113. JH Naismith, SR Sprang. Modularity in the TNF-receptor family (Review). Trends Biochem Sci 23:74–79, 1998.
114. NW Isaacs. Cystine knots. Curr Opin Struct Biol 5:391–395, 1995.
115. JH Naismith, BJ Brandhuber, TQ Devine, SR Sprang. Seeing double: crystal structures of the type I TNF receptor. J Mol Recognit 9:113–117, 1996.
116. JH Naismith, TQ Devine, BJ Brandhuber, SR Sprang. Crystallographic evidence for dimerization of unliganded tumor necrosis factor receptor. J Biol Chem 270:13303–13307, 1995.
117. E Paleolog. Target effector role of vascular endothelium in the inflammatory response: insights from the clinical trial of anti–TNF alpha antibody in rheumatoid arthritis. Mol Pathol 50:225–233, 1997.
118. MI Greene, MS Co. Shared structural features of an internal image bearing monoclonal anti-idiotype and the external antigen. Monogr Allergy 22: 120–125, 1987.
119. E Doring, R Stigler, G Grutz, R von Baehr, J Schneider-Mergener. Identification and characterization of a TNF alpha antagonist derived from a monoclonal antibody. Mol Immunol 31:1059–1067, 1994.
120. G Orfanoudakis, B Karim, D Bourel, E Weiss. Bacterially expressed Fabs of monoclonal antibodies neutralizing tumor necrosis factor alpha in vitro retain full binding and biological activity. Mol Immunol 30:1519–1528, 1993.
121. P Swaminathan, M Hariharan, R Murali, CU Singh. Molecular structure, conformational analysis, and structure-activity studies of dendrotoxin and its homologues using molecular mechanics and molecular dynamics techniques. J Med Chem 39:2141–2155, 1996.
122. BP Schoenborn, A Garcia, R Knott. Hydration in protein crystallography. Prog Biophys Mol Biol 64:105–119, 1995.
123. TA Larsen, AJ Olson, DS Goodsell. Morphology of protein-protein interfaces. Structure 6:421–427, 1998.
124. J Otlewski, W Apostoluk. Structural and energetic aspects of protein-protein recognition. Acta Biochim Pol 44:367–387, 1997.
125. CJ Beverung, CJ Radke, HW Blanch. Adsorption dynamics of L-glutamic acid copolymers at a heptane/water interface. Biophys Chem 70:121–132, 1998.
126. V Prasanna, S Bhattacharjya, P Balaram. Synthetic interface peptides as inactivators of multimeric enzymes: inhibitory and conformational properties of three fragments from Lactobacillus casei thymidylate synthase. Biochemistry 37: 6883–6893, 1998.
127. PL Darke, JR Huff. HIV protease as an inhibitor target for the treatment of AIDS (Review). Adv Pharmacol 25:399–454, 1994.
128. RL DesJarlais, GL Seibel, ID Kuntz, PS Furth, JC Alvarez, PR DeMontellano-

Ortiz, DL DeCamp, LM Babe, CS Craik. Structure-based design of nonpeptide inhibitors specific for the human immunodeficiency virus 1 protease. Proc Natl Acad Sci USA 87:6644–6648, 1990.

129. CW Ward, PA Hoyne, RH Flegg. Insulin and epidermal growth factor receptors contain the cysteine repeat motif found in the tumor necrosis factor receptor. Proteins 22:141–153, 1995.
130. BS Chapman, ID Kuntz. Modeled structure of the 75-kDa neurotrophin receptor. Protein Sci 4:1696–1707, 1995.
131. MC Peitsch, CV Jongeneel. A 3-D model for the CD40 ligand predicts that it is a compact trimer similar to the tumor necrosis factors. Int Immunol 5:233–238, 1993.
132. SV Evans. SETOR: hardware-lighted three-dimensional solid model representations of macromolecules. J Mol Graph 11:134–138, 127–138, 1993.

6

Screening and Design of Tumor Necrosis Factor-α: Converting Enzyme Inhibitors

M. L. Moss
Lineberger Comprehensive Cancer Center, University of North Carolina at Chapel Hill, Chapel Hill, North Carolina

J. D. Becherer, J. G. Conway, J. R. Warner, D. M. Bickett, M. A. Leesnitzer, T. M. Seaton, J. L. Mitchell, R. T. McConnell, T. K. Tippin, L. G. Whitesell, M. C. Rizzolio, K. M. Hedeen, E. J. Beaudet, M. Andersen, M. H. Lambert, R. Austin, J. B. Stanford, D. G. Bubacz, J. H. Chan, L. T. Schaller, M. D. Gaul, D. J. Cowan, V. M. Boncek, M. H. Rabinowitz, D. L. Musso, D. L. McDougald, I. Kaldor, K. Glennon, R. W. Wiethe, Y. Guo, and R. C. Andrews
GlaxoWellcome, Inc., Research Triangle Park, North Carolina

INTRODUCTION

Tumor necrosis factor-α (TNF-α) is the best characterized member of a family of proteins which include, but are not limited to, the lymphotoxins, CD40, CD27, CD30, and Fas ligands (1–3). Like some of the members of its family, TNF-α is a mediator of a diverse set of functions such as immune regulation and host defense (4–8). TNF-α may exist as a membrane-bound precursor of 26 kD or undergo proteolysis to generate a 17-kD soluble form (9). Both the membrane-bound and soluble forms of this cytokine are biologically active. Therefore, research efforts in the areas of TNF-α transcription, secretion, and signal transduction pathways have now been supplemented by drug-discovery efforts centering on TNF-α–neutralizing biomolecules and inhibitors of TNF-α production from cells (10–13).

Although the currently approved therapies utilize large biomolecules, the discovery that certain peptidyl hydroxamate matrix metalloproteases (14–16) inhibit the secretion of TNF-α from monocytes have helped the pharmaceutical industry to aim for the goal of developing small molecule inhibitors that target TNF-α–converting enzyme (TACE). Our efforts (17) as well as those of the Immunex group (18) were successful in characterization, cloning, and expression of TACE. The subject of this discourse is to review how we capitalized on the knowledge obtained from the identification of TACE activity. First, we were able to develop in vitro screens for TACE enzymatic activity such that structure-activity relationships of newly synthesized small molecule inhibitors could be determined. Finally, we have been successful in utilizing in vivo screens to design molecules with good pharmacokinetic properties that can be used for mechanism of action studies and disease efficacy in animal models.

IN VITRO SCREENING OF ENZYME ACTIVITY

Both a source of enzyme and an assay are required in order to screen and prioritize newly synthesized inhibitors of TACE. We chose the monocytic leukemia cell line, Mono Mac 6 as a source of TACE, since the cells produced significantly more TNF-α than other monocytic cell lines. Membrane-bound TACE (mTACE) is prepared by lysis of cells followed by subcellular organelle fractionation by a modification of the method of Fleischer and Kervina (19). The cellular homogenate is centrifuged at two low-speed spins to pellet the mitochondrial, lysosomal, and peroxisomal membranes. The supernatant is further centrifuged to generate a pellet thought to consist of endoplasmic reticulum, Golgi apparatus, and plasma membrane. The pellet is resuspended and placed on a sucrose gradient followed by centrifugation overnight in a swinging bucket rotor. Membrane bands at the interfaces of the discontinuous sucrose gradient (Fig. 1) contain functionally purified TACE activity. This activity has a comparable k_{cat}/K_m to a recombinant soluble form of the enzyme when the peptide substrate Dnp-SPLAQAVRSSSRNH_2 is used. Additionally, inhibition constants of TACE inhibitors for either the recombinant or membrane-bound enzyme are similar (17).

The purified enzyme can be used in several different in vitro assays such as the scintillation proximity assay (SPA) or with an internally quenched fluorescence substrate (20,21). For the TACE SPA assay, a tritiated and biotinylated peptide substrate is synthesized such as biotin-$(CH_2)_6$-SPLAQAVRSSSRTP(^3H)S. The substrate is processed by TACE and then scintil-

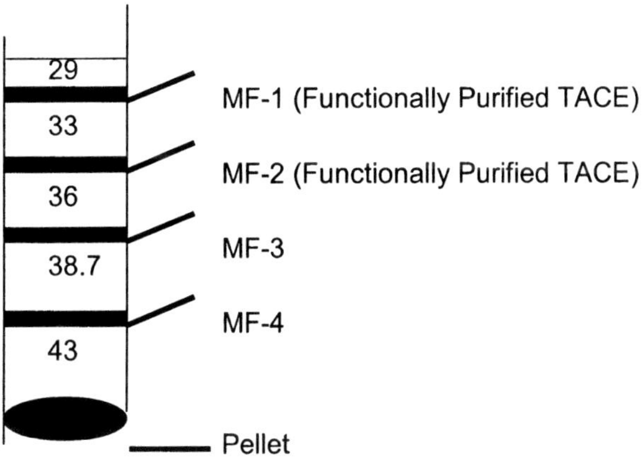

Figure 1 Schematic of the sucrose gradient used to isolate functionally pure TACE. Membranes are passed through a sucrose gradient and separated into different membrane fractions (MF) based on their densities. Functionally purified TACE activity is defined as that activity that cleaves both proTNF and the TNF peptide substrate at the correct site and lacks any other nonspecific protease activities. Typically, the MF-1 and MF-2 fractions contain the TACE activity.

lant-coated streptavidin beads (Amersham) are added to the mixture. In the presence of beads, the biotinylated and tritiated substrate, but not tritiated product, is in close proximity to the scintillant and will send a signal to a scintillation counter.

Alternatively, a fluorescence substrate that relies on internal quenching by resonance energy transfer to provide a low background signal can be used. Two commercially available substrates can be used in place of the SPA assay (20,21). In addition, we developed a substrate with NBD and DMC as the donor/acceptor pair and incorporated these substituents into the substrate, NBD-LAQAVRSK(DMC)NH_2. Once the substrate is cleaved, the released product produces a detectable fluorescence signal. The SPA and fluorescence assays provide inhibition constants of TACE that are comparable.

CELL-BASED ASSAYS

Inhibitors of TACE enzyme activity must also block the release of TNF-α from cells if they are to be effective in vivo. Several cell-based assays are employed in our laboratories that permit TACE inhibition to be correlated

Table 1 Comparison of Enzyme Inhibition Data with Cell-based Assay Results

Compound	mTACE (nM)	PBMC (nM)	MonoMac6 (nM)	Whole blood (nM)
GW 9471	9	140 ± 56	85	3100 ± 1900
GW 9901	9	71 ± 31	35	240 ± 73
CGS27023	60	4400 ± 1300	7500	>10,000
Marimastat	100	6700 ± 4900	1200	7100 ± 2300

PBMC, peripheral blood mononuclear cells.

with inhibition of TNF-α release from cells (Table 1). The first assay utilizes the Mono Mac 6 cell line (MM6) stimulated with lipopolysaccharide (LPS) and phorbal myristate acetate (PMA) (22) as the source of TNF-α. In a similar fashion, isolated peripheral blood mononuclear cells or diluted whole blood can be stimulated with LPS to generate the synthesis and release of TNF-α (14–16). All cell assays are performed using a 96-well format and are relatively short, requiring 4- to 18-h incubations with the aforementioned reagents. At the time of stimulation, varying concentrations of inhibitor are added to produce a dose-response curve that will generate an IC_{50} value.

Table 1 contains a list of inhibitors that have been screened against membrane-bound TACE and in the cell-based assays described above. As expected, the inhibition constants increase as the assay progresses from in vitro screening of native enzyme to the more complex cellular assays. Also noteworthy is that not all inhibitors that exhibit potent IC_{50}s against TACE potently inhibit TNF release from cellular sources. This finding suggests that TNF-α processing does not occur entirely on the cell surface and that cell penetration is necessary for complete inhibition of TNF-α release via TACE inhibition. An alternative hypothesis is that TNF processing may occur on the plasma membrane, but that in order to gain access to the enzyme's active site, the inhibitor is affected by the microenvironment on the cell surface.

IN VIVO TESTING OF TACE INHIBITORS

Once inhibitors are found that possess in vitro potency against membrane-bound and cell-based TACE, they are screened in vivo for oral activity and prevention of TNF-α release by using an LPS-challenge model in mice (14–15). In this model, test compounds are administered and mice are stimulated to release TNF-α by intraperitoneal injection of LPS. Ninety minutes after LPS administration, plasma levels of TNF-α are quantitated by enzyme-linked immunosorbent assay (ELISA) and the percentage of inhibition is determined. This assay provides a quick assessment of a compound's efficacy and oral activity in vivo.

INHIBITORS AND SAR

Utilization of the above assays has allowed us to design potent inhibitors of TACE with good pharmacokinetic properties. Our early leads were structures of compounds, including GW 9471, which inhibit members of the matrix metalloprotease class of enzymes (23). Reviews of the literature have concluded that the great success in inhibition of MMPs have been achieved by compounds formed on the so-called ''right-hand side'' template ZBG—R_1-R_2-R_3-R_4 . . . , where ZBG is a zinc cofactor-binding group and R_1, R_2, R_3, and R_4 represent substituents which might bind in the S_1, S_1', S_2', and S_3' subsites of the enzyme target (24).

Given that the TACE and MMP enzyme catalytic domains are related, coupled with the fact that inhibition of both TACE and MMPs occurs by GW 9471, we felt that similar right-hand side inhibitors would possibly inhibit TACE. Therefore, we screened molecules available to us incorporating different zinc-binding groups (Table 2). Screening against TACE as well as MMPs suggests that TACE discriminates against zinc-binding moieties to a greater degree than do the MMPs. The thiol-based inhibitors are inactive versus TACE yet are adequate MMP inhibitors, and the same is true for the carboxylate inhibitor. The succinyl hydroxamate inhibitor GW 9471 as well as the sulfonamido acetohydroxamate CGS 27023a (25) are good inhibitors of cell-free TACE.

Although it is clear that the potential exists for TACE inhibition with hydroxamate MMP inhibitors, we were intrigued by the possibility that reverse hydroxamates, also called type II hydroxamates, could serve as TACE inhibitors when placed in a left-hand side inhibitor template. Accounts of the study of reverse hydroxamate MMP inhibitors are scarce in the open literature; there exists one study of carboxypeptidase A inhibition by certain N-formylhydroxylamine compounds (26). We felt it would be a worthy challenge to design a reverse hydroxamate-based TACE inhibitor with drug-like properties.

We chose to synthesize our inhibitors based on the template in Figure 2 in which the P_1, P_1', P_2', and P_3' substituents occupy the binding sites for the corresponding amino acid residues in the substrate. A generally applicable

Table 2 Variation in Zinc-binding Group

Structure (entry)	Cell TNF-α inhibition, MM-6 IC_{50}, nM	TACE K_i (nM)	MMP-1 K_i (nM)	MMP-9 K_i (nM)	MMP-3 K_i (nM)
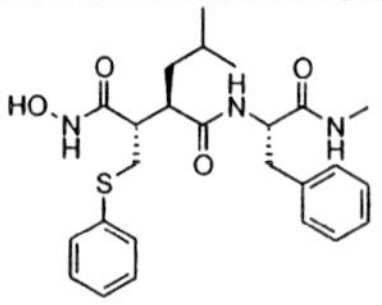 (1) GW9471	100	9	2	1	4
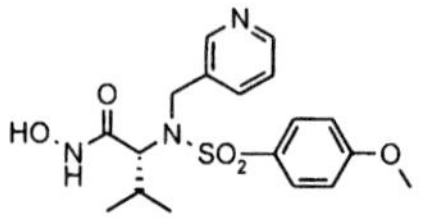 (2) CGS 27023a	7,000	60	13	3	
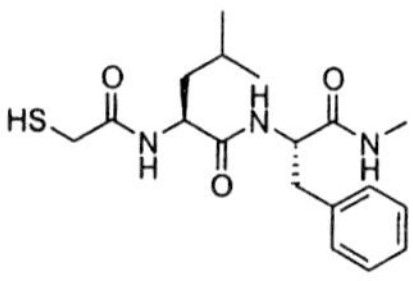 (3)	>10,000	>4,000	73	380	7,500
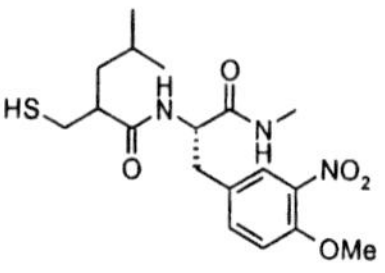 (4)	>10,000	>9,000	190	50	350
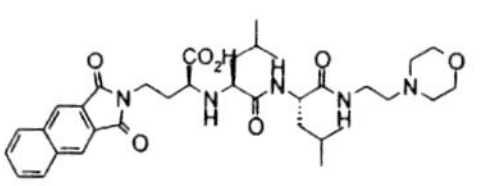 (5)	>10,000	>10,000	40	10	100

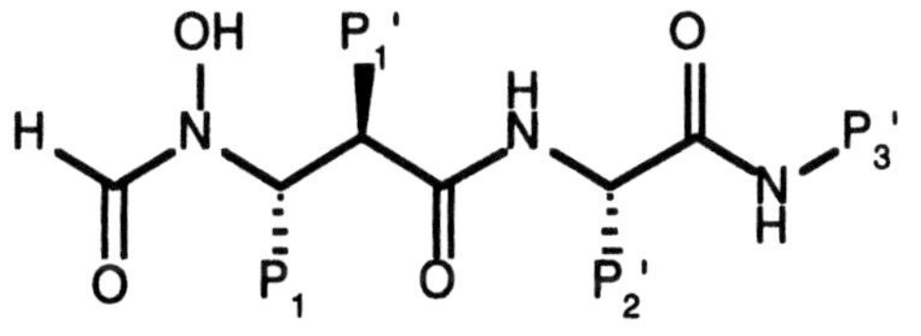

Figure 2 Reverse hydroxamate TACE inhibitor general structure.

synthetic route (Scheme 1) was devised toward compounds in the reverse hydroxamate class to generate the compounds under discussion.

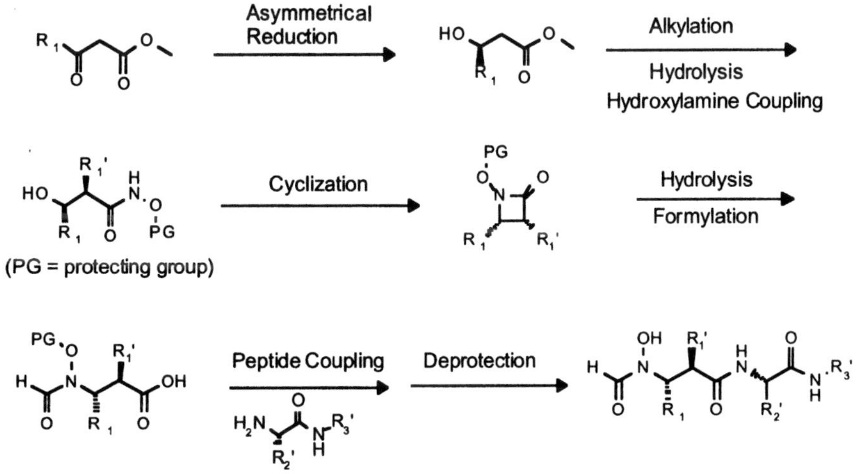

Scheme 1

Clearly there exists the potential for inhibition of TACE within the Figure 2 template, and we sought to define further structure-activity relationships for these compounds as described in Table 3. Primary amides (hydrogen at P_3') are tolerated by TACE. Lipophilic P_2' substituents are preferred (phenylmethyl preferred over methyl). Isobutyl is well tolerated as a P_1' substituent.

TACE seems to be very permissive of the lack of substitution at the P_3' position (Table 4). Moving to a heteroaryl substituent at P_3' improves the whole cell activity of compounds, and in some cases, changes in TACE enzyme K_is are also observed. GW 8536 (entry 4, Table 4) is potent versus TNF release from cells and TACE, whereas also possessing potency versus collagenase-1 and gelatinase B.

Table 3 Variation in P_1'

Structure (entry)	TACE K_i (nM)	MMP-1 K_i (nM)	MMP-9 K_i (nM)	MMP-3 K_i (nM)
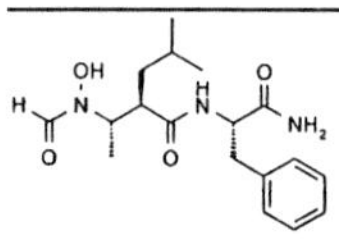 (1)	156	181	33	406
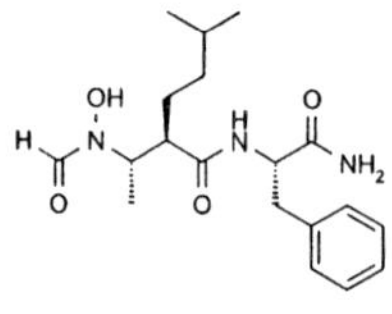 (2)	1,803	390	128	1,148
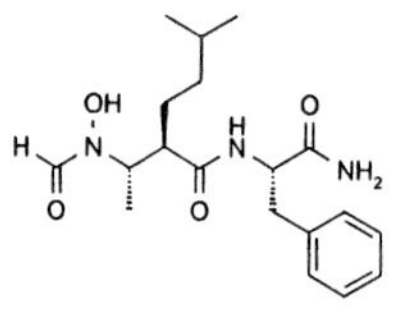 (3)	201	350	113	454
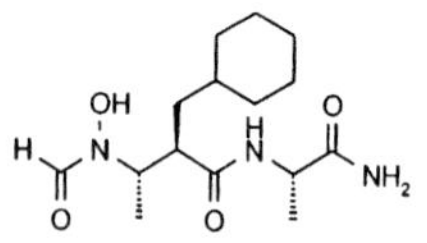 (4)	>10,000	8,158	—	>10,000
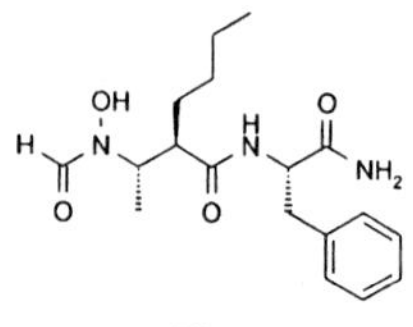 (5)	488	344	158	2,195

Variation of the P_2' substituent (Table 5) in GW 8536 affords compounds which possess potency versus TACE and TNF release from cells. Entries 3 and 4 exhibit selectivity versus stromelysin-1, as does GW 8536. Utilization of a threonine amide derivative as a P_2'-P_3' piece is noteworthy (entry 2) with excellent inhibition of TACE and MMPs.

Table 4 Variation at P_3'

Structure (entry)	Cell TNF-α Inhibition, MM-6 IC_{50}, nM	TACE K_i (nM)	MMP-1 K_i (nM)	MMP-9 K_i (nM)	MMP-3 K_i (nM)
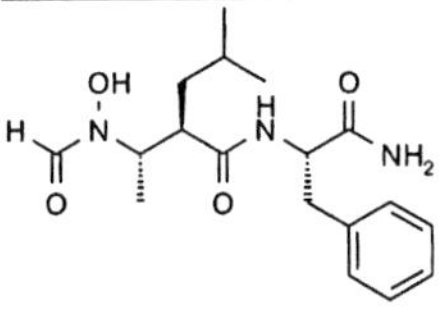 (1)	3073	156	181	33	406
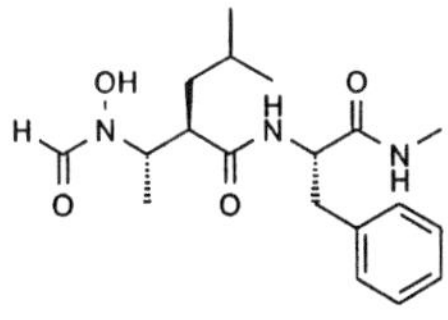 (2)	2270	64	41	31	854
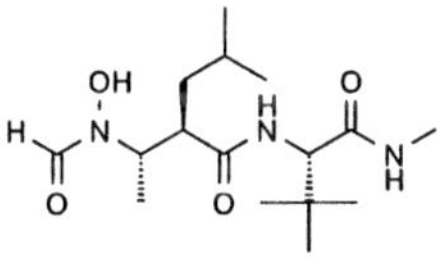 (3)	2954	142	17	21	2323
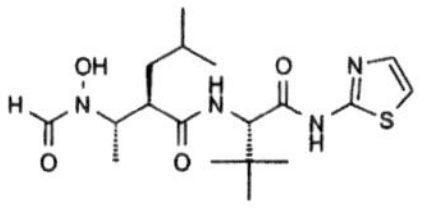 (4) GW 8536	245	20	27	34	437

Change of the P_1 substituent in GW 8536 from methyl to isopropyl (Table 6) afforded GW 3333, which is still a potent inhibitor of TACE. Surprisingly, GW 3333 is particularly active in the murine LPS challenge assay, exhibiting greater than 90% inhibition of plasma TNF when dosed orally at 40 mg/kg.

GW 3333 was studied for pharmacokinetics in the rat and dog (Table 7). Given the activity in the murine LPS-challenge model, one would expect some degree of bioavailability and such is the case. GW 3333 has a good

Table 5 Variation at P_2'

Structure (entry)	Cell TNF-α Inhibition, MM-6 IC_{50}, nM	TACE K_i (nM)	MMP-1 K_i (nM)	MMP-9 K_i (nM)	MMP-3 K_i (nM)
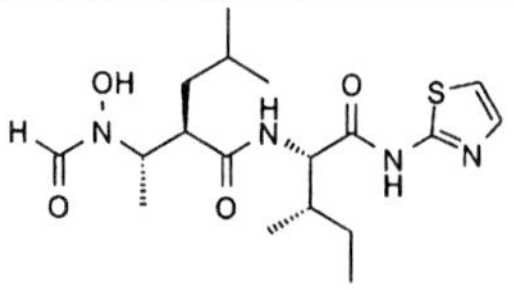 (1)	270	20	40	29	93
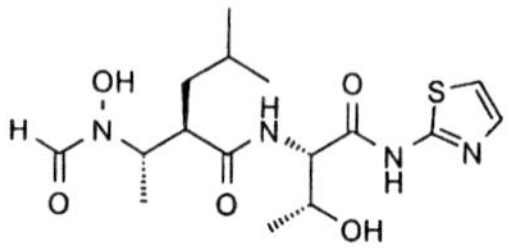 (2)	1136	22	12	8	80
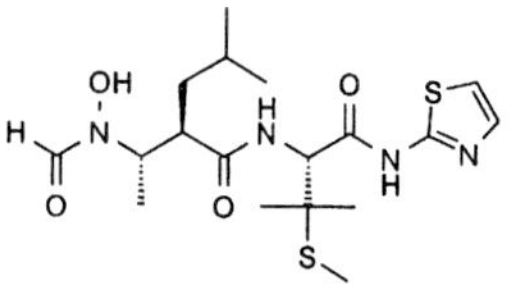 (3)	285	25	30	30	215
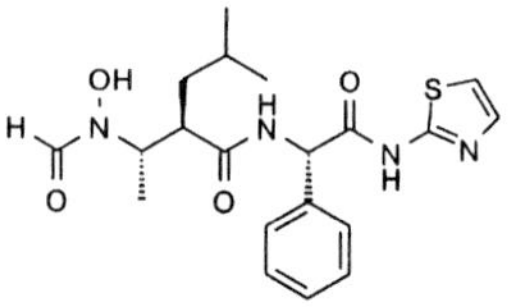 (4)	331	88	353	128	245

pharmacokinetic profile with bioavailabilities in both rat and dog and good half-life in the dog.

CONCLUSIONS

Assays have been devised to screen TACE inhibitors that include measurement of inhibitory activity against membrane bound TACE and cell-based TACE. From the screening effort, reverse hydroxamate inhibitors have been

Table 6 P_1 Isopropyl

Structure (Entry)	Cell TNF-α Inhibition, MM-6 IC_{50}, nM	TACE K_i (nM)	MMP-1 K_i (nM)	MMP-9 K_i (nM)	MMP-3 K_i (nM)	% Inhib. at 40 mg/kg po in mice
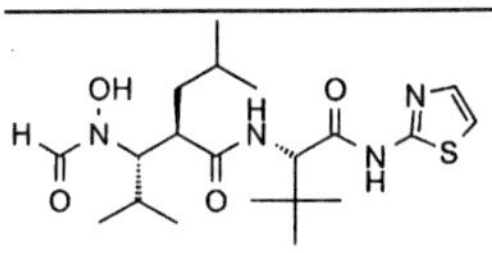 (1)	154	57	13	14	70	67
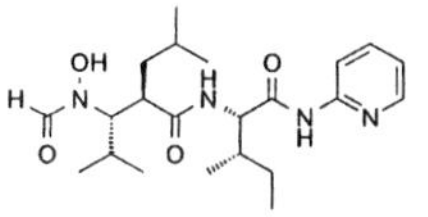 (2) GW 3333	167	42	19	16	20	94

Table 7 GW 8536 and GW 3333

	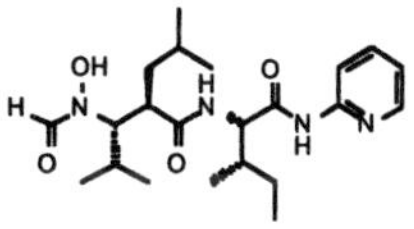 GW 3333
TACE K_i, nM	42
Cell TNF-α Inhibition, MM-6 IC_{50}, nM	167
Mouse LPS %L, 40 mg/kg po	94
$t_{1/2}$, Rat, h	3
F, Rat, %	75
$t_{1/2}$, Dog, h	5
F, Dog, %	100

designed which exhibit potent in vitro activity. A structure-activity relationship was observed based on in vitro screening of TACE inhibitors against cell-free and cell-based TACE. Those molecules that were potent in vitro were further screened with an in vivo assay. This screening process has allowed us to develop reverse hydroxamates, with GW 8536 and GW 3333 as notable examples, with good pharmacokinetic properties and potency against TACE both in vitro and in vivo. As such, they offer the potential to evaluate the effect of TACE inhibition on TNF-α–driven models of human disease.

REFERENCES

1. RJ Armitage. Tumor necrosis factor receptor superfamily members and their ligands. Curr Opin Immunol 6(3):407–413, 1994.
2. HJ Gruss, SK Dower. The TNF ligand superfamily and its relevance for human diseases. Cytokines Mol Ther 1(2):75–105, 1995.
3. D Cosman. A family of ligands for the TNF receptor superfamily. Stem Cells 12(5):440–455, 1994.
4. M Pasparakis, L Alexopoulou, E Douni, G Kollias. Tumor necrosis factors in immune regulation: everything that's interesting is...new!, Cytokine Growth Factor Rev 7(3):223–229, 1996.
5. E Douni, K Akassoglou, L Alexopoulou, S Georgopoulos, S Haralambous, S Hill, G Kassiotis, D Kontoyiannis, M Pasparakis, D Plows, L Probert, G Kollias. Transgenic and knockout analyses of the role of TNF in immune regulation and disease pathogenesis. J Inflamm 47(1–2):27–38, 1995–1996.
6. B Beutler, GE Grau. Tumor necrosis factor in the pathogenesis of infectious diseases. Crit Care Med. 21(10 Suppl):S423–435, 1993.
7. W Fiers. Tumor necrosis factor. Characterization at the molecular, cellular and in vivo level. FEBs Lett 285(2):199–212, 1991.
8. P Vassalli. The pathophysiology of tumor necrosis factors. Annu Rev Immunol 10:411, 1992.
9. M Kriegler, C Perez, Al DeFay, SD Lu. A novel form of TNF/cachectin is a cell surface cytotoxic transmembrane protein: ramifications for the complex physiology of TNF. Cell 53:45, 1988.
10. MG Shire, GW Muller. TNF-α inhibitors and rheumatoid arthritis. Expert Opin Ther Pat 8:531–544, 1998.
11. JR O'Dell. Anticytokine therapy-a new era in the treatment of rheumatoid arthritis? N Engl J Med 340(4):310–2, 1998.
12. M Feldman, P Taylor, E Paleolog, FM Brennan, RN Maini. Anti-TNF alpha therapy is useful in rheumatoid arthritis and Crohn's disease: analysis of the mechanism of action predicts utility in other diseases, Transplant Proc 30(8): 4126–7, 1998.

13. ME Weinblatt, JM Kremer, AD Bankhurst, KJ Bulpitt, RM Fleischmann, RI Fox, CG Jackson, M Lange, DJ Burge. A trial of etanercept, a recombinant tumor necrosis factor receptor: Fc fusion protein, in patients with rheumatoid arthritis receiving methotrexate. N Engl J Med 340:253–259, 1999.
14. GM McGeehan, JD Becherer, RC Bast Jr, CM Boyer, B Champion, KM Connolly, JG Conway, P Furdon, S Karp, S Kidao, A McElroy, J Nichols, K Pryzwansky, F Schoenen, L Sekut, A Truesdale, M Verghese, J Warner, J Ways. Regulation of tumor necrosis factor-α processing by a metalloproteinase inhibitor. Nature 370:558–61, 1994.
15. KM Mohler, PR Sleath, JN Fitzner, DP Cerretti, M Alderson, SS Kerwar, DS Torrance, C Otten-Evans, T Greenstreet, RA Black. Protection against a lethal dose of endotoxin by an inhibitor of tumor necrosis factor processing. Nature 370:218–221, 1994.
16. AJH Gearing, P Beckett, M Christodoulou, M Churchill, J Clements, AH Davidson, AH Drummond, WA Galloway, R Gilbert. Processing of tumor necrosis factor-α precursor by metalloproteases. Nature 370:555–558, 1994.
17. ML Moss, S-LC Jin, ME Milla, DM Bickett, W Burkhart, HL Carter, W-J Chen, WC Clay, JR Didsbury, D Hassler, CR Hoffman, TA Kost, MH Lambert, MA Leesnitzer, P McCauley, G McGeehan, J Mitchell, M Moyer, G Pahel, W Rocque, LK Overton, F Schoenen, T Seaton, J-L Su, J Warner, D Willard, JD Becherer. Cloning of a disintegrin metalloproteinase that processes precursor tumour-necrosis factor-α. Nature 385:733–736, 1997.
18. RA Black, CT Rauch, CJ Kozlosky, JJ Peschon, JL Slack, MF Wolfson, BJ Castner, KL Stocking, P Reddy, S Srinivasan, N Nelson, N Boiani, KA Schooley, M Gerhart, R Davis, JN Fitzner, RS Johnson, RJ Paxton, CJ March, DP Cerretti. A metalloproteinase disintegrin that releases tumour-necrosis factor-α from cells. Nature 385:729–733, 1997.
19. S Fleischer, M Kervina. Subcellular fractionation of rat liver. Methods Enzymol 31:6, 1974.
20. MA Leesnitzer, DM Bickett, ML Moss, JD Becherer. A high-throughtput assay for the TNF converting enzyme. In: M Kahn, ed. High Throughput Screening for Novel Anti-inflammatories. Birkhauser, 1999.
21. DE Van Dyk, P Marchand, RC Bruckner, JW Fox, BD Jaffee, PL Gunyuzlu, GL Davis, S Nurnberg, M Covington, CP Decicco, et al. Comparison of snake venom reprolysin and matrix metalloproteinases as models of TNF-α converting enzyme. Bioorg Med Chem Lett 7:1219–1224, 1997.
22. A Pradines-Figueres, CRH Raetz. Processing and secretion of tumor necrosis factor α in endotoxin-treated mono Mac 6 cells are dependent on phorbol myristate acetate. J Biol Chem 267(32):23261–23268, 1992.
23. RP Beckett, M Whittaker. Matrix metalloproteinase inhibitors. Expert Opin Ther Pat 8:259–282, 1998.
24. RP Beckett, AH Davidson, AH Drummond, P Huxley, M Whittaker. Recent advances in matrix metalloproteinase inhibitor research. Drug Discovery Today 1:16–26, 1996.

25. LJ MacPherson, EK Bayburt, MP Capparelli, BJ Carroll, R Goldstein, MR Justice, L Zhu, IS Hu, RA Melton, L Fryer, R Goldberg, JR Doughty, S Spirito, V Blancuzzi, D Wilson, EM O'Byrne, V Ganu, DT Parker. Discovery of CGS 27023A, a non-peptidic, potent, and orally active stromelysin inhibitor that blocks cartilage degradation in rabbits. J Med Chem 40:2525–2532, 1997.
26. DH Kim, Y Jin. First hydroxamate inhibitors for carboxypeptidase A. N-acyl-N-hydroxy-B-phenylalanines. Bioorg Med Chem Lett 9:691, 1999.

7

Chemokine Receptors: Therapeutic Targets for Human Immunodeficiency Virus Infectivity

Amanda E. I. Proudfoot-Fichard and Timothy N. C. Wells
Serono Pharmaceutical Research Institute S.A., Geneva, Switzerland

Alexandra Trkola
University Hospital Zurich, Zurich, Switzerland

INTRODUCTION

Human immunodeficiency viral (HIV) particles bind to their host cells through specific high-affinity interactions that are mediated through two classes of cell surface proteins on the target cells. The first receptor, CD4, was identified in the early 1980s, and although the evidence for the absolute requirement for CD4 was unequivocal, at least for HIV-1 strains, the evidence that this molecule was not sufficient for viral entry was equally convincing. Throughout the following decade, the search for the coreceptor was carried out in many laboratories using many different approaches, and several molecules were proposed. The identification of the coreceptor as a seven-transmembrane spanning protein with high homology to the chemokine receptor family resulted chemokine biologists and HIV virologists united in a new search for mechanisms to combat the acquired immunodeficiency syndrome (AIDS).

Chemokines and their receptors mediate leukocyte trafficking. Specific cellular recruitment is essential for embryonic development and routine immunosurveillance in the adult, but it also plays a major role in inflammation.

Chemokine receptors belong to the seven-transmembrane G-protein–coupled receptor superfamily, which are the sites of action for innumerable pharmaceutical agents today. This chapter will focus on the validation of their potential as a therapeutic target to prevent the early events necessary for the HIV virus to infect host cells.

HIV ENTRY THROUGH THE CD4 AND CHEMOKINE RECEPTOR COMPLEX

The glycoprotein subunits, gp120 and gp41, of the HIV envelope are the key proteins that the HIV particle uses for entry of host cells. These proteins are covalently associated and assemble to trimers on the HIV viral membrane. Viral entry is achieved by fusion on the cell surface as a result of conformational changes in the envelope glycoproteins on binding to the CD4 molecule (1–3). The evidence for the absolute requirement for CD4 came from human cells which were resistant to certain strains of HIV-1 but were rendered susceptible to infection by the expression of CD4 (1,2). However, nonhuman cells, such as murine cells, were still resistant even if they expressed human CD4. It was therefore concluded that HIV-1 in addition to CD4 utilizes additional surface molecules on human cells to succeed in efficient entry.

A second line of evidence showed that not one but at least two different second receptor molecules are involved in HIV-1 entry. Several groups observed over the past years that HIV isolates have distinct patterns of infection or cellular tropism. Certain primary HIV-1 isolates only replicate in macrophages or activated primary $CD4^+$ T cells, and were hence referred to as macrophage tropic (M-tropic) isolates. M-tropic isolates are unable to replicate in transformed T-cell lines despite the presence of CD4 on these cells. Another group of isolates replicates very efficiently on these cell lines, and are referred to as T-tropic isolates.

IDENTIFICATION OF THE CHEMOKINE RECEPTORS AS THE HIV CORECEPTOR

Before the identification of the essential coreceptor, factors secreted by $CD8^+$ T cells had been shown to have inhibitory properties *on* HIV-1 infection (4). Three candidates for this factor to include three β-chemokines, RANTES (regulated upon activation, normal T-cell expressed, and presumably secreted) and macrophage inflammatory proteins MIP-1α, and MIP-1β which could block infection by M-tropic isolates (5). Shortly after this observation, through an elegant expression cloning approach, a seven-transmembrane G-protein–coupled orphan receptor was identified as the missing cofactor for T-tropic isolates (6). After several name changes and identification of its chemokine ligand, the receptor was finally defined as the CXC chemokine receptor CXCR4 (7,8). Since the three chemokines produced by CD8 cell lines bind to a common receptor, CCR5, this receptor was rapidly identified as being the coreceptor for macrophage-tropic (M-tropic) HIV-1 strains (9–11). These strains infect primary cell types such as macrophages, lymphocytes, and dendritic cells, and it is thus widely accepted that primary infection and viral transmission occurs through CCR5. CXCR4 strains appear to develop as disease progresses from initial infection to fullblown AIDS.

The definition of tropism and syncitia induction is in fact not completely accurate, since the different groups overlap (for more details see ref. 12). A biochemical approach using the classification of HIV stains and isolates based on chemokine receptor usage has recently been adopted (13). Thus, M-tropic or non-syncitia inducing *(NSI)* viruses have been shown to utilize CCR5 for entry, and are hence referred to as R5 viruses. T-tropic viruses use CXCR4 and are referred to as X4 isolates. Dual tropic isolates, which interact with both CCR5 and CXCR4, are now called R5X4 viruses.

CHEMOKINE RECEPTORS EXPLAIN HIV TROPISM

Much of the underlying biology of HIV-1 is linked to differential coreceptor usage. The tropism that was defined early in HIV virology is in fact explained by the pattern of chemokine receptor usage. In addition, HIV viral tropism is closely associated with transmission and pathogenesis. It had been shown early on that sexually transmitted isolates are preferentially R5 (NSI) viruses and stay prevalent early on in infection (14,15). The importance of CCR5 in viral transmission was further underlined by the discovery of a variant in the CCR5 gene, which prevents the formation of a functional receptor molecule, and the finding that individuals who are homozygotic for this deletion are highly resistant to HIV-1 infection (16,18). This observation strengthened the hypothesis that the cell compartments involved in the initial infection might predominantly express CCR5. Macrophages and dendritic cells which are present in the mucosal sites of infection during sexual transmission express CCR5. As time progresses after infection, X4 viruses emerge, and their appearance is closely linked with progression of disease, although this is not observed in all individuals studied. What drives this phenotypic switch, and which factors suppress the emergence of the highly pathogenic X4 isolates in the earlier years of infection is still unclear. Differential expression of chemokine receptors in tissues and on the peripheral $CD4^+$ T cells during progression might be partially responsible. In addition, local overexpression of chemokines might suppress replication of certain groups of viruses.

INTERACTION OF VIRAL ENVELOPE AND THE CHEMOKINE RECEPTOR

The initial interaction between CD4 and the viral envelope drives a conformational change in gp120 which exposes the binding site to allow interaction with the chemokine receptor (19,20). The precise regions of gp120 that interact with the coreceptors are not yet completely identified, but studies with monoclonal antibodies against gp120 epitopes and the solution of the gp120 crystal structure have provided insights into the gp120/CCR5 interaction site. A molecular complex of soluble gp120, membrane-associated CD4, and CXCR4 can be immunoprecipitated, proving the close association of these three elements (21).

Fusion of the viral and cell membrane is then thought to be induced by the conformational change in the gp41/gp120/CD4 complex driving the glycoprotein gp41 into the host membrane. The hypothesis is that this interaction is driven by the hydrophobic amino terminus of gp41, and the model is based on the mechanism of membrane fusion for the influenza virus (22). Chemokine receptors may mediate the conformational change of the HIV envelope proteins required for the insertion of gp41 into the host membrane. The signaling activities of the chemokine receptors mediated by G-proteins or receptor internalization are not required for the chemokine receptor to function as an HIV coreceptor. Infection is not abolished by pertussis toxin (5,23), and mutant receptors that are unable to internalize still support viral entry (24).

Mapping the binding region of the viral envelope protein on the chemokine coreceptors is an area of active research. For R5 viruses, the N-terminus of CCR5 has been shown to be crucial for recognition and coreceptor activity (25,26). Chimeric CCR2b/CCR5 receptors containing only the CCR5 N-terminus are active for R5/NSI viral infection. Yet, monoclonal antibodies that recognize the second extracellular loop (E2) of CCR5 efficiently blocked HIV-1 infection, whereas those specific for the N-terminus were weaker inhibitors (27), suggesting that this loop might not have a specific function in interaction with the envelope but that it nevertheless plays an important role in the fusion process.

For X4 strains, the regions on their coreceptor CXCR4 that are recognized by the envelope are different. Truncation of N-terminal sequences of CXCR4 showed that some X4 viruses required a full-length N-terminus, whereas others were unaffected by removal of most or all N-terminus sequences (28). This different pattern of receptor binding sites between R5 and X4 strains may play an important role in the switch that occurs when the virus mutates from R5 to X4 during the later stages of infection.

THE CHEMOKINE FAMILY

It is perhaps appropriate at this stage to divert the reader from HIV virology to an introduction to the chemokine family and to give a background to their role in immunology and inflammatory diseases. Chemokine biology has recently been the subject of many reviews and the reader is referred to a sample in the references cited (29–32). The main features of the family are outlined here in a brief synopsis. The term *chemokines* was coined from their property to attract leukocytes (a contraction of *chemo*attractant cyto*kines*). Ten years ago, little was known about the factors which controlled basal cell trafficking, as well as recruitment to sites of inflammation. The first member of the chemokine family was discovered just over a decade ago as being the factor responsible for attracting neutrophils, and it was thus appropriately named neutrophil-activating peptide-1 (NAP-1). In the tradition of the cytokine nomenclature, this protein was soon renamed interleukin-8 (IL-8). This protein turned out to be a very special cytokine, since NAP-1, or IL-8, was in fact the first member of a large subfamily of the cytokine family that were renamed chemokines. Around the same time, two new chemokines, monocyte chemotactic protein-1 (MCP-1) and MIP-1α and MIP-β were identified as being able to recruit monocytes-macrophages.

It soon became apparent that this family of proteins had several structural similarities. First, they were all small proteins, approximately 8 kD, and second, they were almost all highly basic—with the obvious exception to this rule being MIP-1α and MIP-1β. Amino acid sequencing revealed two different patterns of four conserved cysteine residues. In IL-8, the two amino terminal cysteines are separated by a single amino acid to form a CXC motif, whereas, in MCP-1, they are adjacent and form a CC motif. These spacings gave rise to the two principal chemokine subclasses, the α, or CXC, chemokines and the β, or CC, chemokines. Early studies of their biology supported this classification, since the α subclass was described to activate neutrophils, whereas the β subclass activates other leukocyte types such as T cells, eosinophils, monocytes-macrophages, basophils, and dendritic cells. The α chemokines were associated with acute inflammation, which is of short duration and is characteristically accompanied by plasma fluid exudates and neutrophil accumulation. Chronic inflammation is characterized by more dense cellular infiltrates which can comprise lymphocytes, eosinophils, mast cells, and monocytes-macrophages and is thus associated with members of the β chemokine subclass. However, it is clear that this classification no longer holds, since the identification of the CXC chemokines IP-10, Mig, and I-TAC have been shown to be responsible for the recruitment of activated $CD4^+$ T cells, thus rendering them intimately associated with all chronic inflammatory disorders.

The chemokine field has exploded in complexity over the last few years. A combination of techniques such as their initial purification from cell culture media or from tissue samples, followed by cDNA cloning by homology, and more recently by bio-informatics using the EST (expressed sequence tag) data bases (33) has resulted in the identification of 36 human chemokines to date. Initially, chemokines were principally associated with inflammatory disease, where they were found at elevated levels in biological fluids or tissues from both patients and from animal models of inflammation. However, more recently, biology has enabled us to identify a second group of chemokines which are constitutively expressed and play a role in development. Others are inducible and play a major role in inflammation. There may, however, be a certain overlap between these two broad categories of chemokines.

The chemokines are further differentiated from other cytokines by the receptors that they activate—they are the only cytokines that act on seven-transmembrane G-protein–coupled receptors. Although the number of distinct human chemokine ligands is approximately 40, the number of receptors is much lower—16 functional receptors have been identified to date. There are two main points to be made: First, there are very few specific receptors that bind only one ligand in in vitro assays, and often an additional ligand is subsequently found to bind and activate a hitherto specific receptor. Second, several ligands bind to more than one receptor. The best example of such ligand promiscuity is RANTES, which binds to CCR1, 3, 4, 5, and 9 (see Figure 1).

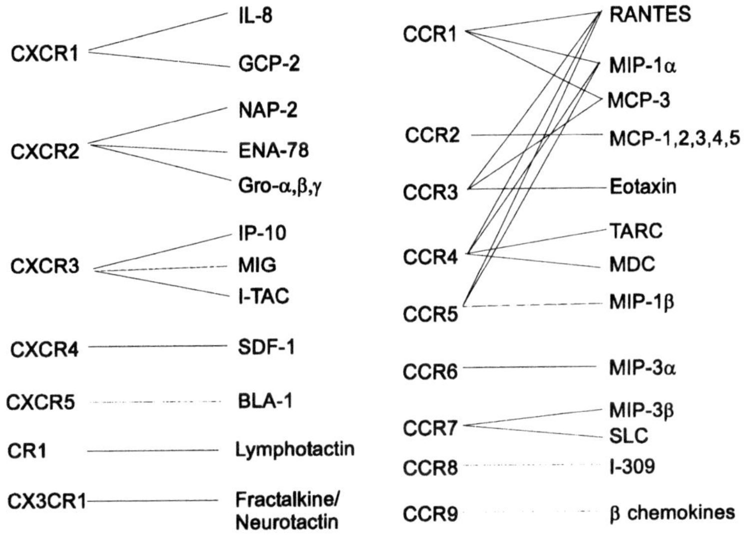

Figure 1 Known receptor-ligand pairs.

WHAT CONTROLS SPECIFICITY IN VIVO?

The discovery of the chemokine family led to the hope that specific recruitment of individual groups of leukocytes could be selectively inhibited, particularly in inflammatory disorders. The complexity of the system has led to the question arising as to whether the redundancy that has been defined allows for therapeutic intervention. However it must be noted that this redundancy has been defined in vitro, and certainly many levels of control exist in vivo which introduce specificity.

Ligand-binding data leading to ligand/receptor pairing from in vitro experiments is obtained where receptors are expressed at very high levels on the surface, often 10–100 times higher than is actually seen in primary cell types. Little data are available as to how this overexpression effects selectivity. Second, the cell context may well be important, since it defines which G-proteins are present, and they in turn may alter the conformation of the receptor. Different cell backgrounds have in fact given conflicting results, as is well demonstrated by the ligand-binding profile of one of the CC receptors, CCR4. When this receptor was originally cloned in our laboratory and expressed in oocytes, it was activated by RANTES, MIP-1α, and MCP-1 (34). RANTES and MIP-1α were shown to bind to CCR4 when expressed in HL60 cells (35). Recently, other laboratories have shown that CCR4 is in fact the receptor for TARC and macrophage-derived chemokine (MDC) (36,37). Third, although a seven-transmembrane receptor may bind several ligands with equal potency in equilibrium-binding assays, there may still be one ligand which is capable of competing away all the others. One explanation of this phenomenon requires a slow conformational change of the receptor-ligand complex. Striking data for these effects have been reported for the virally encoded chemokine receptor, US28, produced by the human cytomegalovirus, where this receptor shows preferential binding for the ligand fraktalkine in heterologous competition experiments (38). Fourth, although in vitro the receptor and ligand may bind, in vivo they have to be expressed in the same place and at the same time. Thus, it is clear that the control of both the ligand and receptor expression is extremely important, and, at least for the receptors, this question is now being addressed by several laboratories. The pioneering work of Loetscher et al. (39) showed that CCR1 and CCR2 were upregulated in the presence of IL-2, clearly demonstrating the importance that cytokines play in the chemokine system. Another example is that the proinflammatory cytokine interferon-γ (IFN-γ), upregulates the CC chemokine receptors CCR1 and CCR3 on neutrophils, a cell type previously thought to be only activated by CXC chemokines (40). Last, it is not well defined in vivo how much presentation by glycosaminoglycans (GAG) alters the pharmacokinetics of the formation and destruction of chemokine gradients, but these are essential questions and are the subject of much active research. We have demonstrated that selectivity does exist in the GAG/chemokine interaction (41) and that, in addition, the binding of chemokines to GAG expressed on the cell surface serves as a mechanism to increase local chemokine concentrations (42).

VALIDATION OF CHEMOKINES AS THERAPEUTIC TARGETS FOR INFLAMMATION

One of the key questions in drug discovery is to make sure that as far as possible the target is validated—and workers in the chemokine area have relied heavily on information derived from animal models of disease. Clearly, the ultimate validation is testing a potential molecule in patients with the disease (the so-called proof of principle in man). However, before attaining this level of validation, it is important to obtain as much information as possible in vivo from the animal models that are relevant to the diseases being studied. The first step is to demonstrate clearly that the chemoattractant effects of chemokines that are observed in vitro can also be seen in vivo. For example, the eosinophil chemoattractant eotaxin, which was originally purified from bronchoalveolar lavage (BAL) of allergen-sensitized animals, causes eosinophil migration in chemotactic assays and also to the site of injection into guinea pig skin in vivo (43). However, although RANTES is clearly an eosinophil chemoattractant in vitro, injection intradermally does not always result in recruitment of eosinophils to the site of injection in nonsensitized animals but requires presensitization of the animals (44). In humans, intradermal injections of RANTES into nonallergic and allergic individuals showed significant eosinophilic recruitment occurring only in allergic patients 6 h after injection (45). Similar results can also be produced using mice that have been genetically manipulated to overexpress a chemokine. Here, the pattern of tissue distribution directed by the expression vector is important in determining the pathology. Mice overexpressing MCP-1 in several organs show specific monocyte infiltration (46). However, when MCP-1 expression was directed to the pancreas, although the mice developed a chronic monocytic infiltrate, the insulitis did not progress (47). MCP-1 on its own appears to be sufficient to elicit cellular migration but not inflammation.

A second line of validation can come from animals in which the gene either for the chemokine or the receptor has been deleted. This type of study addresses the question of redundancy in vivo—the fear that blockade of a particular chemokine/receptor pair will be ineffective, because there are other molecules which can carry the ''inflammatory signal'' so to speak. Deletion of the CXCR2 homolog in mice showed impaired neutrophilic recruitment on infection (48). Deletion of CCR1 again had a normal phenotype, although the mice showed disordered myeloid progenitor cell distribution and trafficking (49). However, these mice showed impaired immune responses when stressed by exposure to pathogens. A human equivalent to the murine deletion experiments exists in approximately 1% of the population, who carry a 32–base pair deletion in the CCR5 gene. This deletion, Δ32-CCR5 prevents any functional surface expression of the receptor, and their ''phenotype'' appears very normal, without any immune defects, and with an added inestimable advantage of being highly resistant to HIV infection (50).

In most fields of cytokine biology, validation of the importance of a given receptor in a disease model may also come from studies using blocking monoclonal antibodies in various animal models of disease. In the chemokine area, there are many reports of blocking an individual ligand. To cite two examples from work in the area of airways inflammation—the neutralization of chemokines whose expression has been demonstrated to be temporally regulated in the ovalbumin sensitization model of murine ''asthma'' has identified the coordinated action of CC chemokines (51). Another study using cockroach antigen to sensitize the mice has again demonstrated the differential recruitment mediated by two eosinophilic chemoattractants, MIP-1α and eotaxin (52). Neutralization of receptors has been limited by the availability of good blocking monoclonal antibodies—seven-transmembrane receptors are notoriously difficult to raise antibodies against, since they have very little protein exposed on the extracellular surface. Such antibodies are now becoming available—so we should expect to see more antibody validation data of receptor blockade over the next couple of years. Antibodies have the main advantage that they mimic therapy much more closely, since they are given as a therapeutic injection. It can always be argued with receptor knock out mice that they have adjusted their chemokine network to make up for the lack of the particular receptor. Clearly, in the future, knock out mice with inducible promoters will be much more important in the study of disease pathology. In the absence of antibodies, much of our understanding of the role of breaking the chemokine network may have on the progression of a given disease model comes from studies with modified chemokines.

MODIFIED CHEMOKINES—AN EFFECTIVE RECEPTOR BLOCKADE STRATEGY

Although a number of studies have produced chemokine receptor antagonists by modifications at the N-terminus, it should be noted that several chemokines have been purified from natural sources as truncated proteins. Truncated forms of IL-8 (53), RANTES (54,55), granulocyte chemotactic protein-1 (GCP-2) (56), and MDC (57), to cite a few examples, have been reported, with the truncations producing different effects on bioactivity. The relevance of these truncations in vivo remains to be elucidated.

The first indication of the importance of the amino-terminus of the chemokines came from the pioneering work of two groups studying IL-8 (53,58). IL-8, as well as several other members of the CXC family, possesses a conserved three–amino acid motif immediately preceding the conserved CXC motif consisting of the residues Glu-Leu-Arg, or ELR, which are essential for receptor binding *and* biological activity. Production of the truncated analog 5-68 IL-8, which retains the Arg of the ELR motif, retains the capacity for binding to both CXCR1 and CXCR2 but does not induce signal transduction (59). It is thus a receptor antagonist, since it is able to prevent biological activities in neutrophils elicited by the full-length IL-8 protein in vitro such as elastase release, chemotaxis, and the respiratory burst.

Although several other CXC receptor antagonists have been generated by similar stategies that show good activity in vitro, there is no information on their effectiveness in preventing inflammation in vivo. One reason may be that although mice are the most widely used animals for inflammatory models, they do not have a direct IL-8 homolog, thus perhaps precluding the use of IL-8 antagonist analogs. Thus, the data on the role of CXCR1 and CXCR2 in inflammation is limited to the studies using monoclonal antibodies directed against the IL-8 protein in rabbit models of sepsis, where neutralizing the activity of IL-8 significantly suppressed neutrophil recruitment and the accompanying damage in this model (60).

Modification in the amino-terminal region of CC chemokines has similarly produced CC chemokine receptor antagonists. Structure-function studies in several laboratories have created amino-terminally truncated RANTES, MCP-1, and MCP-3 (61,62) proteins that antagonize the effects of their full-length parent ligands in vitro. Furthermore, some of these modifications have provided unexpected observations relevant to previously determined receptor-binding properties. The production of chemokines by the total chemical synthesis route has provided a large number of amino-terminally truncated CC chemokines. For instance, deletion of a single residue at the amino-terminus of MCP-1 to produce [2–76] MCP-1 dramatically reduced its activity on basophils, but rendered the protein a potent activator of eosinophils, whereas the full-length protein is inactive on eosinophils (63). This result indicates that both cell types have recognition sites that are very similar but allow a fine tuning—the question can be posed as to whether this mechanism is used in vivo. Removal of the second residue renders the protein inactive on both basophils and eosinophils. Surprisingly, the truncation of the first eight residues of RANTES conferred the ability of RANTES to bind to CCR2, in addition to the usual RANTES receptors, suggesting that the design of high-affinity multispecific CC chemokine antagonists may be feasible (61).

But the most potent chemokine receptor antagonists have in fact been produced by extending the amino-terminal of the chemokine RANTES. The retention of the initiating methionine on RANTES (Met-RANTES) when the recombinant protein is produced in the bacterial host *Escherichia coli* produces a receptor antagonist with nanomolar potency in its ability to inhibit both RANTES and MIP-1α–induced monocyte and T-cell chemotaxis in vitro (64). As determined in vitro, the inhibitory effect appears to be more effective via CCR1 than the other RANTES receptors such as CCR3 (65).

There is a high degree of species cross reactivity between the human and murine chemokine systems, which has allowed the validation of chemokine receptor blockade in preventing inflammation. Met-RANTES has been shown to be efficacious in reducing inflammation in several models. When administered to mice that have been sensitized with collagen in a model which provokes symptoms very closely resembling rheumatoid arthritis in humans, the onset and severity of the arthritic symptoms are reduced in a dose-related manner if the antagonist is administered prior to disease onset (66). In a murine genetic variant, MRL-lpr, mice develop arthritis spontaneously, and the [9–68] MCP-1 analog has been shown to have both prophylactic and therapeutic effects, since its administration both prior to onset as well as during disease is effective in reducing the symptoms (67). Met-RANTES administration to mice that develop crescentic glomerulonephritis induced by nephrotoxic sheep serum results in a reduction of both T-cell and macrophage accumulation in the renal tissues (68). Potent effects of Met-RANTES were observed in an ovalbumin-sensitization model of airways inflammation that results in symptoms closely resembling human asthma (69). In this model, both eosinophilic and T-cell recruitment to the airways were highly reduced, with a concomitant reduction in bronchohyperactivity and inhibition of mucus production. When the effect of Met-RANTES was compared to the blockade of chemokine ligands previously identified as being instrumental in airways inflammation by specific monoclonal antibodies, the strategy of a receptor blocker appeared to be more efficacious (51). Although Met-RANTES has not been shown to be very potent in vitro at blocking CCR3, in a late-phase allergic reaction in mouse skin, it was more effective at preventing eosinophilic recruitment than blocking the activity of the specific CCR3 ligand, eotaxin (70). In another inflammatory scenario where RANTES has been shown to play an important role, the blocking of RANTES receptors by the administration of Met-RANTES has been shown to be very effective in preventing the inflammatory response leading to the rejection of heterologous organ transplants (71). In a mild model, administration of the antagonist alone was capable of significantly inhibiting rejection of the transplanted kidney, and in a more severe model, administration of suboptimal doses of cyclosporin, which alone could not prevent rejection, the synergistic addition of low doses of Met-RANTES significantly abrogated signs of rejection.

Extension of the amino-terminal of RANTES by the chemical coupling of a five-carbon alkyl chain to the oxidized N-terminal Ser residue produces a RANTES analog called aminooxypentane RANTES (AOP-RANTES) (72). This protein was found to have an increased affinity over RANTES and Met-RANTES for CCR5 and was a significantly more potent inhibitor of HIV-1 infection (see below). AOP-RANTES had no activity on freshly isolated monocytes, and it was able to inhibit RANTES and MIP-1β–induced chemotaxis. When tested for its ability to induce calcium mobilization on cells overexpressing CCR5, it is, however, found to be fully active, whereas it had no significant activity on CCR1. Studies of this RANTES analog in vivo are ongoing.

INHIBITION OF HIV INFECTION THROUGH CHEMOKINE RECEPTOR LIGANDS

Coreceptors for which the ligands have been identified can be rendered nonfunctional in their coreceptor capacity in the presence of their ligands. Thus, the CCR5 ligands, RANTES, MIP-1α, and MIP-1β, inhibit infection of R5 strains (5), eotaxin inhibits R3 strains (73), I-309 inhibits R8 using viruses (74), and SDF-1 inhibits infection mediated by CXCR4 (7,8). A number of viruses have been shown to produce their own chemokine proteins, and the implications of this in their propagation in the host is the study of intense research, which is beyond the scope of this chapter. However, one of these needs to be mentioned here in view of its broad spectrum of coreceptor inhibitory properties. Herpesvirus 8 encodes three chemokines which have the closest protein sequence identity to MIP-1α, although this is only in the 40% range. One of these, vMIP-II, has even shown to be very promiscuous as a chemokine in that it is able to bind both CC and CXC chemokine receptors (75,76). It was in fact able to inhibit infection by R5, R3, and X4 HIV-1 strains and had the highest potency on CCR3.

It was, therefore, of great interest to identify whether chemokine receptor antagonists were also capable of inhibiting HIV-1 infection. Chemokines could inhibit viral infection by one of two mechanisms. The first would simply be by steric hindrance where their binding to the receptor would prevent the interactions of the virus (*with the receptor*) required for subsequent fusion. The second would be that the interaction of the chemokine would induce receptor internalization, a common feature of 7TM receptors. Removal of the receptor from the cell surface would obviously abrogate its coreceptor function. An agonist would be able to act through both mechanisms, whereas a true receptor antagonist would only inhibit through steric hindrance, since endocytosis of 7TM receptors is an agonist-driven event.

The three RANTES analogs, [9–68] RANTES (77) Met-RANTES, and AOP-RANTES (72), have all been shown to inhibit peripheral blood mononuclear cell (PBMC) infection by R5 strains of HIV-1. However, both [9–68] RANTES and Met-RANTES are less potent than RANTES itself, whereas AOP-RANTES is significantly more potent than RANTES. Although the inhibition by RANTES of R5 strains of both fusion and infection in recombinant cell lines expressing CD4 and CCR5 as well as that of PBMC has been widely reported by many different groups, the inhibition of macrophage infection by RANTES is still controversial. In fact, enhancement of infection of macrophages in the presence of RANTES has been reported (78). However, Simmons et al. (72) were able to show that AOP-RANTES was very efficient in inhibiting the infection of macrophages by four R5 strains, whereas RANTES showed inhibition of only one of the strains tested, and Met-RANTES showed no inhibition at all. To date, AOP-RANTES remains the most potent inhibitor of R5 strains.

The observation that these three RANTES analogs could inhibit infection lent weight to the mechanism mediated by steric hindrance, since they have been described as receptor antagonists in that they do not induce the signal transduction pathways required for cellular migration. However, although [9–68] RANTES has been reported to be unable to mobilize calcium and induce chemotaxis, it is able to induce CCR5 internalization (24). Downregulation of 7TM receptors is known to be an agonist-mediated event which requires phosphorylation mediated by the GRK family of serine/threonine kinases of the carboxy-terminal region of the receptor (79,80). This is followed by interaction with β-arrestins prior to sequestration into clathrin-coated pits. Similarly, AOP-RANTES efficiently downregulates CCR5 in stably transfected CHO cells as well as PBMC, and in fact is even more efficient than RANTES in this activity (81). Met-RANTES, on the other hand, has been shown to be unable to mediate CCR1 downregulation (82) and is very poor in inducing CCR5 internalization. The ability of chemokines to remove cell surface receptors has been suggested to contribute to their anti-infectivity properties both for X4 and R5 strains, since SDF-1 causes CXCR4 receptor internalization and the modified RANTES proteins [9–68] RANTES and AOP-RANTES cause internalization of CCR5. In the case of AOP-RANTES, the receptor downregulation was in fact greater than that induced by RANTES, possibly explaining its superior potency as an inhibitor of HIV infection.

Furthermore, AOP-RANTES has identified a novel mechanism which may be important in the inhibition of HIV infection. After downregulation and trafficking into early endosomes, 7TM receptors may undergo one of two fates. They are either targeted to late endosomes and then lysosomes where they are degraded, or they are targeted to a recycling compartment where, following dissociation of the ligand, the receptor is dephosphorylated and then recycles to the cell surface. Although the fate of chemokine receptors following internalization has not yet been studied extensively, examples of both pathways have been demonstrated. CCR2b has been shown to follow the degradative pathway through the lysosomal compartment (83), whereas CXCR4 (84) and CCR5 (81) recycle to the cell surface. Removal of RANTES from the culture medium after downregulation of CCR5 allows the internalized receptors to recycle, but surprisingly removal of AOP-RANTES significantly abrogates receptor recycling (81). Under these circumstances, cell surface expression of the receptor is prevented and HIV infection becomes impossible. This conclusion is further supported by the infection of cells with cDNA's coding for CCR5 ligands, including the KDEL motif. This amino acid sequence targets proteins to the endoplasmic reticulum (ER), and it is believed that ligands expressed with this amino acid sequence, named *intrakines*, trap the nascent chemokine receptors in the ER and thus prevent cell surface expression (85). Cells transfected with these cDNA sequences were effectively unable to be infected by HIV-1 viruses.

Studies of the inhibition of HIV-1 infectivity by chemokines, modified chemokines, and truncated receptors has therefore demonstrated that the most important route of inhibition is through steric hindrance. This is further borne out by the efficacy of AMD3100 (see below), a purely competitive inhibitor of SDF-1 for its receptor CXCR4, in inhibiting X4 HIV-1 strains. However, the removal of the coreceptors from the cell surface as demonstrated by AOP-RANTES, which prevents recycling, or by the intervention of the exocytosis pathway, which also prevents their surface expression, contribute to the prevention of host cell infection by the HIV virus.

SMALL MOLECULE INHIBITORS OF CORECEPTORS—THE PHARMACEUTICAL IDEAL

The observations that HIV infection can be inhibited both by the natural ligands as well as by modified chemokines certainly validates chemokine receptors as good targets for therapeutic intervention. Pharmaceutical companies in fact began screening for low molecular weight, orally available inhibitors of coreceptors as soon as they were identified in 1996. This screening effort is now beginning to pay off. Since seven-transmembrane receptors have an exceptionally good track record for being susceptible to low molecular weight therapeutic drugs that can be taken orally, once the worry of crucial interference with the immune system was diminished, these receptors appeared to be an ideal target. However, the majority of these receptors have small ligands of which histamine, epinephrine (adrenaline), serotonin, and dopamine are representative examples. The interaction of chemokines and their receptors involves a larger protein ligand, of about 8 kD, which may explain why small molecule inhibitors were rapidly not found. However, the picture is hopeful, since reports of such compounds are now appearing in the patent and scientific literature, and have recently been well documented in a specialized review (87). Although no compounds have yet been reported for CCR5 in the scientific literature, three classes of compounds were recently described that were identified by their ability to block infection of PBMC by T-tropic HIV-1 strains. These compounds have all been shown to be CXCR4 receptor antagonists in that they block binding of its ligand, SDF-1 as well as inhibiting SDF-1–induced receptor activation (87). The most potent class of compounds is the bicyclam series, where the protoype AMD3100 is 10- to 50-fold more potent in inhibiting HIV infection than SDF-1 itself, but unfortunately does not have the required bioavailability properties. The second type of antagonist is a cyclic 18-mer peptide containing two disulfides that was isolated from *Limulus polyphemus*, whereas the third is pseudo 9-mer peptide.

To date, modified chemokines appear to be the most advanced. However, these proteins are all modifications of RANTES, which, as mentioned above, has been described to enhance HIV-1 infection. Furthermore, at least in vitro, RANTES will enhance infection of several types of virus in a CCR5-independent manner (86), and although these effects were observed at higher concentrations than those which occur in vivo, these results must certainly be taken into consideration when designing a therapeutic approach using modified RANTES proteins.

Obviously, the question arises of what the effect of blocking a receptor involved in the maintenance of a fully competent immune system would be. The answer, at least partially, for the case of CCR5 arises from the existence in the human population of an equivalent to transgenic knock out animals: approximately 1% of the white population are homozygous for a 32–base pair deletion in the CCR5 gene. This results in an absence of functional cell surface CCR5 receptors in these individuals with no apparent immune defects, but with an inestimable advantage in that they are highly resistant to HIV infection. Although CXCR4 appears to be essential for fetal development, since gene deletion of its ligand SDF-1 is neonatal lethal in mice, there is no evidence that blocking this receptor is deleterious in the adult.

It does not, therefore, seem unreasonable to hope that in the near future a low molecular weight coreceptor inhibitor may become an addition to the currently available combination therapies used in the treatment of AIDS. An advantage of this type of drug is that it would not be subject to escape mechanisms of the virus through mutations, since it is directed against a host target. However, if a single coreceptor is targeted, the question of modification of receptor selectivity through selective mutagenesis of the virus must not be overlooked.

REFERENCES

1. AG Dalgleish, PC Beverley, Clapham, DH Crawford, MF Greaves, RA Weiss. The CD4 (T4) antigen is an essential component of the receptor for the AIDS retrovirus. Nature 312:763–767, 1984.
2. PJ Maddon, AG Dalgleish, JS McDougal, PR Clapham, RA Weiss, R Axel. The T4 gene encodes the AIDS virus receptor and is expressed in the immune system and the brain. Cell 47:333–348, 1986.
3. B Chesebro, R Buller, J Portis, K Wehrly. Failure of human immunodeficiency virus entry and infection in CD4-positive human brain and skin cells. J Virol 64:215–221, 1990.
4. JA Levy. HIV pathogenesis and long-term survival (editorial). AIDS 7:1401–1410, 1993.
5. F Cocchi, AL DeVico, A Garzino-Demo, SK Arya, RC Gallo, P LUSSO. Identification of RANTES, MIP-1 alpha, and MIP-1 beta as the major HIV-suppressive factors produced by CD8+ T cells. Science 270:1811–1815, 1995.
6. Y Feng, CC Broder, PE Kennedy, EA Berger. HIV-1 entry cofactor: functional cDNA cloning of a seven-transmembrane, G protein–coupled receptor (see Comments). Science 272:872–877, 1996.
7. CC Bleul, M Farzan, H Choe, C Parolin, I Clark-Lewis, J Sodroski, TA Springer. The lymphocyte chemoattractant SDF-1 is a ligand for LESTR/FUSIN and blocks HIV-1 entry. Nature 382:829–833, 1996.
8. E Oberlin, A Amara, F Bachelerie, C Bessia, JL Virelizier, F Arenzana-Seisdedos, O Schwartz, JM Heard, I Clark-Lewis, DF Legler, M Loetscher, M Baggiolini, B Moser. The CXC chemokine SDF-1 is the ligand for LESTR/fusin and prevents infection by T-cell-line–adapted HIV-1. Nature 382:833–835, 1996.
9. T Dragic, V Litwin, GP Allaway, SR Martin, Y Huang, KA Nagashima, C Cayanan, PJ Maddon, RA Koup, JP Moore, WA Paxton. HIV-1 entry into CD4+ cells is mediated by the chemokine receptor CC-CKR-5 (see Comments). Nature 381:667–673, 1996.
10. G Alkhatib, C Combadiere, CC Broder, Y Feng, PE Kennedy, PM Murphy, EA Berger. CC CKR5: a RANTES, MIP-1alpha, MIP-1beta receptor as a fusion cofactor for macrophage-tropic HIV-1. Science 272:1955–1958, 1996.
11. H Deng, R Liu, W Ellmeier, S Choe, D Unutmaz, M Burkhart, P Di Marzio, S Marmon, RE Sutton, CM Hill, CB Davis, SC Peiper, TJ Schall, DR Littman, NR Landau. Identification of a major co-receptor for primary isolates of HIV-1 (see Comments). Nature 381:661–666, 1996.

12. P Clapham, A McKnight, G Simmons, R Weiss. Is CD4 sufficient for HIV entry? Cell surface molecules involved in HIV infection, (Review). Philosophical Trans R Soc Lond B Biol Sci 342:67–73, 1993.
13. EA Berger, RW Doms, EM Fenyo, BT Korber, DR Littman, JP Moore, QJ Sattentau, H Schuitemaker, J Sodroski, RA Weiss. A new classification for HIV-1 (letter). Nature 391:240, 1998.
14. RI Connor, DD Ho. Transmission and pathogenesis of human immunodeficiency virus type 1 (Review). AIDS Res Hum Retroviruses 10:321–323, 1994.
15. H Schuitemaker, M Koot, NA Kootstra, MW Dercksen, RE de Goede, RP van Steenwijk, JM Lange, JK Schattenkerk, F Miedema, M Tersmette. Biological phenotype of human immunodeficiency virus type 1 clones at different stages of infection: progression of disease is associated with a shift from monocytotropic to T-cell–tropic virus population. J Virol 66:1354–1360, 1992.
16. R Liu, WA Paxton, S Choe, D Ceradini, SR Martin, R Horuk, ME MacDonald, H Stuhlmann, RA Koup, NR Landau. Homozygous defect in HIV-1 coreceptor accounts for resistance of some multiply-exposed individuals to HIV-1 infection. Cell 86:367–377, 1996.
17. M Samson, F Libert, BJ Doranz, J Rucker, C Liesnard, CM Farber, S Saragosti, C Lapoumeroulie, J Cognaux, C Forceille, G Muyldermans, C Verhofstede, G Burtonboy, M Georges, T Imai, S Rana, Y Yi, RJ Smyth, RG Collman, RW Doms, G Vassart, M Parmentier. Resistance to HIV-1 infection in caucasian individuals bearing mutant alleles of the CCR-5 chemokine receptor gene (see Comments). Nature 382:722–725, 1996.
18. M Dean, M Carrington, C Winkler, GA Huttley, MW Smith, R Allikmets, JJ Goedert, SP Buchbinder, E Vittinghoff, E Gomperts, S Donfield, D Vlahov, R Kaslow, A Saah, C Rinaldo, R Detels, SJ O'Brien. Genetic restriction of HIV-1 infection and progression to AIDS by a deletion allele of the CKR5 structural gene. Hemophilia Growth and Development Study, Multicenter AIDS Cohort Study, Multicenter Hemophilia Cohort Study, San Francisco City Cohort, ALIVE Study (see Comments). Science 273:1856–1862, 1996.
19. A Trkola, T Dragic, J Arthos, JM Binley, WC Olson, GP Allaway, C Cheng-Mayer, J Robinson, PJ Maddon, JP Moore. CD4-dependent, antibody-sensitive interactions between HIV-1 and its co-receptor CCR-5 (see Comments). Nature 384:184–187, 1996.
20. L Wu, NP Gerard, R Wyatt, H Choe, C Parolin, N Ruffing, A Borsetti, AA Cardoso, E Desjardin, W Newman, C Gerard, J Sodroski. CD4-induced interaction of primary HIV-1 gp120 glycoproteins with the chemokine receptor CCR-5 (see Comments). Nature 384:179–183, 1996.
21. CK Lapham, J Ouyang, B Chandrasekhar, N Nguyen, DS Dimitrov, H Golding. Evidence for cell-surface association between fusin and the CD4-gp120 complex in human cell lines. Science 274:602–605, 1996.
22. PA Bullough, FM Hughson, JJ Skehel, DC Wiley. Structure of influenza haemagglutinin at the pH of membrane fusion (see Comments) Nature 371:37–43, 1994.

23. I Aramori, J Zhang, SS Ferguson, PD Bieniasz, BR Cullen, M Caron. Molecular mechanism of desensitization of the chemokine receptor CCR-5: receptor signaling and internalization are dissociable from its role as an HIV-1 co-receptor. EMBO J 16:4606–4616, 1997.
24. A Amara, SL Gall, O Schwartz, J Salamero, M Montes, P Loetscher, M Baggiolini, JL Virelizier, SF Arenzana. HIV coreceptor downregulation as antiviral principle: SDF-1alpha–dependent internalization of the chemokine receptor CXCR4 contributes to inhibition of HIV replication. J Exp. Med 186:139–146, 1997.
25. Z Lu, JF Berson, Y Chen, JD Turner, T Zhang, M Sharron, MH Jenks, Z Wang, J Kim, J Rucker, JA Hoxie, SC Peiper, RW Doms. Evolution of HIV-1 coreceptor usage through interactions with distinct CCR5 and CXCR4 domains. Proc Natl Acad Sci USA 94:6426–6431, 1997.
26. T Dragic, A Trkola, SW Lin, KA Nagashima, F Kajumo, L Zhao, WC Olson, L Wu, CR Mackay, GP Allaway, TP Sakmar, JP Moore, PJ Maddon. Amino-terminal substitutions in the CCR5 coreceptor impair gp120 binding and human immunodeficiency virus type 1 entry. J Virol 72:279–285, 1998.
27. L Wu, G LaRosa, N Kassam, CJ Gordon, H Heath, N Ruffing, H Chen, J Humblias, M Samson, M Parmentier, JP Moore, CR Mackay. Interaction of chemokine receptor CCR5 with its ligands: multiple domains for HIV-1 gp120 binding and a single domain for chemokine binding. J Exp Med 186:1373–1381, 1997.
28. L Picard, DA Wilkinson, A McKnight, PW Gray, JA Hoxie, PR Clapham, RA Weiss. Role of the amino-terminal extracellular domain of CXCR-4 in human immunodeficiency virus type 1 entry. Virology 231:105–111, 1997.
29. TNC Wells, CA Power, AEI Proudfoot. Definition, function and pathophysiological significance of chemokine receptors. Trends Pharmacol Sci 19:376–380, 1998.
30. M Baggiolini. Chemokines and leukocyte traffic (Review). Nature 392:565–568, 1998.
31. AD Luster. Chemokines—chemotactic cytokines that mediate inflammation (Review). N Eng J Med 338:436–445, 1998.
32. BJ Rollins. Chemokines. Blood 90:909–928, 1997.
33. TNC Wells, MC Peitsch. The chemokine information source—identification and characterization of novel chemokines using the worldwideweb and expressed sequence tag databases. J Leukoc Biol 61:545–550, 1997.
34. CA Power, A Meyer, K Nemeth, KB Bacon, AJ Hoogewerf, AEI Proudfoot, TNC Wells. Molecular cloning and functional expression of a novel CC chemokine receptor cDNA from a human basophilic cell line. J Biol Chem 270:19495–19500, 1995.
35. A Hoogewerf, D Black, AEI Proudfoot, TNC Wells, CA Power. Molecular cloning of murine CC CKR-4 and high affinity binding of chemokines to murine and human CC CKR-4. Biochem Biophys Res Commun 218:337–343, 1996.
36. T Imai, T Yoshida, M Baba, M Nishimura, M Kakizaki, O Yoshie. Molecular cloning of a novel T cell–directed CC chemokine expressed in thymus by signal

sequence trap using Epstein-Barr virus vector. J Biol Chem 271:21514–21521, 1996.
37. T Imai, D Chantry, CJ Raport, CL Wood, M Nishimura, R Godiska, O Yoshie, PW Gray. Macrophage-derived chemokine is a functional ligand for the CC chemokine receptor 4. J Biol Chem 273:1764–1768, 1998.
38. TN Kledal, MM Rosenkilde, TW Schwartz. Selective recognition of the membrane-bound CX3C chemokine, fractalkine, by the human cytomegalovirus-encoded broad-spectrum receptor US28. FEBS Lett 441:209–214, 1998.
39. P Loetscher, M Seitz, M Baggiolini, B Moser. Interleukin-2 regulates CC chemokine receptor expression and chemotactic responsiveness in T lymphocytes (see Comments). J Exp Med 184:569–577, 1996.
40. R Bonecchi, N Polentarutti, W Luini, A Borsatti, S Bernasconi, AEI Proudfoot, TNC Wells, CR Mackay, A Mantovani, S Sozzani. Upregulation of CCR1 and CCR3 and induction of chemotaxis to CC chemokines by Interferon gamma in human neutrophils. J Immunol 162:474–479, 1999.
41. GS Kuschert, F Coulin, RE Hubbard, CA Power, TNC Wells, AJ Hoogewerf. Chemokines interact selectively with glycosaminoglycans and soluble glycosaminoglycans inhibit 7-transmembrane chemokine receptor binding and cellular responses. Biochemistry (in press).
42. AJ Hoogewerf, GS Kuschert, AEI Proudfoot, F Borlat, LI Clark, CA Power, TNC Wells. Glycosaminoglycans mediate cell surface oligomerization of chemokines. Biochemistry 36:13570–13578, 1997.
43. PJ Jose, DA Griffiths-Johnson, PD Collins, DT Walsh, R Moqbel, NF Totty, O Truong, JJ Hsuan, TJ Williams. Eotaxin: a potent eosinophil chemoattractant cytokine detected in a guinea pig model of allergic airways inflammation. J Exp Med 179:881–887, 1994.
44. R Meurer, G van Riper, W Feeney, P Cunningham, D Hora, Jr, MS Springer, DE MacIntyre, H Rosen. Formation of eosinophilic and monocytic intradermal inflammatory sites in the dog by injection of human RANTES but not human monocyte chemoattractant protein 1, human macrophage inflammatory protein 1 alpha, or human interleukin 8. J Exp Med 178:1913–1921, 1993.
45. LA Beck, S Dalke, KM Leiferman, CA Bickel, R Hamilton, H Rosen, BS Bochner, RP Schleimer. Cutaneous injection of RANTES causes eosinophil recruitment: comparison of nonallergic and allergic human subjects. J Immunol 159:2962–2972, 1997.
46. BJ Rutledge, H Rayburn, R Rosenberg, RJ North, RP Gladue, CL Corless, and BJ Rollins. High level monocyte chemoattractant protein-1 expression in transgenic mice increases their susceptibility to intracellular pathogens. J Immunol 155:4838–4843, 1995.
47. IS Grewal, BJ Rutledge, JA Fiorillo, L Gu, RP Gladue, R Flavell, BJ Rollins. Transgenic monocyte chemoattractant protein-1 (MCP-1) in pancreatic islets produces monocyte-rich insulitis without diabetes: abrogation by a second transgene expressing systemic MCP-1. J Immunol 159:401–408, 1997.
48. G Cacalano, J Lee, K Kikly, AM Ryan, S Pitts-Meek, B Hultgren, WI Wood,

MW Moore. Neutrophil and B cell expansion in mice that lack the murine IL-8 receptor homolog [see comments] (published erratum appears in Science 1995 Oct 20;270[5235]:365). Science 265:682–684, 1994.
49. JL Gao, TA Wynn, Y Chang, EJ Lee, HE Broxmeyer, S Cooper, HL Tiffany, H Westphal, CJ Kwon, PM Murphy. Impaired host defense, hematopoiesis, granulomatous inflammation and type 1-type 2 cytokine balance in mice lacking CC chemokine receptor 1. J Exp Med 185:1959–1968, 1997.
50. WA Paxton, SR Martin, D Tse, TR O'Brien, J Skurnick, NL VanDevanter, N Padian, JF Braun, DP Kotler, SM Wolinsky, RA Koup. Relative resistance to HIV-1 infection of CD4 lymphocytes from persons who uninfected despite multiple high-risk sexual exposure. Nat Med 2:412–417, 1996.
51. JA Gonzalo, CM Lloyd, JP Albar, D Wen, TNC Wells, AEI Proudfoot, C Martinez-A, T Bjerke, AJ Coyle, JC Gutierrez-Ramos. The coordinated action of CC chemokines in the lung orchestrates allergic inflammation and airway hyperresponsiveness. J Exp Med 188:157–167, 1998.
52. EM Campbell, SL Kunkel, RM Strieter, NW Lukacs. Temporal role of chemokines in a murine model of cockroach allergen-induced airway hyperreactivity and eosinophilia. J Immunol 161:7047–7053, 1998.
53. I Clark-Lewis, C Schumacher, M Baggiolini, B Moser. Structure-activity relationships of interleukin-8 determined using chemically synthesized analogs. Critical role of NH2-terminal residues and evidence for uncoupling of neutrophil chemotaxis, exocytosis, and receptor binding activities. J Biol Chem 266:23128–23134, 1991.
54. N Noso, M Sticherling, J Bartels, AI Mallet, E Christophers, JM Schroder. Identification of an N-terminally truncated form of the chemokine RANTES and granulocyte-macrophage colony-stimulating factor as major eosinophil attractants released by cytokine-stimulated dermal fibroblasts. J Immunol 156:1946–1953, 1996.
55. S Struyf, I De Meester, S Scharpe, JP Lenaerts, P Menten, JM Wang, P Proost, J van Damme. Natural truncation of RANTES abolishes signaling through the CC chemokine receptors CCR1 and CCR3, impairs its chemotactic potency and generates a CC chemokine inhibitor. Eur J Immunol 28:1262–1271, 1998.
56. P Proost, I De Meester, D Schols, S Struyf, AM Lambeir, A Wuyts, G Opdenakker, E De Clercq, S Scharpe, J van Damme. Amino-terminal truncation of chemokines by CD26/dipeptidyl-peptidase IV. Conversion of RANTES into a potent inhibitor of monocyte chemotaxis and HIV-1-infection. J Biol Chem 273:7222–7227, 1998.
57. S Struyf, P Proost, S Sozzani, A Mantovani, A Wuyts, E De Clercq, D Schols, J van Damme. Enhanced anti–HIV-1 activity and altered chemotactic potency of NH2-terminally processed macrophage-derived chemokine (MDC) imply an additional MDC receptor. J Immunol 161:2672–2675, 1998.
58. CA Hebert, RV Vitangcol, JB Baker. Scanning mutagenesis of interleukin-8 identifies a cluster of residues required for receptor binding. J Biol Chem 266: 18989–18994, 1991.

59. B Moser, B Dewald, L Barella, C Schumacher, M Baggiolini, I Clark-Lewis. Interleukin-8 antagonists generated by N-terminal modification. J Biol Chem 268:7125–7128, 1993.
60. HG Folkesson, MA Matthay, CA Hebert, VC Broaddus. Acid aspiration–induced lung injury in rabbits is mediated by interleukin-8–dependent mechanisms. J Clin Invest 96:107–116, 1995.
61. JH Gong, M Uguccioni, B Dewald, M Baggiolini, I Clark-Lewis. RANTES and MCP-3 antagonists bind multiple chemokine receptors. J Biol Chem 271:10521–10527, 1996.
62. JH Gong, I. Clark-Lewis. Antagonists of monocyte chemoattractant protein 1 identified by modification of functionally critical NH2-terminal residues. J Exp Med 181:631–640, 1995.
63. M Weber, M Uguccioni, B Ochensberger, M Baggiolini, I Clark-Lewis, CA Dahinden. Monocyte chemotactic protein MCP-2 activates human basophil and eosinophil leukocytes similar to MCP-3 J Immunol 154:4166–4172, 1995.
64. AEI Proudfoot, CA Power, AJ Hoogewerf, MO Montjovent, F Borlat, RE Offord, TNC Wells. Extension of recombinant human RANTES by the retention of the initiating methionine produces a potent antagonist. J Biol Chem 271:2599–2603, 1996.
65. J Elsner, H Petering, R Hochstetter, D Kimmig, TNC Wells, A Kapp, AEI Proudfoot. The CC chemokine antagonist Met-RANTES inhibits eosinophil effector functions through the chemokine receptors CCR1 and CCR3. Eur J Immunol 27:2892–2898, 1997.
66. C Plater-Zyberk, AJ Hoogewerf, AEI Proudfoot, CA Power, TNC Wells. Effect of a CC chemokine receptor antagonist on collagen induced arthritis in DBA/1 mice. Immunol Lett 57:117–120, 1997.
67. JH Gong, LG Ratkay, JD Waterfield, LI Clark. An antagonist of monocyte chemoattractant protein 1 (MCP-1) inhibits arthritis in the MRL-lpr mouse model. J Exp Med 186:131–137, 1997.
68. CM Lloyd, ME Dorf, AEI Proudfoot, DJ Salant, JC Gutierrez-Ramos. Role of MCP-1 and RANTES in inflammation and progression to fibrosis during murine crescentic nephritis. J Leukoc Biol 62:676–680, 1997.
69. AJ Coyle, J Kips, AJ Hoogewerf, R Pauwels, MD Tyers, Y Chvatchko, CA Power, AEI Proudfoot, TNC Wells. Upregulation of chemokine receptors in the airways after allergen provocation:Inhibition of airway inflammation and altered airway responsiveness by Met-RANTES, a chemokine receptor antagonist. J Exp Med (In press).
70. MM Teixeira, TNC Wells, NW Lukacs, AEI Proudfoot, SL Kunkel, TJ Williams, PG Hellewell. Chemokine-induced eosinophil recruitment—evidence of a role for endogenous eotaxin in an in vivo allergy model in mouse skin. J Clin Invest 100:1657–1666, 1997.
71. H-J Gröne, C Weber, KSC Weber, EF Gröne, T Rabelink, CM Klier, TNC Wells, AEI Proudfoot, D Schlondorff, PJ Nelson. Met-RANTES reduces vascular and

tubular damage during acute renal transplant rejection: blocking monocyte arrest and recruitment. FASEB J 13:1371–1383, 1999.
72. G Simmons, PR Clapham, L Picard, RE Offord, MM Rosenkilde, TW Schwartz, R Buser, TNC Wells, AEI Proudfoot. Potent inhibition of HIV-1 infectivity in macrophages and lymphocytes by a novel CCR5 antagonist. Science 276:276–279, 1997.
73. H Choe, M Farzan, Y Sun, N Sullivan, B Rollins, PD Ponath, L Wu, CR Mackay, G LaRosa, W Newman, N Gerard, C Gerard, J Sodroski. The beta-chemokine receptors CCR3 and CCR5 facilitate infection by primary HIV-1 isolates. Cell 85:1135–1148, 1997.
74. R Horuk, J Hesselgesser, Y Zhou, D Faulds, M Halks-Miller, S Harvey, D Taub, M Samson, M Parmentier, J Rucker, BJ Doranz, RW Doms. The CC chemokine I-309 inhibits CCR8-dependent infection by diverse HIV-1 strains. J Biol Chem 273:386–391, 1998.
75. TN Kledal, MM Rosenkilde, F Coulin, G Simmons, AH Johnsen, S Alouani, CA Power, HR Luttichau, J Gerstoft, PR Clapham, I Clarklewis, TNC Wells, TW Schwartz. A broad-spectrum chemokine antagonist encoded by Kaposi's sarcoma–associated herpesvirus. Science 277:1656–1659, 1997.
76. C Boshoff, Y Endo, PD Collins, Y Takeuchi, JD Reeves, VL Schweickart, MA Siani, T Sasaki, TJ Williams, PW Gray, PS Moore, Y Chang, RA Weiss. Angiogenic and HIV-inhibitory functions of KSHV-encoded chemokines. Science 278: 290–294, 1997.
77. F Arenzana-Seisdedos, JL Virelizier, D Rousset, I Clark-Lewis, P Loetscher, B Moser, M Baggiolini. HIV blocked by chemokine antagonist (letter). Nature 383: 400–400, 1996.
78. H Schmidtmayerova, B Sherry, M Bukrinsky. Chemokines and HIV replication. Nature 382:767, 1996.
79. SK Bohm, EF Grady, NW Bunnett. Regulatory mechanisms that modulate signalling by G-protein–coupled receptors (Review). Biochem. J. 322:1–18, 1997.
80. JA Koenig, JM Edwardson. Endocytosis and recycling of g protein–coupled receptors (Review). Trends Pharmacol Sci 18:276–287, 1997.
81. M Mack, B Luckow, PJ Nelson, J Cihak, G Simmons, PR Clapham, N Signoret, M Marsh, M Stangassinger, F Borlat, TNC Wells, D Schlondorff, AEI Proudfoot. Aminooxypentane-RANTES induces CCR5 internalization but inhibits recycling: a novel inhibitory mechanism of HIV infectivity. J Exp Med 187:1215–1224, 1998.
82. R Solari, RE Offord, S Remy, JP Aubry, TNC Wells, E Whitehorn, T Oung, AEI Proudfoot. Receptor-mediated endocytosis of CC-chemokines. J Biol Chem 272:9617–9620, 1997.
83. SG Mueller, WP Schraw, A Richmond. Activation of protein kinase c enhances the phosphorylation of the type b interleukin-8 receptor and stimulates its degradation in non-hematopoietic cells. J Biol Chem 270:10439–10448, 1995.
84. N Signoret, J Oldridge, MA Pelchen, PJ Klasse, T Tran, LF Brass, MM Rosenkilde, TW Schwartz, W Holmes, W Dallas, M Luther, TNC Wells, JA Hoxie,

M Marsh. Phorbol esters and SDF-1 induce rapid endocytosis and down modulation of the chemokine receptor CXCR4. J Cell Biol 139:651–664, 1997.
85. AG Yang, XF Bai, XF Huang, CP Yao, SY Chen. Phenotypic knockout of HIV type 1 chemokine coreceptor CCR-5 by intrakines as potential therapeutic approach for HIV-1 infection. Proc Natl Acad Sci USA 94:11567–11572, 1997.
86. CJ Gordon, MA Muesing, AEI Proudfoot, CA Power, JP Moore, A Trkola. Enhancement of human immunodeficiency virus type 1 infection by the CC-chemokine RANTES is independent of the mechanism of virus-cell fusion. J Virol 73: 684–694, 1999.
87. PD Ponath. Chemokine receptor antagonists: novel therapeutics for inflammation and AIDS. Exp Opin Invest Drugs 7:1–18, 1998.

8
Receptor Antagonists of gp130 Signaling Cytokines

Ann B. Vernallis
Aston University, Birmingham, England

CREATION OF RECEPTOR ANTAGONISTS BY MODIFYING CYTOKINES

Cytokines activate receptors by oligomerizing receptor chains into complexes that bring into proximity the cytoplasmic domains of the receptor subunits. The association of the intracellular domains initiates signaling through second-messenger pathways. Cytokines recruit receptor chains into receptor complexes in an ordered, sequential manner. The clusters of amino acids, or sites, that are used by cytokines to bind each receptor chain are topologically separated. This separation of sites has been exploited in the generation of receptor antagonists; the site that binds the subunit that is recruited first can be preserved, while disrupting sites that bind subunits that are recruited subsequently. The preservation of the first site allows the modified cytokine to sequester the first subunit such that the subunit is no longer available to participate in receptor complexes that signal. Hence, receptor antagonists act as antagonists by sequestering chains that are essential for the assembly of active receptor complexes.

The selective modification of receptor-binding sites was first applied to the generation of receptor antagonists for human growth hormone (GH) receptor. To form a functional receptor complex, each molecule of GH binds two

Figure 1 Schematic representation of a single molecule of growth hormone bound to two molecules of GHR, based on the crystalline structure of the cocomplex (48). The receptor binding sites on GH are topologically separated. The first receptor to bind, shown on the right, occupies site 1. The receptor on the left subsequently occupies site 2. Preserving site 1 while disabling site 2 creates a competitive antagonist (1).

molecules of GH receptor (GHR) (Fig. 1). The two GHR are recruited in a sequential fashion. Mutants of GH that bound the first GHR but not the second acted as competitive antagonists (1). A similar approach has also been applied to prolactin (2), interleukins IL-2 (3), IL-4 (4–6), and IL-5 (7), granulocyte-macrophage colony-stimulating factor (GM-CSF) (8), and to the gp130 family of cytokines (referenced below). A similar approach can be applied to any cytokine or growth factor that drives the stepwise assembly of a multisubunit receptor complex. The development of receptor antagonists for the gp130 family, GM-CSFRα, and IL-4R demonstrates that there has been substantial progress in the creation of receptor antagonists that are potent, specific, and suitable for in vivo applications.

THE gp130 FAMILY OF CYTOKINES

Receptor Assembly

Cytokine receptors often contain more than one kind of subunit, and often one or two of the subunits are shared with other cytokine receptors. The gp130 family of cytokines, which all share gp130 as a signal-transducing subunit, includes IL-6, IL-11, leukemia inhibitory factor (LIF), ciliary neurotrophic factor (CNTF), cardiotrophin (CT-1), and oncostatin M (OSM) (reviewed in ref. 9). Recently, a viral homolog of IL-6 has been added to the family; viral IL-6, (vIL-6), also referred to as Kaposi's sarcoma–associated herpesvirus–IL-6 (KSHV-IL-6), is encoded in the genome of human herpesvirus 8 (10).

IL-6 first binds a ligand-specific receptor, IL-6Rα, before driving the dimerization of gp130 (Fig. 2). The designation of IL-6R as an α subunit emphasizes that IL-6 binds this receptor component first before recruiting additional subunits. The short cytoplasmic domain of IL-6Rα does not participate in signal transduction, which explains why a soluble form of the receptor is capable of binding IL-6 and triggering responses from cells that make gp130 but not transmembrane IL-6Rα (11). Similarly, IL-11 binds IL-11Rα before binding gp130, and soluble IL-11Rα can induce responses to IL-11 in cells that make gp130 but not transmembrane IL-11Rα (12).

Several of the gp130 cytokines also share LIF-R as a common signal-transducing subunit. LIF, CNTF, CT-1, and human OSM all activate gp130 in conjunction with LIF-R. CNTF must first bind a ligand-specific receptor, CNTFRα, before heterodimerizing LIF-R and gp130. CNTFRα is normally tethered to the membrane by a glycosyl-phosphatidylinositol (GPI) linkage, but it can be released, for example, in peripheral nerve injury (13). Unlike CNTF, the soluble CNTF receptor/CNTF complex will potently stimulate cells that express LIF-R and gp130 even if they lack CNTFRα (13). A ligand-specific receptor for CT-1 has also been proposed (14,15). In addition to LIF-R, a third signal-transducing subunit, OSM-R, can also heterodimerize with gp130 (16). Human OSM activates either human LIF-R/gp130 receptors or human OSM-R/gp130 receptors, whereas, murine OSM only activates murine OSM-R/gp130 (17,18). The cross reactivity of human OSM has implications for the application of receptor antagonists (see below).

Almost all cell types express gp130; therefore, whether or not a cell responds to a cytokine in the gp130 family depends on the additional receptors that it expresses. Varying the expression levels of the other signal-transducing subunits, LIF-R and OSM-R, and the ligand-specific subunits IL-6Rα, IL-11Rα, and CNTFRα, allows different cell types to participate in responses to different cytokines. For example, since the expression of CNTFRα is mostly

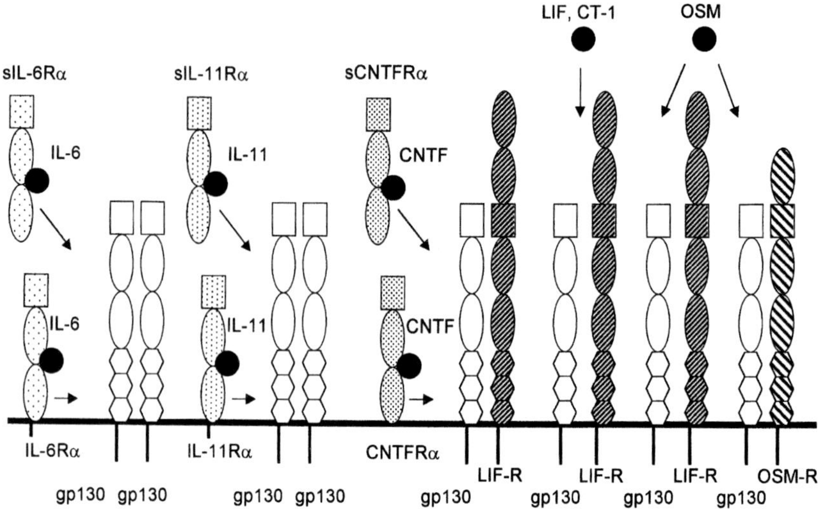

Figure 2 Receptor usage in the gp130 family of cytokines. Receptors are made up of combinations of fibronectin domains that are proximal to the membrane (hexagons), cytokine-binding homology domains (each β barrel is depicted as an oval; two make up a complete CBD), and immunoglobulin domains (squares). IL-6 and IL-11 induce the homodimerization of gp130 via their ligand-specific subunits, IL-6Rα and IL-11Rα, which may be transmembrane or soluble. CNTF, LIF, and OSM induce heterodimerization of LIF-R and gp130. CNTF requires soluble or transmembrane CNTFRα for efficient receptor assembly. OSM also induces the heterodimerization of gp130 and OSM-R. In contrast to human OSM, which can activate both gp130/LIF-R and gp130/OSM-R, murine OSM is restricted to gp130/OSM-R.

restricted to neurons, primarily neurons respond to CNTF in the absence of soluble CNTFRα (19).

Homodimerization of gp130 or heterodimerization of gp130 with LIF-R or OSM-R triggers cellular responses via the activation of the JAK family of cytoplasmic tyrosine kinases. The JAK kinases phosphorylate the cytoplasmic domains of gp130/LIF-R/OSM-R and activate members of the STAT family of transcription factors. Other signaling pathways, including the MAP kinase pathway, are also activated (reviewed in refs. 9 and 20).

Biological Activity of gp130 Cytokines—The Case for Antagonists

The phenotypes of mice deficient in gp130 cytokines and their receptors reveal the importance of gp130 cytokines in a diverse range of physiological systems (reviewed in ref. 21). IL-6−/− mice develop normally; however, their immune response is impaired when challenged by viral and bacterial infections (22–25). Despite the observation that in vitro IL-11 affects most hematopoietic lineages, IL-11Rα−/− mice show normal hematopoiesis (26). However, female mice without IL-11Rα are infertile, because they fail to decidualize properly following implantation of the embryo (27). Interestingly, LIF is also important in the maternal contribution to implantation (28,29). CNTF is prominently expressed in the nervous system; however, CNTF−/− mice exhibit only mild motor neuron degeneration with age (30).

The mild phenotype of CNTF−/− mice contrasts with the severe phenotype of CNTFRα−/− mice. Both CNTFRα and LIF-R knock out mice die at birth from nervous system deficiencies resulting from the premature death of motor neurons (31,32). LIF-R−/− mice also exhibit poor placentation, severe bone abnormalities, metabolic defects, and a reduction in astrocyte number (33). Mice lacking gp130 die even earlier, in mid gestation, probably due to thinning of the ventricular walls of the heart (34). gp130 is not only important in embryogenesis; postnatal inactivation of gp130 results in neurological, cardiac, hematopoietic, immunological, hepatic, and pulmonary defects (35). The large number of systems effected by gp130 deficiency testifies to its widespread importance.

IL-6−/− mice exhibit similar immunological defects to the postnatal gp130−/− mice. For other defects, the mapping to known cytokines is poor suggesting that additional ligands are likely to be found (35). Additional gp130 ligands are also suggested by a comparison between the dramatic phenotypes of the LIF-R−/− and CNTFR−/− mice and the mild phenotypes of LIF−/− and CNTF−/− mice. Understanding receptor function when not all ligands are known requires antagonists that directly target receptors.

The severity of the receptor knock out phenotypes stimulated the development of receptor antagonists as a different approach to studying receptor function in vivo. In principle, antagonists offer flexibility in inhibiting receptor function, since varying the antagonist concentration will vary the degree of inhibition and antagonists can be introduced at selected times. Using an appropriate delivery system, it may also be possible better to localize antagonists.

gp130 Receptor antagonists may be useful in defining the roles of gp130 cytokines in tissue repair such as liver regeneration (36), muscle regeneration (37), and peripheral nerve injury (38,39). Defining the role of gp130 cytokines in repair will contribute to the understanding of repair that is necessary for effective medical intervention aimed at promoting regeneration. For example, in peripheral nerve injury repair the activities of gp130 cytokines are likely to shed light on the importance of trophic support for neuronal and nonneuronal cells and the organization of the inflammatory response.

The development of receptor antagonists for the gp130 family has also been stimulated by an interest in their potential therapeutic applications. gp130 Cytokines contribute to a range of pathologies, including breast cancer (40), multiple myeloma (41), rheumatoid arthritis (42), osteoporosis (43), and inflammation (44). Neutralizing gp130 cytokines in these pathologies may prove to be beneficial (42).

Overview of Receptor Binding Sites on gp130 Cytokines

All of the gp130 cytokines belong to the family of long chain cytokines of which growth hormone is the prototype. They are characterized by a bundle of four antiparallel helices, A, B, C, and D. Although the proteins share limited sequence homology, structural features are conserved. The helices have an up-up-down-down helix orientation with long crossover loops between the first two and last two helices. Crystal structures have been published for IL-6 (45), LIF (46), and CNTF (47). An alignment of IL-6, LIF, and CNTF that highlights the helices is shown in Figure 3.

gp130 Cytokines bind receptor subunits via three topologically conserved sites. A general view of the sites will be presented here, and more

```
                   10        20        30        40        50        60        70
                    |         |         |         |         |         |         |
hIL-6               DVAAPHRQPLTSSERIDKQIRYILDGISALRKETCNKSNMCESSKEALAENNLNLPKMAEKDGCFQSGF--
hLIF           (10) TCAIRHPCHNNLMNQIRSQLAQLNGSANALFILYYTAQGE-------PFPNNLDKLCGPNVTDFPPFHANG
hCNTF       (0)MAFTEHSPLTPHRRDLCSRSIWLARKIRSDLTALTESYVKHQGL-------NKNINLDSADGMPVASTDQWSEL-

            80        90        100                110       120       130       140
            |         |         |                  |         |         |         |
hIL-6       NEETCLVKIITGLLEFEVYLEYLQNRFES--------SEEQARAVQMSTKVLIQFLQKKAKNLDAITTPDPTTNAS
hLIF        TEKAKLVELYRIVVYLGTSLGNITRDQKILN-PS---ALSLHSKLNATADILRGLLSNVLCRLCSKYHVGHVDVT-
hCNTF       TEAERLQENLQAYRTFHVLLARLLEDQQVHFTPTEGDFHQAIHTLLLQVAAFAYQIEELMILLEYKIPRNEADG--

              150       160       170       180       190       200
              |         |         |         |         |
hIL-6       LLTKLQAQNQWLQDMTTHLILRSFKEFLQSSLRALRQM
hLIF        YGPDTSGKDVFQKKLGCQLLGKYKQIIAVLAQAF
hCNTF       MPINVGDGGLFEKKLWGLKVLQELSQWTVRSIHDLRFISSHQTGIPARGSHYIANNKKM
```

Figure 3 Alignment of human IL-6, LIF, and CNTF. Adapted from the structure-based alignment in McDonald et al. (47) to reflect the demarcation of the helices evident in the crystal structure of IL-6 (45). Helices are highlighted. Note that the numbering corresponds to IL-6 and must be adjusted for LIF and CNTF.

specific mappings of the sites will be presented within the context of each receptor antagonist.

The intellectual framework for studying receptor-binding sites on cytokines and the nomenclature for the sites stems from elegant studies of growth hormone and growth hormone receptor. GH provides the simplest and best-understood model of receptor assembly. GH induces the homodimerization of the GH receptor to form functional receptor complexes. The GH-GHR complex, of one GH bound to two GHR, was the first structure of a cytokine-receptor complex to be solved (48). The cocrystalline structure identified the residues of GH at the ligand-receptor interface, allowing the contribution of each residue of GH to be assessed by alanine scanning mutagenesis (49). The mutagenesis revealed that each site is made up of a few important residues surrounded by less important ones. On GH, the first GHR to bind the hormone occupies site 1. Site 1 is centered on the C-terminus of the A- and D-helices and the A-B loop (49). Site 1 binds GHR with higher affinity than site 2, ensuring that site 1 is occupied first. Site 2 on GH includes residues in the N-terminal half of helix A and most of the C-helix (50,51). Site 2 is occupied by the second GHR, which contacts both GH and the first GHR. Receptor-receptor interactions also help stabilize the complex (see Fig. 1).

In gp130 cytokines, site 1 and site 2 are named by homology according to their positions relative to the binding sites on GH. IL-6, IL-11, and CNTF contain a site called site 1 that is topologically similar to site 1 of GH and is occupied by their respective ligand-specific subunits, IL6-Rα, IL-11Rα, and CNTFRα. The usage of site 1 by the ligand-specific receptors may reflect that these receptors are also the first subunits to bind the cytokines. All gp130 cytokines use a homologous site to site 2 of GH on the A- and C-helices to bind gp130.

The occupation of site 1 by nonsignaling subunits means that IL-6, IL-11, and CNTF must form more complex receptor associations in order to include at least two signal-transducing receptor subunits. An additional site, site 3, which is not present on GH, facilitates the increase in receptor complexity. Site 3 is centered at the N-terminus of the D-helix. On cytokines that dimerize gp130 such as IL-6 and IL-11, site 3 is occupied by gp130, whereas LIF-R or OSM-R occupies site 3 on cytokines that bind these receptors (see below) (Table 1).

Receptor Structure

The receptors for gp130 cytokines share both sequence and structural homology. The receptors are constructed from combinations of cytokine-binding

Table 1 Receptor Binding Sites of gp130 Cytokines

	Site 1	Site 2	Site 3
IL-6	IL-6Rα	gp130	gp130
IL-11	IL-11Rα	gp130	gp130
LIF		gp130	LIF-R
OSM		gp130	LIF-R/OSM-R
CNTF	CNTFRα	gp130	LIF-R

References are cited in the text.

homology domains (CBD), immunoglobulin domains, and fibronectin domains (see Fig. 2). CBD are about 220 amino acids in length and contain several conserved elements in the primary structure, including four cysteine residues and the motif Trp-Ser-X-Trp-Ser (52). The crystalline structures of GHR (48) and prolactin receptor (53) first showed that the domain is made up of two β barrels each consisting of seven antiparallel β strands. Subsequently, the crystalline structure of the CBD of gp130 revealed a similar fold (54). By analogy with GH/GHR, sites 1 and 2 of the gp130 cytokines are predicted to bind CBD. Consistent with this hypothesis, the CBD of IL-6Rα is sufficient to mediate IL-6 binding (55) and the CBD of gp130 is sufficient to mediate binding to site 2 of OSM (K.R. Hudson and J.K. Health, personal communication, 1999) (56). Although the GH/GHR complex provides a model for site 1 and site 2 interactions, a model for site 3 interactions awaits a cocrystalline structure of a cytokine engaging a receptor through site 3.

RATIONAL DESIGN OF RECEPTOR ANTAGONISTS

Starting from a crystalline structure or a molecular model, residues known or predicted to be solvent exposed can be rationally selected for mutagenesis. By analogy with GH, residues in homologous positions to site 1 and site 2 can also be preferentially targeted. However, a mutagenesis strategy that is appropriate for the identification of receptor-binding sites may not be equally effective for the generation of antagonists. For a residue to participate directly in a binding site, it must be surface exposed and present at the ligand-receptor interface. Identifying the contribution of individual residues is best achieved by alanine substitutions, because they are least likely to disrupt the conformation of the protein. In contrast, to create an antagonist, the complete disruption of the conformation of the binding site is desirable as long as the binding site to the sequestered subunit is preserved. Thus, simultaneous substitutions of multiple residues and nonconservative substitutions may be more effective than single alanine substitutions. Substitutions of buried residues may also be effective by indirectly disabling sites. The criteria for evaluating substitutions in an antagonist is whether the antagonist exhibits the desired potency and specificity—not whether the substitutions are surface exposed and in the direct binding site.

DEVELOPMENT OF IL-6Rα ANTAGONISTS

Assembly of IL-6 Receptor Complexes

IL-6 binds the 80-kD IL-6Rα with relatively low affinity (1000 pM) (57), whereas the IL-6/IL–6Rα complex associates with gp130 with 100-fold higher affinity (10 pM) (58). The binding of IL-6 to IL-6Rα is mediated via site 1, composed of residues on the carboxyl end of the D-helix and residues of the A-B loop (59–61) (Fig 4). Although IL-6 has little intrinsic affinity for gp130, when IL-6 is bound to IL-6Rα, it is capable of dimerizing gp130 (62). Sites 2 and 3 mediate the binding of IL-6 to gp 130. Site 2 is on the A- and C-helices (63–65), whereas site 3 is formed by residues of the beginning of the D-helix spatially flanked by residues on the initial part of the A-B loop (65–67) (Fig. 4). Each gp130-binding site is independently capable of binding a single molecule of gp130, but it is incapable on its own of dimerizing gp130 (68). Therefore, sites 2 and 3 on IL-6 are both required for the dimerization gp130. How occupation of sites 2 and 3 causes dimerization of gp130 can be explained by the formation of hexameric receptor complexes made up of two molecules of IL-6, two molecules of IL-6Rα, and two molecules of gp130 (68,69). In the model, an individual molecule of gp130 contacts one molecule of IL-6 via site 2 and another molecule of IL-6 via site 3 such that two trimers of IL-6–IL-6R/gp130 are brought together to form a hexamer. The hexameric complex is stabilized both by cytokine-receptor interactions and receptor-receptor interactions (68,69).

The receptor-binding sites on IL-6 were originally mapped by mutagenesis studies based on molecular models of IL-6. A reassessment of the mutagenesis data in light of the surface accessibility of the mutated residues in the crystalline structure of IL-6 confirms that sites 1, 2, and 3 are topologically separate (45).

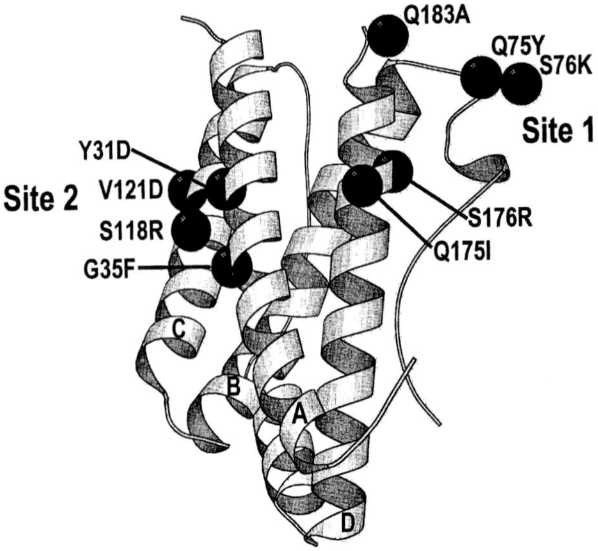
Q183A
Q75Y
S76K
Site 1
Y31D
Site 2
V121D
S118R
G35F
S176R
Q175I
C
B
A
D

(a)

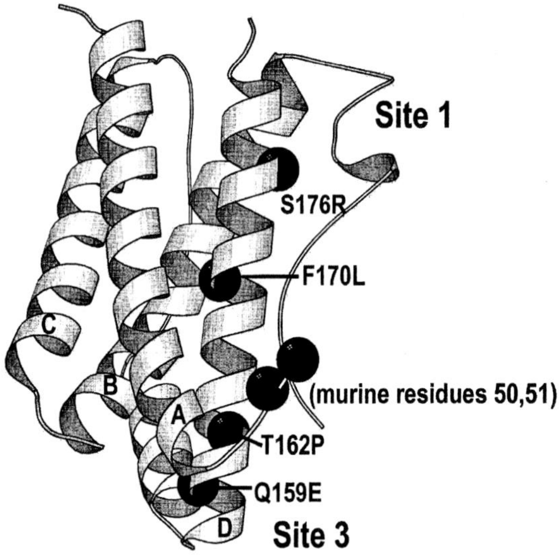
Site 1
S176R
F170L
C
B
A
(murine residues 50,51)
T162P
Q159E
D
Site 3

(b)

Creation of IL-6Rα Antagonists by Disruption of Site 2

The first site of IL-6 binding to gp130 targeted by Savino, Ciliberto, and colleagues was site 2 on the A- and C-helices. An IL-6 mutant containing the combined point mutations in the A-helix, Y31D/G35F, exhibited normal binding to IL-6Rα, but behaved as a partial agonist on Hep3B cells as assayed by the acute phase response (for the physical location of these mutations on IL-6, see Fig 4). As expected for a partial agonist, Y31D/G35F could only partially inhibit wild-type (wt) IL-6 responses (63). The performance of Y31D/G35F was subsequently enhanced by incorporating two additional mutations in the C-helix, S118R/V121D (64). The resulting IL-6 site 2 mutant, named DFRD, bound normally to IL-6Rα, and bound a single molecule of gp130 but did not efficiently dimerize gp130 (68). DFRD did not stimulate Hep3B cells at concentrations as high as 4 μg/mL and required a 44-fold molar excess over wt IL-6 for 50% inhibition. Although promising on Hep3B cells, when DRFD was tested on XG-1, a myeloma cell line that was extremely sensitive to IL-6, a molar excess of 1860-fold was required for 50% inhibition of proliferation (64). Three conclusions can be drawn from the experiments with DFRD. First, progressing from a partial agonist to a molecule with no detectable agonist activity required further disruption of site 2. Second, the potency of DFRD is highly cell type dependent; cells that are very sensitive to IL-6 require high ratios of DFRD to wt IL-6. Nonconservative substitutions on the

---

Figure 4 Schematic representation of the hIL-6 crystal structure (45) with the substitutions in the IL6-Rα antagonists imposed upon the structure. The A- and B-helices are labeled on the C-terminus and the C- and D-helices are labeled on the N-terminus. (a) Substitutions in Sant 7 developed by Savino, Ciliberto, and colleagues (70). Mutations in site 1 that enhance binding to IL-6Rα include Q1751 and S176R in the D-helix and the first residue following the C-terminal end of the D-helix, Q183A, plus Q75Y and S76K in the A-B loop. Substitutions in site 2 that reduce gp130 binding include Y31D and G35F on the A-helix and S118R and V121D on the C-helix. The residues in the A-B loop that when substituted enhance IL6-Rα binding and simultaneously reduce gp130 binding at site 3 (L57D, E59F, N60W) do not appear in the representation since their positions were not defined in the crystal structure (45). (b) The distribution of substitutions in an IL-6Rα antagonist developed by Rose-John, Brakenhoff, and colleagues (74). Substitutions in site 1 that enhance binding to IL-6Rα include F170L and S176R in the D-helix. Of the murinized residues in the A-B loop that reduce gp130 binding at site 3, C50-K55, only the first two are resolved in the crystalline structure (45). Substitutions in site 3 that reduce gp130 binding include Q159E and T162P in the D-helix (note that T162 is not solvent exposed).

surface of IL-6 can effectively disrupt site 2, as all of the residues that are mutated in DFRD are now known to be surface exposed (45) (see Fig 4).

Creation of Superantagonists with Enhanced Binding to IL-6Rα

Since efficacy on XG-1 cells was considered an appropriate criterion for a potent IL-6Rα antagonist, DFRD required further refinements. The potency of DFRD was improved by incorporating mutations that strengthen the binding of IL6 to IL6Rα (64). Tighter binding of DFRD to IL-6Rα boosts the ability of DRFD to bind and sequester IL-6Rα. Three mutations in the D-helix, S176R/Q175I/Q183A, together provide a fivefold affinity increase in binding to IL-6Rα (for the location of these residues, see Fig 4). When introduced into DFRD, the resulting mutant, called Sant1 (for superantagonist), was more potent than DFRD on XG-1 cells, requiring about a 200-fold molar excess for 50% inhibition of wt IL-6 responses compared to the 1860-fold molar excess required by DFRD. Similarly, the concentration required for complete growth inhibition dropped from 4 μg/mL for DFRD to 1 μg/mL for Sant1. Thus the fivefold affinity increase in binding to IL-6Rα translated to about a fivefold increase in potency.

Strengthening the binding of the antagonists to IL-6Rα even further resulted in additional gains in potency. Sant5 incorporates the mutations of DFRD, plus the additional mutations included in Sant1, S176R/Q175I/Q183A, plus two new mutations, Q75Y/S76K, that also strengthen the binding to IL-6Rα (see Fig. 4). Sant5 showed increased IL-6Rα binding relative to Sant1 of about 10-fold and displayed a similar 10-fold increase in potency over Sant1 on XG-1 cells (70). The mutations, Q75Y/S76K, were first identified in a screen for IL-6 variants with high affinity for IL-6Rα. In the screen, phase display was used to select high-affinity variants with randomized mutations in A-B loop (61). The potency of Sant5, which requires only about 20-fold molar excess over wt IL-6 on X-G1 cells, demonstrates the utility of phage display in identifying mutations that enhance receptor binding.

The binding of Sant5 to IL-6Rα was further strengthened by additional mutations in the A-B loop that influence both site 1 and site 3. The mutations, L57D/E59F/N60W, have the unusual property that they both strengthen the binding of site 1 to IL-6Rα by about 12-fold and simultaneously disrupt site 3 (61). The residues lie adjacent on the A-B loop to the residues in site 3 identified by Ehler et al. (65) and Brakenhoff et al. (71); perhaps explaining their indirect effects on site 3 (since the A-B loop is poorly defined in the crystalline structure of IL-6, the residues do not appear in the schematic representation in Fig. 4). The inclusion of the mutations, L57D/E59F/N60W, in Sant7 resulted in a small increase in the binding to IL-6Rα relative to Sant5 and the complete loss of gp130 binding. Interestingly, mutations designed to target more directly site 3 such as W157R/D160R were no more effective at disrupting site 3 than these indirect ones (45,68). Additional site 3 mutants are discussed below.

To review, Sant7 contains substitutions in all three receptor-binding sites. Sant7 includes three mutations in the D-helix and two mutations in the A-B loop of site 1 that enhance IL-6Rα binding. Sant7 also includes four mutations in the A- and C-helices in site 2 (DFRD) that abolish binding of gp130 to site 2. Sant7 further includes three mutations in the A-B loop that enhance the binding of site 1 to IL-6Rα, whereas abolishing the binding of site 3 to gp130. Sant7 is a potent antagonist, exhibiting about a 30-fold increase in potency over DFRD and a 4-fold increase in potency over Sant1 on XG-1 cells and requiring about a 60-fold molar excess to reach 50% inhibition (Table 2).

Table 2 Antagonist Potency of IL-6 Mutants on XG-1 Cells

Mutant	Site 1 enhanced	Site 2 disrupted	Site 3 disrupted	sIL-6-Rα binding compared to wt IL-6	Molar excess needed to reach 50% inhibition on XG-1 cells
DFRD	No	Yes	No	1	2,033 ± 578
Sant1	Yes	Yes	No	4.5	230 ± 60
Sant5	Yes	Yes	No	40	19 ± 4
Sant7	Yes	Yes	Yes	65	61 ± 29
Murine K41-A56/Q159E/T162P	No	No	Yes	0.2	90,000
Murine K41-A56/Q159E/T162P/F171L/S177R	Yes	No	Yes	1	9000
Murine C50-E55/Q159E/T162P/F171L/S177R	Yes	No	Yes	1	3000

Enhancing the binding of IL6-Rα antagonists to IL-6Rα dramatically improves their potency on XG-1 cells. The amino acid composition of DFRD and the Sant mutants are described in the text. Values were taken with permission from the experiments presented by Sporeno et al. (70). Values for murine K41-A56/Q159E/T162P and murine K41-A56/Q159E/T162P/F171L/S177R were adapted from de Hon et al. (75). Values for murine C50-E55/Q159E/T162P/F171L/S177R were adapted from Ehlers et al. (74).

The application of the superantagonists to a variety of myeloma cell lines revealed important differences in their biological activity. The rank order of potencies of the antagonists varied in a cell type–specific manner. For example, Sant7 is more potent than Sant5 on XG-4 cells, equally potent on XG-6 cells, but less potent on XG-1 cells (70). Understanding the shifts in potency may shed light on the factors controlling the responsiveness of myeloma cell lines to gp130 cytokines. Sant7 was selected as the most promising antagonist, because on one of the cell lines, XG-2, Sant7 was the only one that showed no agonist activity, whereas Sant1, DFRD, and the others were partial agonists (70). The lack of stimulation by Sant7, a site 2 and site 3 mutant, and partial stimulation by those with an intact site 3, argues that disrupting site 2 may not be sufficient to remove residual agonist activity on all cell lines. In applications where the responsiveness of individual cell types cannot be verified, a combined site 2 + 3 receptor antagonist might be more appropriate than a simple site 2 antagonist.

A more thorough study of the effects of Sant7 on XG-1 cells revealed that Sant7 not only inhibited the IL-6–dependent growth of XG-1 cells, but it also promoted apoptosis (72). In high-density cultures in which IL-6 is produced by XG-1 cells at high levels, Sant7 is more effective at inducing apoptosis than withdrawal of IL-6 (72). The apoptosis probably results from the blockade of autocrine stimulation of IL-6 receptors. The ability to induce apoptosis was restricted to antagonists that have no residual gp130 binding such as Sant7 and correlated with an ability to inhibit STAT activation over an extended period (72).

Creation of IL-6Rα Antagonists by Disruption of Site 3

Concurrently as Savino and Ciliberto and colleagues targeted site 2 on the A- and C-helices, Brakenhoff and Rose-John and colleagues led the way at targeting what would subsequently be called site 3. Whereas site 2 of IL-6 had been identified by homology to site 2 of GH, site 3 was identified, because it mapped to a putative gp130-binding region on IL-6 that was defined by a monoclonal antibody, mAb 16(II). The mAb neutralized IL-6 activity in biological assays but did not interfere with IL-6Rα binding (66). One of the IL-6 mutants, Q159E/T162P (referred to as Q160E/T163P at the time), interfered with antibody binding but bound IL-6Rα normally and showed no activity on CESS cells, whereas acting as a partial agonist on HepG2 cells. The mutant, Q159E/T162P, antagonized wt IL-6 on CESS cells and partially antagonized wt IL-6 on HepG2 cells. However, when challenged with XG-1 cells, Q159E/T162P proved to be a full agonist with a reduction in specific activity relative to wt IL-6 of about 1000-fold (66). Neutralizing mAb against gp130 inhibited the remaining activity, suggesting that residual gp130 binding was the source of the residual activity (73). The cell type dependency in the effects of residual gp130 binding is reminiscent of the behavior of superantagonists described above. Interestingly, the threonine residue mutated in Q159E/T162P turns out to be buried, suggesting an indirect effect on gp130 binding (45).

Inserting murine sequences into human IL-6 (hIL-6) subsequently identified additional residues that influence site 3. Murine sequences were chosen because mIL-6 does not have biological activity on human cells. The introduction of only six murine residues to replace human residues C50-E55 resulted in a large reduction in biological activity on XG-1 cells (1000-fold) with no significant change in IL-6Rα binding (74). Combining Q159E/T162P with the murinized residues resulted in a more effective disruption of site 3 (74). In the combined site 3 mutant, stimulation of XG-1 cells was not observed. This is probably a stronger reduction in biological activity than that obtained with the mutations in the A-B loop, L57D/E59F/N60W, incorporated into Sant7, described above. L57D/E59F/N60W by itself still exhibited a small amount of activity on Ba/F3/hgp130 cells (Ba/F3 cells stably transfected with human gp130) costimulated with soluble IL-6Rα (61). The reduction in biological activity of the murinized C50-E55/Q159E/T162P is also probably a stronger reduction than that observed with W157R/D160R, which maintains a small amount of activity on X-G1 cells (68). Since the site 3 mutant, murinized C50-E55/Q159E/T162P, was not applied to the same panel of myeloma cell lines as the superantagonists described earlier, it is unknown whether fully disrupting site 3 is sufficient to remove residual agonism on all myeloma cell lines.

Fully disrupting site 3 binding to gp130 is not, however, sufficient to generate a potent antagonist. Hence, additional mutations were introduced into the murinized C50–E55/Q159E/T162P antagonist at site 1, F170L/S176R, to strengthen IL6Rα binding. The combined mutant succeeded in antagonizing wt IL-6 on X-G1 cells, requiring about a 3000-fold molar excess to achieve 50% inhibition and about 3 μg/mL to achieve full inhibition (74) (Table 2). Achieving greater potency will require further strengthening of the binding to IL-6Rα via site 1. Similar results to introducing murine residues C50-E55 were obtained when residues K41-A56 were substituted instead (75). A single substitution, L57A (referred to as L58A at the time), was subsequently identified that, when combined with Q159E/T162P and F170L/S176R, was as potent an antagonist as if murine residues K41-A56 had been similarly combined (71). This last mutation, L57A, is in a residue of the A-B loop that is also mutated in Sant7 (see above). Subsequently, mutating K54 on its own was shown to be as effective as substituting positions C50-E55 with murine residues (76). Substituting fewer residues may decrease the antigenicity of the receptor antagonists, making them easier to apply in vivo.

SUMMARY OF THE GENERATION OF IL-6Rα ANTAGONISTS

IL-6 mutants with diminished binding of site 2 or site 3 to gp130 antagonized wt IL-6. They bound a single molecule of gp130 but failed to dimerize gp130. Eliminating binding to gp130 at both sites 2 and 3 may be necessary to reduce residual agonism. Antagonists with mutations in both sites completely failed to bind gp130 and thus failed to dimerize gp130. Incorporating mutations that enhance the binding of the IL-6 mutants to IL-6Rα increased their potency. Such superantagonists were effective on the very sensitive myeloma cell line, XG-1, which displays an EC_{50} for IL-6 of about 1 pM. The molar excess that was required to obtain 50% inhibition ranged between 15- and 230-fold. For full inhibition, 30–1000 ng/mL was required, depending on the antagonist. The large gains in potency achieved by the superantagonists illustrate the enormous benefits of creating antagonists that bind IL-6Rα with higher affinity than wt IL-6 (see Table 2).

The development of IL-6 antagonists demonstrates that making potent antagonists depends on identifying all of the receptor-binding sites of a cytokine and carefully modifying those sites. Refinement of an antagonist requires both selections for complete inhibition of binding to the targeted receptor subunits and enhancement of the binding to the receptor subunit to which binding is preserved.

SPECIFICITY OF IL-6Rα ANTAGONISTS

Initial experiments suggested that IL-6Rα antagonists were strictly specific for IL-6Rα. For example, on HepG2 cells, the site 2 antagonist DFRD inhibited the induction of the C-reactive protein (CRP) promoter in response to IL-6 but not in response to OSM (64). Similarly, DFRD inhibited the activation of STAT complexes in HepG2 cells in response to IL-6 but not to OSM or to LIF (64). The site 3 antagonist L57A/Q159E/T162P/F170L/S176R also inhibited the proliferation of the erythroleukemic TF-1 cells in response to IL-6 but not in response to OSM (71). All of the IL-6Rα antagonists discussed so far can be described as competitive inhibitors: Excess IL-6 will overcome the antagonists.

More recently, some IL-6Rα antagonists have been shown to inhibit responses to IL-11 as well as to IL-6. In two myeloma cell lines, XG-4CNTF and XG-6, antagonists with mutations in site 2, such as DFRD and Sant5, inhibited IL-6 and IL-11 responses. CNTF, LIF, and OSM were not inhibited even at concentrations of antagonist that were high enough fully to occupy IL-6Rα (77). However, the combined site 2 and site 3 mutant, Sant7, inhibited IL-6 but had no effect on IL-11, CNTF, LIF, and OSM. The difference in behavior between the single-site and the combined-site mutants suggests that the inhibition of IL-11 responses is dependent on the preservation of gp130 binding through an individual site. Site 2 or site 3 antagonists sequester gp130, because they are still capable of binding a single molecule of gp130 even though they are not capable of triggering gp130 dimerization. Antagonists with mutations in both site 2 and site 3 do not bind gp130 and so cannot sequester the subunit. Similar widening of specificity due to sequestering of gp130 has been observed in CNTFRα antagonists (see below).

The unique susceptibility of IL-11 to IL-6Rα antagonists, whereas CNTF, LIF, and OSM were spared merits speculation. IL-11 may compete poorly for gp130 because of low numbers of IL-11Rα relative to IL-6Rα and LIF-R. Alternatively the affinity of the IL-11/IL-11Rα complex for gp130 may be low relative to the other competing subcomplexes. Last, the outcome of competition with the IL-6Rα antagonist may differ because of the way the cytokines interact with gp130. IL-6 and IL-11 may compete for similar sites on gp130, whereas the ligands that engage LIF-R may use slightly different binding sites on gp130 (78).

Thus, for applications in which the specificity of the IL6-Rα antagonist for IL6-Rα is important, only antagonists that completely lack binding to gp130 will be appropriate.

gp130 ANTAGONISTS

A broad gp130 antagonist might be useful when several members of the gp130 family of cytokines are active or when their identities are unknown. For example, some multiple myeloma cells proliferate in response to several gp130 cytokines (79). The generation of a gp130 antagonist entails modifying a ligand such that it binds to gp130 but fails to drive the homodimerization or heterodimerization of gp130. High doses of wt hOSM will inhibit IL-6 responses on Hep3B cells, because the cells lack functional LIF-R and OSM-R capable of dimerizing with gp130 (80). To be useful on a cell that expresses these receptors, the binding of OSM to LIF-R and OSM-R would have to be disrupted. To make the antagonist potent, the binding of OSM to gp130 would have to be substantially enhanced.

Another approach, taken by Rose-John and colleagues, is to start with a complex ligand such as IL-6 bound to soluble IL-6Rα (IL-6 needs to be bound to IL-6Rα, because IL-6 on its own has little native affinity for gp130). The interaction of IL-6 and soluble IL-6Rα with gp130 can be enhanced by covalently linking IL-6 to IL-6Rα. Hyper–IL-6 is a fusion protein made from linking IL-6 to IL-6Rα via a flexible protein linker. Hyper–IL-6 is 100- to 1000-fold more active than the separate proteins on cells that only express gp130 (81) and is about 10-fold more active than IL-6 on cells that express gp130 and IL-6Rα (82).

In the same way that a fusion protein of soluble IL-6Rα and IL-6 behaves as an enhanced agonist, a fusion protein of IL-6Rα and an IL-6Rα receptor antagonist behaves as an enhanced antagonist. Broad gp130 antagonists were generated by covalently coupling a site 2 defective antagonist (DFRD, described above) or a site 3 defective antagonist (murine residues C50-A58/Q159E/T162P) to soluble IL-6Rα by a peptide linker (83). The fusion antagonists bound gp130 as predicted by their retention of a single gp130-binding site, and they showed no agonist activity on BAF/3 cells stably transfected with gp130. On BA/F3 cells transfected with gp130 and IL-6Rα or gp130, LIF-R, and CNTFRα, they inhibited responses to IL-6, OSM, LIF, and CNTF (83). The inhibition of OSM, LIF, and CNTF by the fusion antagonists suggests a broad effect on gp130 cytokines. Previously, the site 2 antagonists on their own had only antagonized IL-6 and, in some situations, IL-11 (see above). The broader inhibitory activity of the fusion proteins probably reflects an increased stability in the fusion protein–gp130 complex, which allows more effective sequestration of gp130. The fusion antagonists are potent, requiring 100-fold molar excess over hLIF for 50% inhibition of proliferation of BA/F3/hgp130/hLIF-R cells. Applying these antagonists in vivo will, however, require careful monitoring, because gp130 plays critical roles in adult physiology (see above).

IL-11Rα ANTAGONISTS

IL-11 binds to IL-11Rα with about 10-fold lower affinity than IL-6 binds to IL-6Rα (10 vs 1 nM). The IL-11–IL-11Rα complex subsequently associates with gp130 with high affinity (300–800 pM), but the affinity is still lower than that observed with IL6/IL6Rα in association with gp130 (10 pM) (84). By analogy with IL-6, IL-11 is predicted to form hexameric complexes of two IL-11 molecules, two IL-11Rα, and two gp130s. However, pentamers containing only one gp130 were observed when human ligand-receptor complexes were assembled in solution (85). In contrast, murine IL-11, IL-11Rα, and gp130 form hexameric receptors in solution (M.A. Hall, K.R. Hudson, and J.K. Heath, personal communication, 1998).

Receptor-binding sites on murine IL-11 have been recently mapped by site-directed mutagenesis (86). Residues were targeted for substitution if they mapped to topologically similar sites to the receptor-binding sites of IL-6, according to a model of IL-11 that was based on the IL-6 crystalline structure. IL-11 binds to IL-11Rα through residues in site 1. Site 1 of IL-11 is dominated by R169 and is predicted to lie within the C-terminus end of the D-helix. Binding to gp130 is reduced by substitutions in the site 2 region of the C-helix, whereas binding to IL-11Rα is preserved. The site 2 mutant, R111A/L115A, however, retains significant biological activity on Ba/F3/mgp130/mIL-11Rα cells, making it unsuitable as an antagonist without further modification. Binding to gp130 is also reduced by substitutions in site 3 within the CD loop or the N-terminal end of the D-helix. In contrast to the site 2 mutant, W147A in site 3 shows no detectable agonist activity. Preliminary experiments indicate that W147A is a receptor antagonist for IL-11Rα (V.A. Barton, K.R. Hudson, and J.K. Heath, personal communication, 1999). Many of the approaches used to refine IL-6Rα antagonists are likely to be applicable to antagonists for IL-11Rα.

LIF-R ANTAGONISTS

Assembly of LIF-R into Receptor Complexes

LIF-R binds several gp130 cytokines as a shared signal-transducing subunit. LIF, CNTF, and CT-1 all require LIF-R for signaling, and in some human cell types, OSM also utilizes LIF-R. All currently known LIF-R ligands signal by heterodimerizing LIF-R and gp130. In the simplest example, LIF forms a complex with LIF-R and gp130; the formation of complexes in solution suggests that the complexes are heterotrimers of LIF, LIF-R, and gp130 in a ratio of 1 : 1 : 1 (87). In a more complex example, CNTF must bind CNTFRα before binding to LIF-R and gp130 to assemble a hexameric complex (88).

gp130 Cytokines Bind to LIF-R Through Residues in Site 3

hLIF first binds to hLIF-R with low affinity (K_D = 1 nM (89,90)) and then associates with gp130 in a higher affinity complex (K_D = 10 pM (91)). Based on the crystalline structure of murine LIF, residues likely to be surface exposed on hLIF were subjected to alanine scanning mutagenesis to identify receptor-binding sites. Specific regions were chosen for analysis based on the site usage of GH and on previous analyses of mouse-human chimeras that had indicated a role for the C-D loop (92) and the D-helix (46) in species-specific binding to LIF-R. The alanine scanning mutagenesis demonstrated that the primary binding of LIF-R is mediated via site 3 of hLIF, including residues at the amino-terminus of the D-helix, carboxyl-terminus of the B-helix, and C-D loop. Site 3 is dominated by residues Phe-156 and Lys-159 at the top of the D-helix (93). These residues are spatially close together and have their side chains prominently exposed to the solvent in the mLIF structure (46). Substitution of either amino acid resulted in more than a 100-fold reduction in LIF-R binding and a substantial reduction in biological activity, measured in a proliferation assay of Ba/F3 cells stably transfected with hLIF-R and hgp130. The importance of Phe-156 and Lys-159 in site 3 is underlined by their conservation in all known ligands for LIF-R, including mLIF, hOSM, CNTF, and CT-1 (46,94). IL-6 and IL-11, cytokines that use site 3 but do not bind LIF-R, do not have phenylalanine and lysine in these positions.

In contrast to the profound effects of changes in site 3, there is evidence for at best weak binding of hLIF to LIF-R via site 1 at the carboxyl-terminus of the D-helix (93). A cocrystalline structure for LIF and LIF-R would be helpful in determining if LIF-R occupies site 1 on cytokines that do not bind ligand-specific receptors.

Creation of a LIF-R Antagonist by Disruption of Site 2 of hLIF

As the affinity of hLIF for gp130 is probably 100- to 1000-fold lower than the affinity for LIF-R, multiple simultaneous mutations were used in place of single mutations to identify the gp130-binding site (93). Both alanine substitutions and nonconservative substitutions in the beginning of the A- and C-helices (site 2) reduced gp130 binding to undetectable levels. Perhaps because of limitations of the gp130 binding assay, mutants that showed equal losses in gp130 binding showed varying amounts of biological activity in stimulating the proliferation of Ba/F3-hLIF-R/hgp130 cells. Activities ranged from a 250-fold reduction in proliferation activity to a complete absence of agonist activity. In all cases, LIF-R binding was equivalent to wt hLIF (93). The preservation of LIF-R binding confirms the independence of sites involved in LIF-R and gp130 binding.

Several of the site 2 mutants were able to antagonize hLIF in proliferation assays on Ba/F3-hLIF-R/hgp130 cells. An hLIF site 2 mutant with no detectable agonist activity, hLIF-05, was chosen for further characterization (95). hLIF-05 contains the nonconservative mutations A117E, D120R, I121K, G124N, S127L, Q25L, S28E, Q32A, S36K (for the placement of the residues in the LIF structure, see Fig. 5). The antagonist is modestly potent on cells that are highly sensitive to LIF, requiring about 15 nM or a 10,000-fold excess over wt hLIF to inhibit by 50% the proliferation Ba/F3-hLIF-R/hgp130 cells. However, on HepG2 cells, a ratio of only 200-fold hLIF-05 to wt LIF was sufficient for 50% inhibition of hLIF (95). The modest potency is not surprising given that hLIF-05, which only binds LIF-R, must compete with wt LIF, which also assembles high-affinity LIF-R/gp130 complexes. Improving the potency of hLIF-05 by strengthening its binding to LIF-R has not yet been attempted.

As predicted for an antagonist that sequesters LIF-R, hLIF-05 inhibited the biological response to not just LIF, but to all known LIF-R ligands, including LIF, CNTF, CT-1, and OSM (95). hLIF-05 is best thought of as a broad

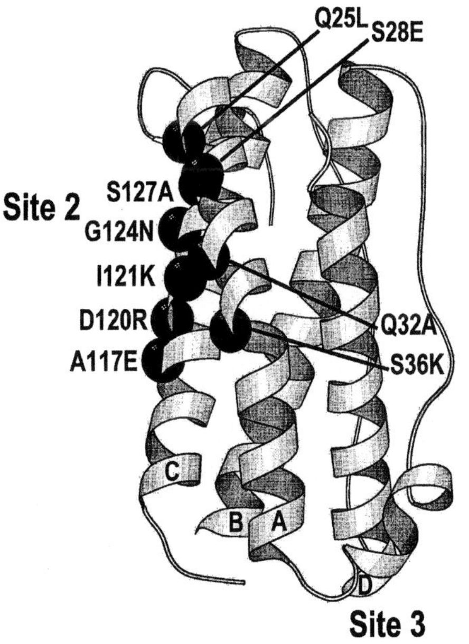

Figure 5 Schematic representation of hLIF, modeled on the mLIF crystalline structure, showing the distribution of substitutions in hLIF-05 (93). The substitutions in site 2 reduce binding to gp130. In the A-helix, the mutations include Q25L, S28E, Q32A, and S36K. In the C-helix, they include A117E, D120R, I121K, G124N, and S127L.

LIF-R antagonist and not as an antagonist particular to hLIF. hLIF-05 was shown to be a competitive antagonist: An excess of wt ligand overcomes the antagonism by hLIF-05. The antagonism is specific, since hLIF-05 did not inhibit IL-6Rα–mediated responses and hLIF-05–inhibited hOSM responses only when they were mediated by LIF-R and not when they were mediated by the related OSM-R (95). The ability to discriminate between LIF-R and OSM-R may be useful in dissecting the behavior of hOSM in pathologies in which hOSM is elevated, such as rheumatoid arthritis (96).

To examine events closer to ligand binding, receptor phosphorylation was measured on cells that make all three receptor subunits, LIF-R, OSM-R, and gp130. Tyrosine phosphorylation of LIF-R and gp130 in response to hLIF was blocked by hLIF-05. The phosphorylation of LIF-R in response to OSM was also blocked. However, the phosphorylation of gp130 was not blocked by hLIF-05 in response to OSM, indicating that OSM-R was able to associate with gp130 in the presence of hLIF-05 (95). The inhibition of receptor phosphorylation argues that hLIF-05 is acting at an early step in signal transduction and is consistent with the simple model that hLIF-05 works by sequestering LIF-R.

ANTAGONISTS FOR OSMR?

In a reversal of the normal family pattern, hOSM binds to gp130 with fairly high affinity (10 nM) but only weakly binds hLIF-R (93) and does not significantly bind hOSM-R (16). Binding of mOSM to the recently cloned mOSMR has not yet been quantified (18). Reducing the binding of hOSM to gp130 is unlikely to generate an effective antagonist for LIF-R or OSM-R, since the remaining binding to these subunits is so weak. Reducing the binding of OSM to LIF-R and OSM-R should generate a gp130 antagonist and not an antagonist of LIF-R and OSM-R. Consequently, although the placement of receptor binding sites on hOSM resembles the placement of binding sites on LIF (K.R. Hudson and J.K. Heath, personal communication, 1998) (56), one cannot make antagonists in the same fashion.

CNTFRα ANTAGONISTS

Assembly of CNTF Receptor Complexes

CNTF assembles receptor complexes by first binding to CNTFRα and then heterodimerizing gp130 and LIF-R (97). CNTF can activate LIF-R and gp130 without its ligand-specific receptor; however, formation of a CNTF-CNTFRα complex increases the ability of CNTF to activate LIF-R and gp130 by 10,000-fold (98). In vivo, CNTF concentrations are likely to be too low to activate LIF-R and gp130 in the absence of CNTFRα. Hence neutralizing CNTFRα is probably sufficient to block CNTF responses in vivo.

Three receptor binding sites have been identified on CNTF. CNTF binds to CNTFRα with moderate affinity via site 1 on the A-B loop and the D-helix (99–102). CNTF binds to gp130 via residues in site 2 on the A-helix (99) and binds to LIF-R via site 3 at the boundary region of the C-D loop and the D-helix (103,104). A hexameric complex of two CNTF, two CNTFRα, one gp130, and one LIF-R has been proposed based on studies of the complex formation of the extracellular domains in solution (88).

Creation of CNTFRα Antagonist by Disruption of Site 3

As in the case of IL-6Rα antagonists, CNTFRα antagonists have been progressively refined. Site 3 was the first site to be targeted to develop CNTFRα antagonists. Inoue et al. substituted K155 with alanine in site 3 of hCNTF, abolishing the survival activity of hCNTF on chick dorsal root ganglion (DRG) neurons (103). The absence of agonist activity on chick DRG neurons could be demonstrated at concentrations as high as 20 μg/mL. On chick DRG neurons, K155A was a modestly potent antagonist, requiring 1000- to 10,000-fold excess to inhibit wt CNTF. However, on rat DRG neurons, K155A was a partial agonist (101). Introducing a bulkier substitution in the form of a tryptophan instead of an alanine (K155W) resulted in a lowering of the agonistic activity on rat cells but did not create an antagonist for rat cells. Therefore, to obtain an antagonist for rat cells, further modifications were required. The simultaneous substitution of both F152 and K155 was subsequently shown to broaden the species specificity. F152S/K155A and F152D/K155A antagonized CNTF on both chick and rat cells (101) (for the placement of the residues on the CNTF structure, see Fig. 6). The species specificity of the early CNTFRα antagonists must stem from subtle differences in how site 3 is engaged in the receptor complexes of the two species that are not apparent with wt hCNTF.

In a similar manner, Di Marco et al. enhanced the performance of the CNTF mutant, K155A, by simultaneously substituting F152 with alanine (104). Although a partial reduction in agonist activity occurred with the individual mutations, the combined mutations exhibited a complete absence of agonist activity on HepG2 cells. The reduction in agonist activity by F152A/K155A occurred without changes in the binding to CNTFRα or in the binding of the F152A/K155A–CNTFRα complex to gp130. In contrast, LIF-R binding

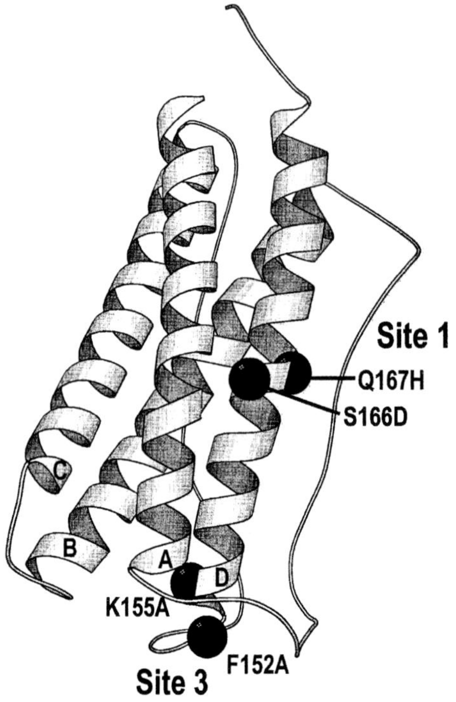

Figure 6 Schematic representation of the hCNTF crystalline structure, showing the distribution of substitutions in the CNTFRα antagonist developed by Di Marco et al. (104). Substitutions in site 1 that enhance binding to CNTFRα include S166D and Q167H in the D-helix. Substitutions in site 3 that reduce binding to LIF-R include F152A and K155A at the N-terminus of the D-helix.

by F152A/K155A was completely lost, indicating that site 3 of CNTF is a crucial site for LIF-R binding (104). The residues F152 and K155 in CNTF are homologous to the residues F156 and K159 that are crucial for the binding of LIF to LIF-R. The use of homologous residues to bind LIF-R suggests conservation in the way LIF-R ligands engage LIF-R. F152A/K155A was a moderately potent antagonist in assays on human IMR32 cells that express endogenous transmembrane CNTFRα (about 3000-fold excess required) but had no antagonistic activity on HepG2 cells when copresented with soluble CNTFRα (104). The preferential activity of the antagonist on cell lines that makes transmembrane receptor probably reflects the undirectional capture of CNTF by membrane-bound CNTFRα, which makes the formation of receptor complexes relatively insensitive to changes in CNTF affinity for LIF-R and gp130. The lack of sensitivity to changes in the affinity for LIF-R enables the antagonist to compete effectively with wt CNTF for CNTFRα.

When the mutations in site 3 were combined with mutations in site 1 (S166D/Q167H) that enhance the affinity of CNTF for CNTFRα by 30- to 50-fold, the enhanced antagonist was effective on HepG2 cells when copresented with soluble CNTFRα (104) (see Fig. 6). Combining site 1 and site 3 mutations also greatly improved the potency of the CNTFRα antagonist; CNTF responses were inhibited by 50% at a ratio of 30- to 100-fold excess over wt CNTF on IMR32 cells (and HepG2 cells) compared to the 3000-fold excess observed previously. On HepG2 cells, the EC_{50} of CNTF was shifted in the presence of the antagonist up to the value observed in the absence of CNTFRα, suggesting that the antagonist worked simply by competing for CNTFRα binding (104).

The observation that site 3 antagonists for CNTFRα behaved differently in assays dependent on membrane-bound CNTFRα than in assays dependent on soluble CNTFRα subsequently led to a more careful study of the specificity of CNTFRα antagonists. On cells with high concentrations of membrane-bound CNTFRα, such as IMR32 cells, the CNTF mutant with increased site 1 binding and decreased site 3 binding, F152A/K155A/S166D/Q167H, inhibited not only CNTF, but also LIF (102). The investigators suggested that the antagonist was able to trap gp130 in a nonproductive complex with CNTFRα. In contrast, on HepG2 cells, when soluble CNTFRα was made limiting, the antagonist inhibited CNTF, whereas sparing LIF and IL-6. The prediction is that in vivo, the antagonist will inhibit CNTF responses that are mediated by soluble CNTFRα, whereas acting as a general gp130 antagonist on cells that express high levels of CNTFRα.

A more specific CNTFRα antagonist could be created by simultaneously reducing the affinity of CNTF for both LIF-R and gp130. Existing mutations in site 2 of CNTF do not fully reduce gp130 binding, indicating a need for further delineation of site 2. Reducing the binding of the antagonist to gp130 would presumably widen the gap between the affinity of the antagonist for CNTFRα and the high affinity of wt CNTF for the receptor complex of CNTFRα, LIF-R, and gp130. To regain potency, the binding to CNTFRα may have to be further enhanced to bridge the gap in affinity.

MECHANISM: DO RECEPTOR ANTAGONISTS ALWAYS WORK AS COMPETITIVE ANTAGONISTS?

The evidence that receptor antagonists work as competitive antagonists derives from the observation that the antagonists can be overcome by increasing concentrations of wt cytokine. Furthermore, the block in signal transduction can be observed as early as JAK activation, STAT activation, and receptor phosphorylation. One prediction of this model is that if the wt cytokine is not present to activate the receptor, then the presence of the antagonist should not matter. An intriguing series of experiments on a GM-CSF receptor antagonist may suggest otherwise.

GM-CSF stimulates the function of mature neutrophils and eosinophils and promotes their viability as well as the viability of leukemic cells (105,106). The GM-CSF mutant, E21R, binds to the GM-CSF receptor α chain (GM-CSFR-α), a ligand-specific signal-transducing subunit, but does not bind to βc, the common signaling chain shared with IL-3 and IL-5 receptor complexes (8). Surprisingly, E21R induces the apoptosis of hematopoietic cells, including eosinophils, in the absence of detectable GM-CSF (107,108). The E21R-triggered apoptosis can be recreated in Jurkat T cells that have been transfected with both GM-CSF α and βc, indicating that both subunits are required. In the doubly transfected cells, treatment with E21R decreased the activity of the mitogen-activated protein kinase ERK (109). The mechanism is not yet understood. Distinguishing between an apoptotic effect secondary to blocking the GM-CSF receptor and a direct apoptotic effect has proven to be difficult. One possibility is that GM-CSF is produced by hematopoietic cells in an autocrine fashion at levels below current detection but sufficient for cell survival. Another possibility is that since a portion of the GM-CSF α and βc on myeloid cells has been shown to exist in preformed complexes in the absence of GM-CSF (110), the preformed complexes may provide a trickle of signal that is blocked by E21R (111). If receptor subunits prove to have activities that are independent from ligand-mediated oligomerization, receptor antagonists may cause unexpected effects.

IN VITRO APPLICATIONS OF IL-6Rα AND LIF-R ANTAGONISTS

Receptor antagonists can be used to dissect the hierarchical relationship among cytokines. For example, IL-6, tumor necrosis factor (TNF), IL-1, parathyroid hormone–related protein, and 1,25-dihyroxyvitamin D_3 all induce human osteoclast-like cell formation in human bone marrow culture. TNF and IL-1 induce IL-6, suggesting that their effects may be mediated by IL-6. Application of one of the IL-6Rα superantagonists, Sant5, revealed that the stimulation of osteoclast-like multinucleated cells (MNC) by IL-1 and TNF was dependent on IL-6, but the stimulation by parathyroid hormone–related protein and 1,25-dihyroxyvitamin was not (112).

Inhibiting IL-6–induced osteoclast formation may be therapeutically beneficial in situations in which bone resorption is undesirably high such as postmenopausal osteoporosis. The inhibition of IL-6 would have to be maintained over a long period of time. Constitutively high levels of Sant5 were generated by transfecting Sant5 cDNA into a stromal cell line (PSV10), which normally expresses IL-6 and stimulates osteoclast formation. Sant5 expression inhibited the stimulatory effects of the transfected cells in cocultures with normal human bone marrow. Conditioned media from the transfected cells also inhibited the stimulatory effects of the parental cell line in similar cocultures (112). The successful expression of Sant5 in transfected cells suggests that gene therapy might be a promising method for delivering IL-6Rα antagonists. So far, the in vivo analysis of the superantagonists has been restricted by their species specificity. Although they are antagonists on human cells, the superantagonists are agonists on murine cells (112).

Another important use for receptor antagonists has been to determine the receptor usage of newly identified cytokines. For example, LIF-R antagonists were used to define the receptor usage of CT-1 when CT-1 was first characterized. A LIF-R antagonist was able to block the induction of c-fos by CT-1 in murine cardiac myotubes, indicating that CT-1 responses in heart were in fact dependent on LIF-R for signal transduction (113).

LIF-R antagonists have also been useful in dissecting the receptor dependence of activities exhibited by cells in culture. A LIF-R antagonist was used to demonstrate that LIF-R activation is required for the arrest of rod differentiation exerted by Müller glial cells in cultures of mouse retinal cells (114). In situations where multiple LIF-R ligands may be present and the identities of the ligands are uncertain, LIF-R antagonists may be more useful than neutralizing antibodies for individual LIF-R ligands.

IN VIVO APPLICATIONS OF RECEPTOR ANTAGONISTS

Difficulties to Overcome

The in vivo applications of receptor antagonists for research purposes or as therapeutic agents present several challenges. Both wt cytokines and cytokines that have been modified to be receptor antagonists are rapidly cleared from the body. Hence, if the receptor antagonists are injected as a bolus, they must be repeatedly administered at frequent intervals. In several diseases, such as multiple myeloma, cytokine synthesis is continuously elevated; thus, for treatment, the appropriate antagonist would need to be administered continuously and at high levels. Each of the examples of in vivo applications described below addresses the problem of applying a sustained level of receptor antagonist.

Another concern in the application of receptor antagonists is the possibility of eliciting an immune response. Repeated applications of antagonists over a long period may result in a neutralizing response to the antagonist if the amino acid substitutions create new antigenic regions on the cytokine. One way around the problems of both antigenicity and the need for continuous treatment might be gene therapy, which might also allow for the targeting of specific tissues. An alternative method to achieve long-term blockade of a cytokine might be to use antagonists deliberately to induce a neutralizing response to the wt cytokine. Both gene therapy approaches and vaccination approaches are discussed in the examples below.

Application of IL-4R Antagonists

IL-4 receptor antagonists are interesting molecules to clinicians because IL-4 has a central role in T-cell (Th2)–dominated diseases such as type 1 allergic responses. An antagonist based on murine IL-4 (Q116D/Y119D) disrupts binding to γc, whereas preserving binding to the IL-4R. In a variety of murine cell lines, the antagonist inhibited IL-4 responses by 50% at a 100- to 500-fold molar excess of antagonist (6). Both IL-4 and IL-13 responses are inhibited, because the two receptor complexes share IL4-R as a subunit (115). During immunization with ovalbumin (OVA), BALB/c mice were administered the IL4R antagonist (given in two intraperitoneal injections per day for 8 days). The antagonist prevented immediate cutaneous hypersensitivity in response to intradermal rechallenge with antigen and prevented anaphylactic shock in response to intravenous rechallenge with antigen. Treatment completely inhibited the synthesis of OVA-specific IgE and IgG1. Interestingly, unlike in the IL4−/− mice or the IL-4R−/− mice, the antagonist also inhibited the production of IgG2a, IgG2b, and IgG3 (115). The different effects on immunoglobulin synthesis highlight the utility of applying receptor antagonists even when receptor knock outs are available for study. Explaining the differences between applying IL-4R antagonists and genetic deletion of IL4-R will enrich our understanding of IL4R in allergy.

GM-CSFR Antagonist in Juvenile Myelomonocytic Leukemia

The GM-CSFRα antagonist, E21R, induces apoptosis in a variety of hematopoietic cells including juvenile myelomonocytic leukemia (JMML) cells

(116). Cells from JMML patients and normal donors were engrafted into immunodeficient mice and a continuous supply of E21R was administered via minipump at the time of transplantation or 4 weeks later. At both time points, E21R profoundly reduced the JMML cell load in the mouse bone marrow, whereas sparing cells from normal donors (117). The sparing of normal cells, presumably because they are less sensitive to both GM-CSF and E21R, suggests that E21R might be a good therapeutic candidate for JMML.

Gene Therapy May Allow a Steady Supply of Receptor Antagonist

A recombinant adenovirus was constructed by inserting an IL-6 receptor superantagonist (Sant1, described above) under the control of a RSV promoter into a replication-incompetent adenoviral vector (118). Supernatants of cells infected in vitro contained active antagonist. After intravenous injection of the virus into mice, 1–2 ng/mL of antagonist could be detected in the serum. In vivo blockade of receptors could not be measured, because the IL-6 superantagonist does not block murine IL-6 receptors. However, the antagonist was active, since serum from infected mice could inhibit wt IL-6 responses on human cell lines. (118).

Vaccination of Mice with an IL-6Rα Antagonist to Neutralize IL-6

Exploiting the potential antigenic nature of a receptor antagonist is the basis of an unusual approach to inhibiting cytokines. Vaccinating mice with an IL-6Rα antagonist, Sant1, in the presence of either complete Freund's adjuvant or aluminum hydroxide resulted in the production of autoantibodies against IL-6. When the procedure was carried out in transgenic mice that had been engineered to express high levels of hIL-6, a strong antibody response to both the receptor antagonist and wt hIL-6 was generated, suggesting that the vaccination overcame tolerance to the transgenic hIL-6. Although vaccination with wt IL-6 also elicited autoantibodies, it was much less potent than the antagonist. The antibodies elicited by vaccination completely masked the circulating IL-6. Mice injected with hIL-6 following vaccination with the receptor antagonist did not show a normal acute phase response in response to an injection of hIL6. (119).

CONCLUSIONS

Many of the characteristic properties of receptor antagonists, including residual agonism, potency, and specificity, are cell-type dependent. In the example of IL-6Rα antagonists, testing a variety of cell lines revealed that in order to remove all residual agonism, both gp130-binding sites may need to be disrupted. Similar behavior is likely to occur in all receptor antagonists based on cytokines that bind multiple subunits.

The high potency of the IL-6Rα superantagonists on cell lines that are very sensitive to IL-6 demonstrated that enhancing the binding of IL-6 antagonists to IL-6Rα greatly improved their potency. More generally, the generation of a potent antagonist depends on the antagonist binding to the sequestered receptor subunit with higher affinity than the antagonist binds to the other receptor subunits and with higher affinity than the wt cytokine. The creation of receptor antagonists by inhibiting cytokine-receptor interactions may be unsuited to receptor complexes in which the cytokine binds to its ligand-specific subunit with low affinity. For example, OSM binds to its specific signal-transducing subunit, OSM-R, too weakly to generate an antagonist by disrupting gp130 binding. Cytokines may require substantial site modifications to become antagonists when the stability of the cytokine-receptor interactions that are preserved is substantially lower than the stability of the native receptor complex.

Testing different cell lines with receptor antagonists revealed subtle differences in the specificity of the antagonists. A survey of myeloma cell lines demonstrated that IL-6Rα antagonists with mutations in site 2 binding to gp130 inhibited both IL-6 and IL-11 responses. Only antagonists with mutations in both sites 2 and 3 were strictly specific for IL-6Rα. As a general rule, for complete specificity, only binding to the subunit targeted for sequestration should be preserved. The behavior of CNTFRα antagonists is also influenced by their preservation of gp130 binding. The CNTFRα antagonists inhibit CNTF responses that depend on soluble CNTFRα, whereas inhibiting all gp130 responses on cells that express transmembrane CNTFRα. The CNTFRα antagonists with their unusual mix of targets might be useful as a general gp130 antagonist that is restricted to the nervous system. As a general gp130 antagonist effective in all tissues, the fusion protein–antagonists are promising. When IL-6Rα antagonists containing a single gp130-binding site were fused to soluble IL-6Rα, they inhibited all gp130 cytokines. In turn, as a general LIF-R antagonist, hLIF-05 can inhibit all of the LIF-R–dependent cytokines, CNTF, CT-1, OSM, and LIF. Ideally, researchers and clinicians can benefit from a panel of receptor antagonists with different specificity.

The unexpected proapoptotic activity of the GM-CSFRα antagonist, E21R, is potentially very interesting. If the apoptosis induced by E21R derives from a blockade of signaling by a preformed complex of GMR-CSFRα and βc, then E21R will have important activities in the absence of wt GM-CSF. Whether similar phenomena occur in other receptor systems remains to be seen. Regardless of the detailed mechanism, the heightened sensitivity of leukemic cells to E21R-induced apoptosis makes the therapeutic application of E21R an attractive possibility.

The therapeutic application of IL-4R antagonists and IL-6Rα antagonists are also becoming attractive. The IL-6Rα superantagonists are now sufficiently potent that, at low concentrations, they are likely to be effective against endogenous IL-6 in vivo. The low concentrations of antagonists required make the development of controlled release formulations or gene therapy a justifiable goal.

ACKNOWLEDGMENTS

I thank V.A. Barton and M.A. Hall for their critical review of the manuscript.

REFERENCES

1. G Fuh, BC Cunningham, R Fukunaga, S Nagata, DV Goeddel, JA Wells. Rational design of potent antagonists to the human growth hormone receptor. Science 256:1677–1680, 1992.
2. G Fuh, JA Wells. Prolactin receptor antagonists that inhibit the growth of breast cancer cell lines. J Biol Chem 270:13133–13137, 1995.
3. SM Zurawaski, G Zurawski. Receptor antagonist and selective agonist derivatives of mouse interleukin-2. EMBO J 11:3905–3910, 1992.
4. N Kruse, H-P Tony, W Sebald. Conversion of human interleukin-4 into a high affinity antagonist by a single amino acid replacement. EMBO J 11:3237–3244, 1992.
5. H-P Tony, B-J Shen, P Reusch, W Sebald. Design of human interleukin-4 antagonists inhibiting interleukin-4–dependent and interleukin-13–dependent responses in T-cells and B-cells with high efficiency. Eur J Biochem 225:659–665, 1994.
6. SM Grunewald, S Kunzmann, B Schnarr, J Ezernieks, W Sebald, A Duschl. A murine interleukin-4 antagonistic mutant protein completely inhibits interleukin-4–induced cell proliferation, differentiation, and signal transduction. J Biol Chem 272:1480–1483, 1997.

7. J Tavernier, T Tuypens, A Verhee, G Plaetinck, R Devos, J Van der Heyden, Y Guisez, C Oefner. Identification of receptor-binding domains on human interleukin 5 and design of an interleukin 5-derived receptor antagonist. Proc Natl Acad Sci USA 92:5194–5198, 1995.
8. TR Hercus, CJ Bagley, B Cambareri, M Dottore, JM Woodcock, MA Vadas, MF Shannon, AF Lopez. Specific human granulocyte-macrophage colony-stimulating factor antagonists. Proc Natl Acad Sci USA 91:5838–5842, 1994.
9. T Taga. gp130, a shared signal transducing receptor component for hematopoietic and neuropoietic cytokines. J of Neurochem 67:1–10, 1996.
10. F Neipel, JC Albrecht, A Ensser, YQ Huang, JJ Li, AE Friedman-Kien, B Fleckenstein. Human herpesvirus 8 encodes a homolog of interleukin-6. J Virol 71:839–842, 1997.
11. T Taga, M Hibi, Y Hirata, K Yamasaki, K Yasukawa, T Matsuda, T Hirano, T Kishimoto. Interleukin-6 triggers the association of its receptor with a possible signal transducer, gp130. Cell 58:573–581, 1989.
12. J Karow, KR Hudson, MA Hall, AB Vernallis, JA Taylor, A Gossler, JK Heath. Mediation of interleukin-11-dependent biological responses by a soluble form of the interleukin-11 receptor. Biochem J 318:489–495, 1996.
13. S Davis, TH Aldrich, NY Ip, N Stahl, S Scherer, T Farruggella, PS DiStefano, R Curtis, N Panayotatos, H Gascan, S Chevalier, GD Yancopoulos. Released form of CNTF receptor α component as a soluble mediator of CNTF responses. Science 259:1736–1739, 1993.
14. D Pennica, V Arce, TA Swanson, R Vejsada, RA Pollock, M Armanini, K Dudley, HS Phillips, A Rosenthal, AC Kato, CE Henderson. Cardiotrophin-1, a cytokine present in embryonic muscle, supports long-term survival of spinal motoneurons. Neuron 17:63–74, 1996.
15. O Robledo, M Fourcin, S Chevalier, C Guillet, P Auguste, A Pouplard-Barthelaix, D Pennica, H Gasgan. Signaling of the cardiotrophin-1 receptor. Evidence for a third receptor component. J Biol Chem 272:4855–4863, 1997.
16. B Mosley, C De Imus, D Friend, N Boiani, B Thoma, LS Park, D Cosman. Dual oncostatin M (OSM) receptors. Cloning and characterization of an alternative signaling subunit conferring OSM-specific receptor activation. J Biol Chem 271:32635–32643, 1996.
17. M Ichihara, T Hara, H Kim, T Murate, A Miyajima. Oncostatin M and leukemia inhibitory factor do not use the same functional receptor in mice. Blood 90: 165–173, 1997.
18. RA Lindberg, TS-C Juan, AA Welcher, Y Sun, R Cupples, B Guthrie, FA Fletcher. Cloning and characterization of a specific receptor for mouse oncostatin M. Molecular and Cell Biol 18:3357–3367, 1998.
19. NY Ip, J McClain, NX Barrezueta, TH Aldrich, L Pan, Y Li, SJ Wiegand, B Friedman, S Davis, GD Yancopoulos. The α component of the CNTF receptor is required for signaling and defines potential CNTF targets in the adult and during development. Neuron 10:89–102, 1993.
20. PC Heinrich, I Behrmann, G Muller-Newen, F Schaper, L Graeve. Interleukin-

6–type cytokine signalling through the gp130/Jak/STAT pathway. Biochem J 334:297–314, 1998.
21. T Taga, T Kishimoto. gp130 and the interleukin-6 family of cytokines. Annu Rev Immunol 15:797–819, 1997.
22. M Kopf, H Baumann, G Freer, M Freudenberg, M Lamers, T Kishimoto, R Zinkernagel, H Bluethmann, G Kohler. Impaired immune and acute-phase responses in interleukin-6–deficient mice. Nature 368:339–342, 1994.
23. CH Ladel, C Blum, A Dreher, K Reifenberg, M Kopf, SH Kaufmann. Lethal tuberculosis in interleukin-6–deficient mutant mice. Infect Immun 65:4843–4849, 1997.
24. T van der Poll, CV Keogh, X Guirao, WA Buurman, M Kopf, S Lowry. Interleukin-6 gene–deficient mice show impaired defense against pneumococcal pneumonia. J Infect Dis 176:439–444, 1997.
25. DM Williams, BG Grubbs, T Darville, K Kelly, RG Rank. A role for interleukin-6 in host defense against murine Chlamydia trachomatis infection. Infect Immun 66:4564–4567, 1998.
26. HH Nandurkar, L Robb, D Tarlinton, L Barnett, F Kontgen, CG Begley. Adult mice with targeted mutation of the interleukin-11 receptor (IL11Ra) display normal hematopoiesis. Blood 90:2148–2159, 1997.
27. L Robb, R Li, L Hartley, HH Nandurkar, F Koentgen, CG Begley. Infertility in female mice lacking the receptor for interleukin 11 is due to a defective uterine response to implantation. Nat Med 4:303–308, 1998.
28. JL Escary, J Perreau, D Dumenil, S Ezine, P Brulet. Leukaemia inhibitory factor is necessary for maintenance of haematopoietic stem cells and thymocyte stimulation. Nature 363:361–364, 1993.
29. CL Stewart, P Kaspar, LJ Brunet, H Bhatt, I Gadi, F Kontgen, SJ Abbondanzo. Blastocyst implantation depends on maternal expression of leukaemia inhibitory factor (see Comments). Nature 359:76–79, 1992.
30. Y Masu, E Wolf, B Holtmann, M Sendtner, G Brem, H Thoenen. Disruption of the CNTF gene results in motor neuron degeneration. Nature 365:27–32, 1993.
31. TM DeChiara, R Vejsada, WT Poueymirou, A Acheson, C Suri, JC Conover, B Friedman, J McClain, L Pan, N Stahl, NY IP, A Kato, GD Yancopoulos. Mice lacking the CNTF receptor, unlike mice lacking CNTF, exhibit profound motor neuron deficits at birth. Cell 83:313–322, 1995.
32. M Li, M Sendtner, A Smith. Essential function of LIF receptor in motor neurons. Nature 378:724–727, 1995.
33. CB Ware, MC Horowitz, BR Renshaw, JS Hunt, D Liggitt, SA Koblar, BC Gliniak, HJ McKenna, T Papayannopoulou, B Thoma, L Cheng, PJ Donovan, JJ Peschon, PF Bartlett, CR Willis, BD Wright, MK Carpenter, BL Davison, DP Gearing. Targeted disruption of the low-affinity leukemia inhibitory factor receptor gene causes placental, skeletal, neural and metabolic defects and results in perinatal death. Development 121:1283–1299, 1995.
34. K Yoshida, T Taga, M Saito, S Suematsu, A Kumanogoh, T Tanaka, H Fuji-

wara, M Hirata, T Yamagami, T Nakahata, T Hirabayashi, Y Yoneda, K Tanaka, WZ Wang, C Mori, K Shiota, N Yoshida, T Kishimoto. Targeted disruption of gp130, a common signal transducer for the interleukin 6 family of cytokines, leads to myocardial and hematological disorders. Proc Natl Acad Sci USA 93:407–411, 1996.
35. UAK Betz, W Block, M van den Broek, K Yoshida, T Taga, T Kishimoto, K Addicks, K Rajewsky, W Muller. Postnatally induced inactivation of gp130 in mice results in neurological, cardiac, hematopoietic, immunological, hepatic, and pulmonary defects. J Exp Med 188: 1955–1965, 1998.
36. N Omori, RP Evarts, M Omori, Z Hu, ER Marsden, SS Thorgeirsson. Expression of leukemia inhibitory factor and its receptor during liver regeneration in the adult rat. Lab Invest 75:15–24, 1996.
37. JB Kurek, S Nouri, G Kannourakis, M Murphy, L Austin. Leukemia inhibitory factor and interleukin-6 are produced by diseased and regenerating skeletal muscle. Muscle Nerve 19:1291–1301, 1996.
38. Y Sun, RE Zigmond. Leukaemia inhibitory factor induced in the sciatic nerve after axotomy is involved in the induction of galanin in sensory neurons. Eur J Neurosci 8:2213–2220, 1996.
39. SWN Thompson, AA Majithia. Leukemia inhibitory factor induces sympathetic sprouting in intact dorsal root ganglia in the adult rat in vivo. J Physiol London 506:809–816, 1998.
40. MB Crichton, JE Nichols, Y Zhao, SE Bulun, ER Simpson. Expression of transcripts of interleukin-6 and related cytokines by human breast tumors, breast cancer cells, and adipose stromal cells. Mol Cell Endocrinol 118:215–220, 1996.
41. B Klein. Update of gp130 cytokines in multiple myeloma. Curr Opin Hematol 5:186–191, 1998.
42. G Carroll, M Bell, H Wang, H Chapman, J Mills. Antagonism of the IL-6 cytokine subfamily—a potential strategy for more effective therapy in rheumatoid arthritis. Inflamm Res 47:1–7, 1998.
43. S Manolagas. The role of IL-6 type cytokines and their receptors in bone. Ann NY Acad Sci 840:194–204, 1998.
44. MA Brown, D Metcalf, NM Gough. Leukaemia inhibitory factor and interleukin 6 are expressed at very low levels in the normal adult mouse and are induced by inflammation. Cytokine 6:300–309, 1994.
45. W Somers, M Stahl, J Seehra. 1.9 A crystal structure of interleukin 6: implications for a novel mode of receptor dimerization and signaling. EMBO J 16: 989–997, 1997.
46. RC Robinson, LM Grey, D Staunton, H Vankelecom, AB Vernallis, J-F Moreau, DI Stuart, JK Heath, EY Jones. The crystal structure and biological function of leukemia inhibitory factor: implications for receptor binding. Cell 77: 1101–1116, 1994.
47. NQ McDonald, N Panayotatos, WA Hendrickson. Crystal structure of dimeric

human ciliary neurotrophic factor determined by MAD phasing. EMBO J 14: 2689–2699, 1995.
48. AM De Vos, M Ultsch, AA Kossiakoff. Human growth hormone and extracellular domain of its receptor: crystal structure of the complex. Science 255:306–312, 1992.
49. BC Cunningham, JA Wells. Comparison of a structural and a functional epitope. J Mol Biol 234:554–563, 1993.
50. BC Cunningham, M Ultsch, AM De Vos, MG Mulkerrin, KR Clauser, JA Wells. Dimerization of the extracellular domain of the human growth hormone receptor by a single hormone molecule. Science 254:821–825, 1991.
51. JA Wells, AM de Vos. Hematopoietic receptor complexes. Ann Rev Biochem 65:609–634, 1996.
52. JF Bazan. Haemopoietic receptors and helical cytokines. Immunol Today 11: 350–354, 1990.
53. W Somers, M Ultsch, AM De Vos, AA Kossiakoff. The X-ray structure of a growth hormone–prolactin receptor complex. Nature 372:478–481, 1994.
54. J Bravo, D Staunton, JK Health, EY Jones. Crystal structure of a cytokine-binding region of gp130. EMBO 17:1665–1674, 1998.
55. H Yawata, K Yasukawa, S Natsuka, M Murakami, K Yamasaki, M Hibi, T Taga, T Kishimoto. Structure-function analysis of human IL-6 receptor: dissociation of amino acid residues required for IL-6 binding and for IL-6 signal transduction through gp130. EMBO J 12:1705–1712, 1993.
56. D Staunton, KR Hudson, JK Heath. The interactions of the cytokine-binding homology region and immunoglobulin-like domains of gp130 with oncostatin M: implications for receptor complex formation. Protein Eng 11:1093–1102, 1998.
57. K Yamasaki, T Taga, Y Hirata, H Yawata, Y Kawanishi, B Seed, T Taniguchi, T Hirano, T Kishimoto. Cloning and expression of the human interleukin-6 (BSF-2/IFN beta 2) receptor. Science 241:825–828, 1988.
58. M Hibi, M Murakami, M Saito, T Hirano, T Taga, T Kishimoto. Molecular cloning and expression of an IL-6 signal transducer, gp130. Cell 63:1149–1157, 1990.
59. R Savino, A Lahm, M Giorgio, A Cabibbo, A Tramontano, G Ciliberto. Saturation mutagenesis of the human interleukin 6 receptor-binding site: implications for its three-dimensional structure. Proc Natl Acad Sci USA 90:4067–4071, 1993.
60. A Hammacher, LD Ward, J Weinstock, H Treutlein, K Yasukawa, R Simpson. Structure-function analysis of human IL-6: identification of two distinct regions that are important for receptor binding. Protein Sci 12:2280–2293, 1994.
61. C Toniatti, A Cabibbo, E Sporeno, AL Salvati, M Cerretani, S Serafini, A Lahm, R Cortese, G Ciliberto. Engineering human interleukin-6 to obtain variants with strongly enhanced bioactivity. EMBO J 15:2726–2737, 1996.
62. M Murakami, M Hibi, N Nakagawa, T Nakagawa, K Yasukawa, K Yamanishi,

T Taga, T Kishimoto. IL-6-induced homodimerization of gp130 and associated activation of a tyrosine kinase. Science 260:1808–1810, 1993.
63. R Savino, A Lahm, AL Salvati, L Ciapponi, E Sporeno, S Altamura, G Paonessa, C Toniatti, G Ciliberto. Generation of interleukin-6 receptor antagonists by molecular-modeling guided mutagenesis of residues important for gp130 activation. EMBO J 13:1357–1367, 1994.
64. R Savino, L Ciapponi, A Lahm, A Demartis, A Cabibbo, C Toniatti, P Delmastro, S Altamura, G Ciliberto. Rational design of a receptor super-antagonist of human interleukin-6. EMBO J 13:5863–5870, 1994.
65. M Ehlers, J Grotzinger, FD deHon, J Mullberg, JPJ Brakenhoff, J Liu, A Wollmer, S Rose-John. Identification of two novel regions of human IL-6 responsible for receptor binding and signal transduction. J Immunol 153:1744–1753, 1994.
66. JPJ Brakenhoff, FD de Hon, V Fontaine, E ten Boekel, H Schooltink, S Rose-John, PC Heinrich, J Content, LA Aarden. Development of a human interleukin-6 receptor antagonist. J Biol Chem 269:86–93, 1994.
67. L Ciapponi, R Graziani, G Paonessa, A Lahm, G Ciliberto, R Savino. Definition of a composite binding site for gp130 in human interleukin-6. J Biol Chem 270:31249–31254, 1995.
68. G Paonessa, R Graziani, A De Serio, R Savino, L Ciapponi, A Lahm, AL Salvati, C Toniatti, G Ciliberto. Two distinct and independent sites on IL-6 trigger gp130 dimer formation and signalling. EMBO J 14:1942–1951, 1995.
69. LD Ward, GJ Howlett, G Discolo, K Yasukawa, A Hammacher, RL Moritz, RJ Simpson. High affinity interleukin-6 receptor is a hexameric complex consisting of two molecules each of interleukin-6, interleukin-6 receptor, and gp130. J Biol Chem 269:23286–23289, 1994.
70. E Sporeno, R Savino, L Ciapponi, G Paonessa, A Cabibbo, A Lahm, K Pulkki, R-X Sun, C Toniatti, B Klein, G Ciliberto. Human interleukin-6 receptor superantagonists with high potency and wide spectrum on multiple myeloma cells. Blood 87:4510–4519, 1996.
71. JPJ Brakenhoff, HK Bos, J Grotzinger, S Rose-John, LA Aarden. Identification of residues in the putative 5th helical region of human interleukin-6, important for activation of the IL-6 signal transducer, gp130. FEBS Lett 395:235–240, 1996.
72. A Demartis, F Bernassola, R Savino, G Melino, G Ciliberto. Interleukin 6 receptor superantagonists are potent inducers of human multiple myeloma cell death. Cancer Res 56:4213–4218, 1996.
73. FD de Hon, E ten Boekel, J Herrman, C Clement, M Ehlers, T Taga, K Yasukawa, Y Ohsugi, T Kishimoto, S Rose-John, J Widjenes, R Kastelein, LA Aarden, JPJ Brakenhoff. Functional distinction of two regions of human interleukin 6 important for signal transduction via gp130. Cytokine 7:398–407, 1995.
74. M Ehlers, FD de Hon, HK Bos, U Horsten, G Kurapkat, HS van De Leur, J Grotzinger, A Wollmer, JPJ Brakenhoff, S Rose-John. Combining two muta-

tions of human interleukin-6 that affect gp130 activation results in a potent interleukin-6 receptor antagonist on human myeloma cells. J Biol Chem 270: 8158–8163, 1995.
75. FD de Hon, M Ehlers, S Rose-John, SB Ebeling, HK Bos, LA Aarden, JPJ Brakenhoff. Development of an interleukin (IL) 6 receptor antagonist that inhibits IL-6-dependent growth of human myeloma cells. J Exp Med 180:2395–2400, 1994.
76. M Ehlers, J Grotzinger, M Fischer, HK Bos, JPJ Brakenhoff, S Rose-John. Identification of single amino acid residues of human IL-6 involved in receptor binding and signal initiation. J Interferon Cytokine Res 16:569–576, 1996.
77. R-X Sun, G Ciliberto, S Rocco, Z-J Gu, B Klein. Interleukin-6 receptor antagonists inhibit interleukin-11 biological activity. Eur Cytokine Netw 8:51–56, 1997.
78. S Chevalier, M Fourcin, O Robledo, J Wijdenes, A Pouplard-Barthelaix, H Gascan. Interleukin-6 family of cytokines induced activation of different functional sites expressed by gp130 transducing protein. J Biol Chem 271:14764–14772, 1996.
79. X-G Zhang, J-J Gu, Z-Y Lu, K Yasukawa, GD Yancopoulos, K Turner, M Shoyab, T Taga, T Kishimoto, R Bataille, B Klein. Ciliary neurotropic factor, interleukin 11, leukemia inhibitory factor, and oncostatin M are growth factors for human myeloma cell lines using the interleukin 6 signal transducer gp130. J Exp Med 177:1337–1342, 1994.
80. E Sporeno, G Paonessa, AL Salvati, R Graziani, P Delmastro, G Ciliberto, C Toniatti. Oncostatin M binds directly to gp130 and behaves as interleukin-6 antagonist on a cell line expressing gp130 but lacking functional oncostatin M receptors. J Biol Chem 269:10991–10995, 1994.
81. M Fischer, J Goldschmitt, C Peschel, JP Brakenhoff, KJ Kallen, A Wollmer, J Grotzinger, S Rose-John. I. A bioactive designer cytokine for human hematopoietic progenitor cell expansion. Nat Biotechnol 15:142–145, 1997.
82. M Peters, G Blinn, F Solem, M Fischer, KH Meyer zum Buschenfelde, S Rose-John. In vivo and in vitro activities of the gp130-stimulating designer cytokine Hyper-IL6. J Immunol 161:3575–3581, 1998.
83. C Renne, K-J Kallen, J Mullberg, T Jostock, J Grotzinger, S Rose-John. A new type of cytokine receptor antagonist directly targeting gp130. J Biol Chem 273: 27213–27219, 1998.
84. DJ Hilton, AA Hilton, A Raicevic, S Rakar, M Harrison-Smith, NM Gough, CG Begley, D Metcalf, NA Nicola, TA Wilson. Cloning of a murine IL-11 receptor α-chain; requirement for gp130 for high affinity binding and signal transduction. EMBO J 13:4765–4775, 1994.
85. P Neddermann, R Graziani, G Ciliberto, G Paonessa. Functional expression of soluble human interleukin-11 (IL-11) receptor α and stoichiometry of in vitro IL-11 receptor complexes with gp130. J Biol Chem 271:30986–30991, 1996.
86. VA Barton, KR Hudson, JK Heath. Identification of three distinct receptor binding sites of murine interleukin-11. J Biol Chem 274:5755–5761, 1999.

87. J-G Zhang, CM Owczarek, LD Ward, GJ Howlett, LJ Fabri, BA Roberts, NA Nicola. Evidence for the formation of a heterotrimeric complex of leukaemia inhibitory factor with its receptor subunits in solution. Biochem J 325:693–700, 1997.
88. A De Serio, R Graziani, R Laufer, G Ciliberto, G Paonessa. In vitro binding of ciliary neurotrophic factor to its receptors: evidence for the formation of an IL-6-type hexameric complex. J Mol Biol 254:795–800, 1995.
89. MJ Layton, P Lock, D Metcalf, NA Nicola. Cross-species receptor binding characteristics of human and mouse leukemia inhibitory factor suggest a complex binding interaction. J Biol Chem 269:17048–17055, 1994.
90. DP Gearing, CJ Thut, T VandenBos, SD Gimpel, PB Delaney, J King, V Price, D Cosman, MP Beckmann. Leukemia inhibitory factor receptor is structurally related to the IL-6 signal transducer, gp130. EMBO J 10:2839–2848, 1991.
91. DP Gearing, MR Comeau, DJ Friend, SD Gimpel, CJ Thut, J McGourty, KK Brasher, JA King, S Gillis, B Mosley, SF Zeigler, D Cosman. The IL-6 signal transducer, gp130: an oncostatin M receptor and affinity converter for the LIF receptor. Science 255:1434–1437, 1992.
92. CM Owczarek, MJ Layton, D Metcalf, P Lock, TA Willson, NM Gough, NA Nicola. Inter-species chimeras of leukaemia inhibitory factor define a major human receptor-binding determinant. EMBO J 12:3487–3495, 1993.
93. KR Hudson, AB Vernallis, JK Heath. Characterization of the receptor binding sites of human leukemia inhibitory factor and creation of antagonists. J Biol Chem 271:11971–11978, 1996.
94. D Pennica, KL King, KJ Shaw, E Luis, J Rullamas, S-M Luoh, WC Darbonne, DS Knutzon, R Yen, KR Chien, JB Baker, WI Wood. Expression cloning of cardiotrophin 1, a cytokine that induces cardiac myocyte hypertrophy. Proc Natl Acad Sci USA 92:1142–1146, 1995.
95. AB Vernallis, KR Hudson, JK Heath. An antagonist for the leukemia inhibitory factor receptor inhibits leukemia inhibitory factor, cardiotrophin-1, ciliary neurotrophic factor, and oncostatin M. J Biol Chem 272:26947–26952, 1997.
96. W Hui, M Bell, G Carroll. Detection of oncostatin M in synovial fluid from patients with rheumatoid arthritis. Ann Rheum Dis 56:184–187, 1997.
97. N Stahl, S Davis, V Wong, T Taga, T Kishimoto, NY Ip, GD Yancopoulos. Cross-linking identifies leukemia inhibitory factor-binding protein as a ciliary neurotrophic factor receptor component. J Biol Chem 268:7628–7631, 1993.
98. C Piquet-Pellorce, L Grey, A Mereau, JK Heath. Are LIF and related cytokines functionally equivalent? Exp Cell Res 213:340–347, 1994.
99. N Panayotatos, E Radziejewska, A Acheson, R Somogyi, A Thadani, WA Hendrickson, NQ McDonald. Localization of functional receptor epitopes on the structure of ciliary neurotrophic factor indicates a conserved, function-related epitope topography among helical cytokines. J Biol Chem 270:14007–14014, 1995.
100. I Saggio, I Gloaguen, G Poiana, R Laufer. CNTF variants with increased biolog-

ical potency and receptor selectivity define a functional site of receptor interaction. EMBO J 14:3045–3054, 1995.
101. M Inoue, H Karita, C Nakayama, H Noguchi. Construction and characterization of ciliary neurotrophic factor (CNTF) antagonists: microenvironmental difference in the CNTF receptor between rat and chicken cells for recognizing the D1 cap region. J Neurochem 69:95–101, 1997.
102. A Di Marco, I Gloaguen, A Demartis, I Saggio, R Graziani, G Paonessa, R Laufer. Agonistic and antagonistic variants of ciliary neurotrophic factor (CNTF) reveal functional differences between membrane-bound and soluble CNTF α-receptor. J Biol Chem 272:23069–23075, 1997.
103. M Inoue, C Nakayama, K Kikuchi, T Kimura, Y Ishige, A Ito, M Kanaoka, H Noguchi. D1 cap region involved in the receptor recognition and neural cell survival activity of human ciliary neurotrophic factor. Proc Natl Acad Sci USA 92:8579–8583, 1995.
104. A Di Marco, I Gloaguen, R Graziani, G Paonessa, I Saggio, KR Hudson, R Laufer. Identification of ciliary neurotrophic factor (CNTF) residues essential for leukemia inhibitory factor receptor binding and generation of CNTF receptor antagonists. Proc Natl Acad Sci USA 93:9247–9252, 1996.
105. AF Lopez, DJ Williamson, JR Gamble, CG Begley, JM Harlan, SS Klebanoff, A Waltersdorph, G Wong, SC Clark, MA Vadas. Recombinant human granulocyte-macrophage colony-stimulating factor stimulates in vitro mature human neutrophil and eosinophil function, surface receptor expression, and survival. J Clin Invest 78:1220–1228, 1986.
106. PD Emanuel, LJ Bates, RP Castleberry, RJ Gualtieri, KS Zuckerman. Selective hypersensitivity to granulocyte-macrophage colony-stimulating factor by juvenile chronic myeloid leukemia hematopoietic progenitors. Blood 77:925–929, 1991.
107. PO Iverson, D Robinson, S Ying, Q Meng, AB Kay, I Clark-Lewis, AF Lopez. The GM-CSF analogue E21R induces apoptosis of normal and activated eosinophils. Am J Respir Crit Care Med 156:1628–1632, 1997.
108. PO Iversen, LB To, AF Lopez. Apoptosis of hemopoietic cells by the human granulocyte-macrophage colony-stimulating factor mutant E21R. Proc Natl Acad Sci USA 93:2785–2789, 1996.
109. PO Iversen, TR Hercus, B Zacharakis, JM Woodcock, FC Stomski, S Kumar, BH Nelson, A Miyajima, AF Lopez. The apoptosis-inducing granulocyte-macrophage colony-stimulating factor (GM-CSF) analog E21R functions through specific regions of the heterodimeric GM-CSF receptor and requires interleukin-1β-converting enzyme-like proteases. J Biol Chem 272:9877–9883, 1997.
110. JM Woodcock, BJ McClure, FC Stomski, MJ Elliott, CJ Bagley, AF Lopez. The human granulocyte-macrophage colony-stimulating factor (GM-CSF) receptor exists as a performed receptor complex that can be activated by GM-CSF, Interleukin-3 or Interleukin-5. Blood 90:3005–3017, 1997.
111. CJ Bagley, JM Woodcock, FC Stomski, AF Lopez. The structural and functional basis of cytokine receptor activation: lessons from the common β subunit

of the granulocyte-macrophage colony-stimulating factor, interleukin-3 (IL-3), and IL-5 receptors. Blood 89:1471–1482, 1997.

112. RD Devlin, SV Reddy, R Savino, G Ciliberto, GD Roodman. IL-6 mediates the effects of IL-1 or TNF, but not PTHrP or 1,25$(OH)_2D_3$, on osteoclast-like cell formation in normal human bone marrow cultures. J Bone Miner Res 13: 393–399, 1998.
113. KC Wollert, T Taga, M Saito, M Narazaki, T Kishimoto, CC Glembotski, AB Vernallis, JK Heath, D Pennica, WI Wood, KR Chien. Cardiotrophin-1 activates a distinct form of cardiac muscle cell hypertrophy. Assembly of sarcomeric units in series via gp130/leukemia inhibitory factor receptor–dependent pathways. J Biol Chem 271:9535–9545, 1996.
114. C Neophytou, AB Vernallis, A Smith, MC Raff. Muller-cell-derived leukaemia inhibitory factor arrests rod photoreceptor differentiation at a postmitotic prerod stage of development. Development 124:2345–2354, 1997.
115. SM Grunewald, A Werthmann, B Schnarr, CE Klein, EB Brocker, M Mohrs, F Brombacher, W Sebald, A Duschl. An antagonistic IL-4 mutant prevents type 1 allergy in the mouse: inhibition of the IL-4/IL-13 receptor system completely abrogates humoral immune response to allergen and development of allergic symptoms in vivo. J Immunol 160:4004–4009, 1998.
116. PO Iversen, RL Rodwell, L Pitcher, KM Taylor, AF Lopez. Inhibition of proliferation and induction of apoptosis in juvenile myelomonocytic leukemic cells by the granulocyte-macrophage colony-stimulating factor analogue E21R. Blood 88:2634–2639, 1996.
117. PO Iversen, ID Lewis, S Turczynowicz, H Hasle, C Niemeyer, K Schmiegelow, S Bastiras, A Biondi, TP Hughes, AF Lopez. Inhibition of granulocyte-macrophage colony-stimulating factor prevents dissemination and induces remission of juvenile myelomonocytic leukemia in engrafted immunodeficient mice. Blood 90:4910–4917, 1997.
118. I Saggio, L Ciapponi, R Savino, G Ciliberto, M Perricaudet. Adenovirus-mediated gene transfer of a human IL-6 antagonist. Gene Ther 4:839–845, 1997.
119. L Ciapponi, D Maione, A Scoumanne, P Costa, MB Hansen, M Svenson, K Bendtzen, T Alonzi, G Paonessa, R Cortese, G Ciliberto, R Savino. Induction of interleukin-6 (IL-6) autoantibodies through vaccination with an engineered IL-6 receptor antagonist. Nat Biotechnol 15:997–1001, 1997.

9
Negative-Feedback Regulations of Cytokine Signals

Tetsuji Naka, Masashi Narazaki, and Tadamitsu Kishimoto
Osaka University, Suita City, Osaka, Japan

INTRODUCTION

Advancements in molecular biology during the past several years have resulted in significant clarification of signal-transduction systems for cytokines. As a result, the mechanisms of cell proliferation and differentiation induced by cytokines at the molecular level have also become clearer (1–3). However, much remains unknown about negative control systems which keep the effects of cytokine in a transient state, and studies about such systems have been started only recently. In particular, analyses of negative control systems, including negative feedback mechanisms, are expected to make an important contribution to the treatment of chronic inflammatory or proliferative diseases such as rheumatoid arthritis, Castleman's disease, and cancers, in which the overproduction of cytokines or the abnormal transduction of cytokine signals is considered to be pathogenetic (4,5).

Cytokines play pivotal roles not only in immune systems, including their defensive functions or inflammatory reactions, but also in homeostasis, which is regulated by cell differentiation, proliferation, and apoptosis. They are essential for the maintenance and regulation of biological functions at the cell, tissue, organ, and organism level. In the living body, however, a negative mechanism to inhibit signal transduction is also present, so that the physiological activity of cytokines does not continue unchecked. In other words, the effects of cytokines are transient, and homeostasis of the living body is maintained when the positive and negative control systems are working in a well-balanced manner. In some diseases, however, disturbance of this balance induces production of cytokines at abnormal level. In rheumatoid arthritis, for example, excess production and signal transduction of cytokines such as interleukins IL-1β and IL-6 and tumor necrosis factor-α (TNF-α) have been reported, and inflammatory reactions by these cytokines and the subsequent induction of proteases are thought to be part of the mechanisms involved in the destruction of joint architecture (4). In addition, it has been established that continuous signaling of IL-6 performs an essential function in the malignant proliferation of multiple myeloma cells (5). Under normal circumstances, a negative control system would put an end to such cytokine reactions, but undesirable inflammatory reactions may spread when the positive signal pathway continues to function as a result of the collapse of the negative control system. Clarification of the character, function and mechanism of the negative control system can therefore be expected to contribute to the treatment of various diseases.

SIGNAL TRANSDUCTION MEDIATED BY JANUS KINASE

Cytokines transmit their signals through binding to their cognate receptors on target cells. Cytokine receptors belong to a cytokine receptor family which is characterized by four conserved cysteine residues and a WSXWS motif in the extracellular domain. Structurally, these receptors are also related to interferon (IFN) receptors. Although most cytokine receptors do not contain a kinase motif in their cytoplasmic regions, cytokines have been shown to activate intracellular tyrosine kinases and to induce the tyrosine phosphorylation of cellular proteins, including their receptors. Through extensive studies of non-receptor tyrosine kinases, the essential role of the Janus kinase (JAK) family of tyrosine kinases was first described in connection with the signal transduction of IFN, and it has now been established that a large number of cytokines and growth factors activate JAK for the first steps of their signal transductions (1,2).

Four members of the JAK family, JAK1, JAK2, JAK3, and TYK2, have been identified up to now, and it is thought that a particular JAK is bound to the cytoplasmic membrane proximal region (called Box1) of the respective cytokine receptor. With the binding of cytokines to their receptors, cytokine receptors form dimers or oligomers, and JAKs, which associate with receptors in a constitutive manner, come close enough to one another to be activated.

This results in the phosphorylation of tyrosine residues in the cytoplasmic region of the receptor. Subsequently, members of the STAT (signal transducers and activators of the transcription) family of transcription factors with an SH2 domain bind to tyrosine-phosphorylated receptors and are phosphorylated by activated JAK and then dimerize and translocate into the nuclei. The STAT dimer bound to DNA in the nucleus promotes the expression of various target genes (1–3). The JAK-STAT signaling pathway is the most common of the cytokine-signaling pathways. The Ras-MAPK pathway is also activated by JAK in response to cytokine stimulation, possibly mediated by SHP-2 (the SH2 domain containing protein tyrosine phosphatase-2) and/or Shc, which binds to the Grb2-SOS complex (6). Furthermore, the STAM (signal-transducing adaptor molecule) is known to be a substrate for JAK in the transduction of the IL-2 signal (7).

The negative control systems which inhibit cytokine signal pathways are not as well known as the positive signaling cascades which are being identified rapidly as a result of the recent discovery of JAK-STAT and Ras-MAPK cascades. In 1997, we discovered a novel molecule which is induced in response to IL-6 and the leukemia inhibitory factor (LIF) via STAT3 and inhibits JAK resulting in the inactivation of STAT3 (8). It therefore acts in a negative-feedback loop to inhibit signal transduction of IL-6 and LIF, and in view of its nature, we named it SSI-1 (STAT-induced STAT inhibitor-1). Since SSI-1 was discovered independently by three groups, it is also known as SOCS-1 (suppressor of cytokine signaling-1) (9) or JAB (JAK binding protein) (10). The SSI, SOCS, and JAB nomenclatures all have their respective merits to explain the nature of this protein, but in this chapter, this protein is called SSI and we describe characteristics of the SSI-family members and the results which we obtained recently from a study with SSI-1–deficient mice.

MECHANISMS OF NEGATIVE CYTOKINE REGULATION

One of the characteristics of cytokines is that their effects are transient. This means that some regulatory mechanism must be present in the signal transduction by cytokines. The following section describes the negative regulatory mechanisms of cytokine effects, and mainly focuses on SSI family proteins. However, there are still many unanswered questions about these negative regulatory mechanisms: which mechanisms are the most important, how they are related to one another and how they are induced and regulated.

Negative Regulation by Soluble Cytokine Receptors

Soluble cytokine receptors are generated by proteolytic cleavage or by alternative mRNA splicing. In many cases, they compete with membrane-bound receptors for ligand binding and inhibit the function of ligands. For example, soluble receptors for epidermal growth factor (EGF), IL-1, and TNF inhibit the activity of their membrane-bound receptors in a dose-dependent manner (11). In the case of IL-6, which transduces signals via multisubunit receptors, the soluble IL-6 receptor has been shown to have an agonistic effect, whereas the soluble form of the common signaling subunit (gp130) inhibits IL-6–mediated signaling (12).

Endocytosis of Receptors

The intracellular region of gp130 contains a sequence consisting of 10 amino acids with a di-leucine motif. Deletion of this part results in injury to IL-6–induced endocytosis of the IL-6 receptor system. Therefore, it is highly possible that this sequence plays a role in the endocytosis of the IL-6 receptor and induces downregulation to prevent excessive ligand stimulation (13,14).

Inhibition of Signal Pathway by Tyrosine Phosphatase

Erythropoietin (EPO) stimulation induces activation of JAK2 and phosphorylation of the tyrosine residues (e.g., Y429) in the cytoplasmic domain of the EPO receptor, SHP-1 (SH2-containing protein-tyrosine phosphatase, also known as SHPTP-1) binds to the phosphorylated tyrosine residue in the EPO receptor via the SH2 domain. After binding, dephosphorylation and inactivation of JAK2 occurs as a result of the phosphatase activity of SHP-1, and EPO signaling is inhibited (15). It has also been reported that SHP-1 binds to the β chain, which is the common signal-transducing subunit of IL-3, IL-5, and the granulocyte-macrophage colony-stimulating factor (GM-CSF), and is also thought to inhibit the signaling which mediates the common β chain (16). Furthermore, inhibition of signaling as a result of the direct binding of SHP-1 and JAK has also been reported as an alternative mechanism (17). The motheaten mouse, which is known to show abnormal corpuscular proliferation and immunodeficiency, has been demonstrated to have a mutation in the SHP-1 gene. This finding indicates the importance of the SHP-1 function (18).

Protein Inhibitor of Activated STAT

The protein inhibitor of activated STAT3 (PIAS3) was cloned as a STAT-binding protein. It consists of 583 amino acids and is expressed in various human tissues (19). PIAS3 binds specifically to the active form of STAT3 and inhibits binding of STAT3 homo- and heterodimers to DNA. It has also been reported that PIAS1 binds specifically to STAT1 and inhibits the DNA-binding activity of STAT1 (20). It thus seems that PIAS genes form a family and that there may be a specific PIAS for each member of the STAT family.

Ubiquitin-Proteasome System

Ubiquitinized proteins become the targets of proteasomes and undergoes proteolysis. This ubiquitin-proteasome system regulates cellular function via intracellular protein degradation.

This system participates in the endocytosis of growth hormone receptors and prolonged JAK activation by proteasome inhibitors has been reported (21,22). In addition, it was found that INF-γ–activated STAT1 is conjugated to ubiquitin, and degradation of the STAT1 is reduced by treatment of the cells with a proteasome inhibitor (23). These findings suggest that the ubiquitin-proteasome system participates not only in the receptor level, but also at the STAT transcription factor level.

Negative Crosstalk Among Cytokines

When a certain cytokine signaling system is activated, stimulation by other cytokines is inhibited. For example, cells stimulated by GM-CSF become less responsive to other cytokines such as IL-6 and INF-γ (24). Although the function of SSI-1, which is a negative-feedback molecule, can explain this phenomenon satisfactorily, so far no data have been produced which indicate a direct relationship between these systems.

SSI Family

Up to the present, eight members of the SSI family have been identified by their structural characteristics. Each member of this family has a different name, since three groups of investigators have independently cloned the family members, resulting in CIS1 (CIS), SSI-1/SOCS-1/JAB, SSI-2/SOCS-2/CIS2, SSI-3/SOCS-3/CIS3, SOCS-4/CIS7, SOCS-5/CIS6, SOCS-6/CIS4, and SOCS-7/CIS5/NAP (8–10, 25–28). All of these proteins contain one SH2 domain, and at the COOH-terminal side of the SH2 domain, they possess the well-conserved, highly homologous domain which we designated as the SC motif (the SSI COOH-terminal motif consists of two conserved motifs, the

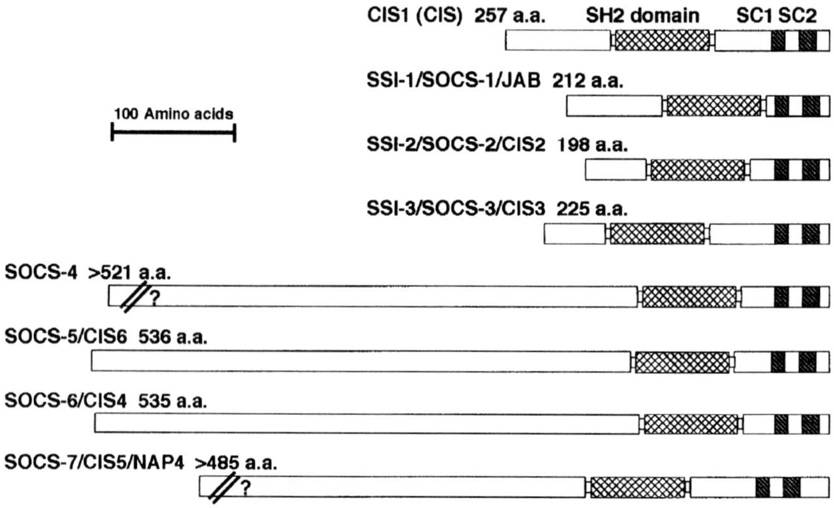

Figure 1 Schematic structure of SSI family proteins. The SH2 domain and the SC-motif (SC-motif1 and 2) are indicated. The full-length cDNAs of SOCS-4/CIS7 and SOCS-7/CIS5/NAP4 have not been isolated.

SC-motif1 and SC-motif2) (26) (Fig. 1). This domain is also called the SOCS box or CH domain (CIS-homology domain) (9,27). It is interesting to note that, in addition to the SSI family proteins, four groups of proteins containing the SC motif at the C-terminal side without the SH2 domain have been identified by DNA database searches: (1) WSB-1 and WSB-2 with WD-40 repeats and SC motif; (2) SSB-1 to SSB-3 with the SPRY domain and SC motif; (3) ASB-1 to ASB-3 with ankyrin repeats and SC motif; and (4) the RAR and RAR-like protein with the guanosine triphosphatase (GTPase) domains and SC motif (28). However, the functions of these molecules have yet to be determined.

SSI FAMILY PROTEINS

CIS1 (Cytokine-Inducible SH2-Containing Protein 1)

CIS1 (CIS) was described in 1995 as an early gene induced by STAT5-activating cytokines such as IL-2, IL-3, and EPO (25). It is the first member of the SSI family and consists of 257 amino acids with the SH2 domain in its central region. The mRNA of CIS1 has been detected in a wide range of organs, and at particularly high levels in kidney, lung, and liver and at low levels in heart and stomach but not in brain or spleen. Results obtained with the mutant IL-2 receptor, which does not activate STAT5 but activates JAK1 and JAK3, suggested that activation of STAT5 is essential for induction of the CIS1 gene. The 5′-flanking region of the CIS1 gene contains four potential STAT5 binding sites (MGF boxes), and the essential role of STAT5 in expression of the CIS1 gene was confirmed by the finding that STAT5 knock out mice fail to induce the CIS1 gene (29). CIS1 must therefore be a target gene of the JAK-STAT5 pathway. The CIS1 protein associates with the tyrosine-phosphorylated IL-3 receptor β chain and EPO receptor, and the forced expression of this protein suppresses EPO-induced tyrosine phosphorylation and activation of STAT5. Because CIS1 is induced by STAT5 and inhibits STAT5, it functions as a negative-feedback molecule of the JAK/STAT5 pathway (25,29). The inhibitory effect of CIS1 on cytokine signaling can be explained in two ways. One explanation is that binding of CIS1 to the tyrosine phosphorylated receptor may function as a ''mask'' and thus competitively suppresses interaction between STAT5 and the receptor. This hypothesis is based on the observation that CIS1 associates with Tyr-401 of the cytoplasmic region of the EPO receptor, and that Tyr-401 is one of the activation sites for STAT5 (30). Another possibility is that CIS1 is a scavenger of tyrosine phosphorylated receptors. Indeed, CIS1 is ubiquitinated in cells, and the proteasome inhibitors increase the amount of CIS1-EPO receptor complexes and inhibit the inactivation of the EPO receptor and STAT5. The target proteins of CIS1 may thus become a substrate of proteolytic enzymes (30). The main mechanisms for the inhibitory function of CIS1 warrant further examination.

The CIS1 gene is located in the distal region of mouse chromosome 9, which is linked to *Trf*, *Gnai2*, and *Col7a1* (25). This region shares homology with the short arm of human chromosome 3 (3p21), so that the putative CIS1 locus in human is thought to be 3p21. It is known that the human 3p21 locus is frequently lost or rearranged in renal cell carcinoma and lung cancer (31,32). In view of the function of CIS1 and its tissue distribution, loss of the CIS1 gene may thus constitute one of the mechanisms involved in unregulated cell proliferation.

SSI-1

SSI-1 was cloned on the basis of the antigenic similarity of its SH2 domain to that of STAT3 (8). In monocytic leukemia M1 cells, whose differentiation and apoptosis are induced by IL-6 or LIF via STAT3 (33,34), SSI-1 is induced in response to IL-6 or LIF. When STAT3 activity is inhibited by the forced expression of dominant negative STAT3 in M1 cells, SSI-1 is not induced even if the cells are stimulated by IL-6, which suggests that STAT3 is important for the induction of SSI-1. In M1 cells with forced expression of SSI-1, differentiation and apoptosis induced by stimulation of IL-6 or LIF are completely inhibited. These findings demonstrate that SSI-1 is a negative-feedback molecule. In view of its character, we named it SSI-1 (STAT-induced STAT inhibitor-1). SSI-1 was cloned independently by another group of investigators, who named it JAB (JAK binding protein) because of its ability to interact with JAK (10). It was also cloned as an inhibitor of IL-6–induced growth arrest of M1 cells, and was therefore referred to as SOCS-1 (suppressor of cytokine signaling-1) (9).

The SSI-1 protein consists of 212 amino acids. Its mRNA is detected most strongly in lymph tissues such as the thymus and spleen followed by the lungs and testis. It is also been identified in some cell lines, including MH60 cells, whose proliferation is dependent on IL-6, in IL-4–dependent CT4S cells, and in G-CSF–dependent NFS60 cells after 30 min of stimulation by their respective cytokines.

Stimulation of IL-6 induces tyrosine phosphorylation of the IL-6 signal-transducing receptor component gp130, which in turn results in the activation of STAT3 by JAK tyrosine kinases. How does SSI-1 exert its inhibitory effect on the gp130-mediated signaling pathway? Tyrosine phosphorylation of gp130 and STAT3 was found to be greatly reduced in M1 cells expressing SSI-1 compared with that in control M1 cells. It has also been established that JAK family tyrosine kinases are activated in response to cytokine stimulation, which suggests that the primary site of SSI-1 is found in JAK kinases. In fact, by using three different systems, COS cell expression, recombinant protein, and yeast two-hybrid, it was demonstrated that SSI-1 associates with JAK1, JAK2, JAK3, and TYK2 and reduces the activities of these kinases (8–10). However, binding of SSI-1 with JAK was decreased when JAK was dephosphorylated while SSI-1 could not bind the kinase-negative mutant of JAK, which suggests that interaction may occur with a tyrosine phosphorylated JAK (10,35).

Mutational analyses of the SSI-1 protein revealed that at least two domains, the pre-SH2 domain (the 24 amino acids, 52–76, in front of the SH2 domain) and the SH2 domain, are essential for the inhibition of JAK activity by SSI-1, which associates with JAK via its SH2 domain (35). Studies of the ability of SSI-1 to bind to various phosphopeptides of JAK2 showed that SSI-1 specifically binds to phosphopeptide containing a phosphorylated Y1007 (36). This tyrosine residue (Y1007) is located in the activation loop of JAK2 and autophosphorylated by JAK itself, while phosphorylation of this residue is known to be critical for the activation of JAK kinase. Although SSI-1 associates with JAK via its SH2 domain, binding of the SH2 domain to JAK is not sufficient to inhibit kinase activity, so that the pre-SH2 domain is also required for such inhibition. The C-terminal half of the pre-SH2 domain shows sequence homology among the eight members of the SSI family, whereas the N-terminal half shows homology between SSI-1 and SSI-3 and between SOCS-4 and SOCS-5 (Fig. 2). Moreover, the SSI-1 and SSI-3 pre-SH2 domains have been found to be functionally interchangeable (37). Detailed mutational analyses of the pre-SH2 domain showed that the C-terminal half of this domain contributes to the binding to the phosphorylated tyrosine residue of JAK as a part of the SH2 domain, and that the N-terminal half of the pre-SH2 domain is required for inhibition of JAK kinase activity. The first half of the pre-SH2 domain is thought to function as a pseudosubstrate and/or adenosine triphosphate (ATP) for the catalytic pocket of JAK kinases (36) (Fig. 3). The molecular mechanism for the inhibition by the pre-SH2 domain should become clear as a result of crystallographic analyses of SSI-1 and the kinase domain of JAK.

The C-terminal conserved domain of the SSI family, which we call the SC-motif (also referred to as the SOCS box or CH domain), is composed of two conserved motifs, SC-motif1 and SC-motif2. Although neither of these two is directly required for inhibition of JAK by SSI-1, deletion of the SC-motif markedly reduces expression of SSI-1 protein in M1 cells. This reduction is reversed by treatment with proteasome inhibitors, so that the SC-motif appears to function as a protector from proteolytic degradation (35). The N-terminal half of the SC-motif, the SC-motif1, includes the consensus elongin BC complex binding sequence (T, S) (L, M) XXX (C, S) XXX (V, L, I) (see Fig. 2). This complex binds not only to members of the SSI family but also to proteins such as WSB-1, ASB-2, and RAR-1 which contain the SC-motif

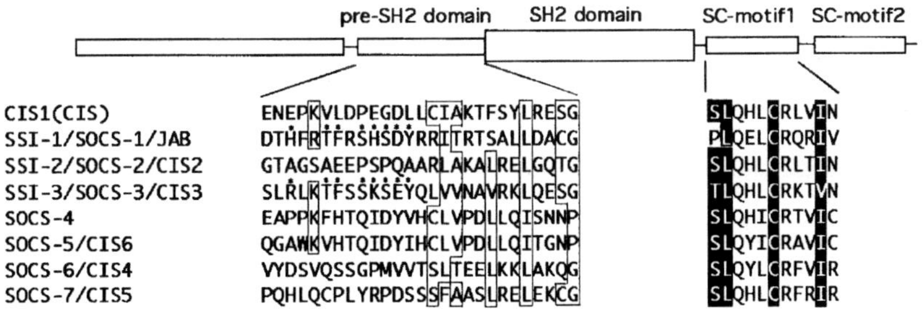

Figure 2 Amino acid sequence of the pre-SH2 domain and the SC-motif1 of the eight SSI family proteins. Boxes in the pre-SH2 domain indicate areas where five or more amino acids are identical in terms of their characters. Asterisks indicate homology between SSI-1 and SSI-3 of the pre-SH2 domain. Black boxes in the SC-motif1 indicate the elongin BC complex binding sequence motif.

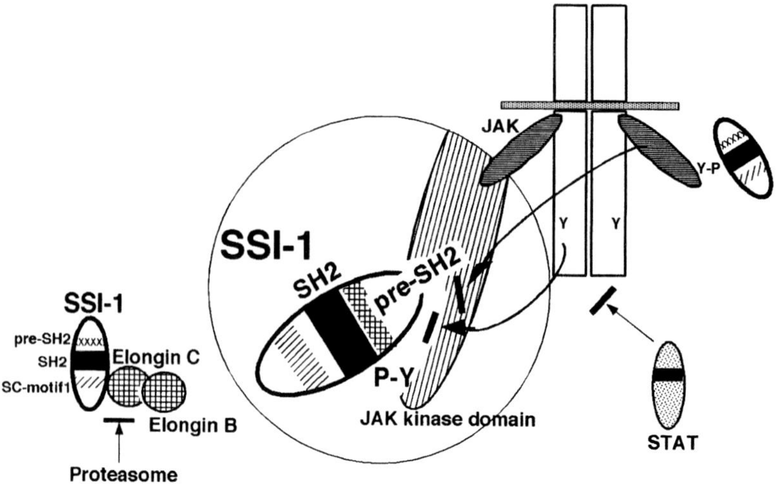

Figure 3 Schematic model of the inhibitory mechanism by the SSI-1 protein. SSI-1 binds to the tyrosine phosphorylated kinase domain of JAK via the latter's SH2 domain and inhibits kinase activity via its pre-SH2 domain. The SSI-1 protein is protected from protein degradation by binding to the elongin BC complex via the latter's SC-motif1.

(38,39). Elongin B is an ubiquitin-like protein, and elongin C is a Skp-1-like protein which binds to elongin A or the von Hippel-Lindau (VHL) tumor suppressor protein. By interacting with the SC-motif, the elongin BC complex enhances expression of the SSI-1 protein, probably by preventing it from degradation (38). Interestingly, elongin C also appears to interact with Cullin-2 (a putative E3 ubiquitin ligase), and Cullin-2 may induce ubiquitination of substrates and subsequent degradation of them. It is unknown, however, how the elongin BC complex protects proteins containing the SC-motif from degradation, and whether SSI-1/elongin BC complex induces ubiquitination of JAK.

It has also been reported that SSI-1 binds with tyrosine kinases other than those in JAK-family, such as Tec (40), PYK2, and tyrosine kinase receptors, including c-Kit, fibroblast growth factor receptor (FGFR) (27), and insulin-like growth factor-I receptor (IGF-IR) (41), but its specific physiological characteristics are not known. Moreover, induction of SSI-1 by various cytokines such as IL-3, IL-4, IL-13, GM-CSF, EPO, and IFN-γ (9), has been reported and analysis of its physiological functions is drawing great interest.

SSI-2

With the aid of an expression sequence tag (EST) data base search, SSI-2 was cloned as a protein related to SSI-1 (9,26,27) and consists of 198 amino acids. The mRNA of SSI-2 is detected in lung, placenta, prostate, and kidney. It is induced in bone marrow in response to a number of cytokines such as EPO, IL-3, IFNγ, G-CSF, LIF and IL-1, and in liver by IL-6 (9). Although SSI-2 is structurally similar to SSI-1 and CIS1, it does not bind to JAK2 nor inhibits EPO signaling (27). SSI-2 was also cloned on the basis of its ability to interact with the insulin-like growth factor I receptor (IGF-IR), but it remains unclear whether SSI-2 affects IGF-IR signaling (41). Since SSI-2 is a cytokine-inducible gene and structurally resembles SSI-1 and CIS1, it may be a negative-feedback molecule. Further studies are needed to determine the precise functions of SSI-2.

SSI-3

SSI-3 was also cloned with the aid of an EST data base search (9,26,27). It consists of 225 amino acids and is induced in bone marrow by IL-1, IL-2, IL-3, IL-4, IL-6, IL-7, IL-12, IL-13, LIF, G-CSF, GM-CSF, EPO, TPO, INF-γ, and TNF-α. Although stimulation of 3T3-442A cell and mouse liver cells with GH leads to expression of SSI-1 to SSI-3, expression of SSI-3 is most strongly induced. SSI-3, as well as SSI-1, can inhibit signal transduction of GH, whereas SSI-2 cannot (42). Expression of SSI-1 to SSI-3 has been detected in the mouse hypothalamus, but only SSI-3 expression can be induced by in vivo stimulation with leptin. SSI-3 can also inhibit signal transmission from leptin and phosphorylation of leptin receptors (43). Further, in mouse pituitary glands, SSI-3 is induced in vivo by stimulation of LIF and IL-1β. It has been reported that in the ArT-20 cell SSI-3 inhibits the secretion of adrenocorticotropic hormone (ACTH) and the induction of proopimelanocortin (POMC) mRNA after stimulation of LIF (44). SSI-3 is thus thought to have physiological significance in both the endocrine and metabolic systems.

Forced expression of SSI-3 in M1 cells, as in the case of SSI-1, inhibits both IL-6–and LIF–induced differentiation and tyrosine phosphorylation of STAT3. Again as in the case of SSI-1, SSI-3 binds to JAK2 and inhibits its kinase activity, while both the SH2 domain and the pre-SH2 domain of SSI-3 are required for its inhibitory effect. However, the inhibitory effect of SSI-3 on JAK kinase activity is weaker than that of SSI-1, suggesting that SSI-3 has a lower affinity to JAK kinases than does SSI-1 (45).

Other Members of the SSI Family

SOCS-4/CIS7, SOCS-5/CIS6, SOCS-6/CIS4, and SOCS-7/CIS5 were also cloned with the aid of an EST data base search (27,28). These four proteins have a more extensive N-terminal region preceding their SH2 domain than do CIS1 and SSI-1 to SSI-3. SOCS-7/CIS5 is the same molecule as NAP4, an adapter molecule (46), which can bind to the SH3 domain of Nck and Ash via the N-terminal proline-rich region and bind to EGF receptors via the SH2 domain of SOCS-7/CIS5. SOCS-5/CIS6 has been shown to be induced by GM-CSF and EPO (27).

The specific functions of these proteins remain to be identified.

SSI-1–Deficient Mice

SSI-1–deficient (SSI-1 (−/−)) mice were generated to clarify physiological roles (47,48). SSI-1 (−/−) mice are born normally in accordance with Mendel's law and no differences are detected between them and control littermate (SSI-1 (+/+) mice). However, a body weight loss of about 40% compared to SSI-1 (+/+) mice is observed from postnatal day 9 and all the animals die within 3 weeks. When the thymocytes of the 10-day-old mice were stimulated with IL-4, and tyrosine phosphorylation of STAT 6 was examined at 0, 1, and 3 h, phosphorylation of STAT6 was transient in SSI-1 (+/+) mice, whereas the tyrosine phenomenon was observed in SSI-1 (−/−) mice as late as 3 h after stimulation, so that thymocytes from SSI-1 (−/−) mice showed prolonged signaling in response to IL-4. This finding suggests that SSI-1 acts also in vivo as a negative-feedback molecule for the JAK-STAT signal pathway.

The numbered thymocytes and splenic lymphoid cell of SSI-1 (−/−) mice did not differ much from those in SSI-1 (+/+) mice on the first postnatal day, but the number of lymphocytes had decreased by 75–80% on postnatal day 10. Moreover, SSI-1 (−/−) mice exhibited thymic and splenic atrophy with aging and the boundary between the cortex and medulla of the thymus became unclear. Irregularity and atrophy of the splenic white pulp were also observed. In terms of differentiation of lymphocytes lacking SSI-1, although SSI-1 (−/−) mice at 10 days of age exhibited a remarkable decrease in the number of thymocytes and splenocytes, there was no distinct difference in the populations of T cells, monocytes, and granulocytes, except for a slight decrease in CD4-single positive T cells in the thymus and a disappearance of activated B cells ($IgM^+/CD23^+$ fraction) in the spleen. According to Starr, et al. (48), SSI-1 (−/−) mice exhibited a decrease in the number of lymphocytes from bone marrow cells, and slight decrease in pre-B cells and mature

B cells ($B220^+/IgM^+$) without changes in pro-B cells. They concluded that there was no abnormality in the lymphocytes themselves, since in vitro culture these bone marrow cells differentiated into mature B ($B220^+$ IgM^+) cells by in vitro culture.

Immunohistological staining using the TUNEL method of the thymus and spleen of 10-day-old SSI-1 (−/−) mice indicated that the number of cells stained by TUNEL increased remarkably. Analysis with an electron microscope found numerous cells showing typical apoptotic features such as nuclear chromatin aggregation as well as apoptotic microbodies in the thymus and spleen of SSI-1 (−/−) mice.

In immunohistological staining using Bcl-2 and Bax antibodies, which are apoptosis-related molecules, expression of Bcl-2 showed no difference between SSI-1 (+/+) mice and SSI-1 (−/−) mice, but expression of Bax increased greatly in the thymus and spleen of SSI-1 (−/−) mice. Increased expression of Bax was thus considered to be one of the causes of acceleration of apoptosis in SSI-1 (−/−) mice resulting in the decrease in the number of lymphocytes.

Analysis for other organs demonstrated myocardial degeneration and lymphocyte infiltration in the heart and fatty degeneration, cell death of hepatocytes, and lymphocyte infiltration in the liver. However, the causes of these pathologies are not known at present.

The JAK-STAT signal from cytokines such as IL-2 has been shown in many studies (49–54) to act as an antiapoptosis signal by inducing expression of antiapoptosis molecules such as Bcl-2 and Bcl-xL, but the result of our analysis of SSI-1–(−/−) mice contradict these findings.

The following are considered to be possible causes of accelerated apoptosis of lymphocytes observed in SSI-1 (−/−) mice. First, there is a possibility that SSI-1 suppresses not only the JAK-STAT signal transduction pathway, but also other signals that promote apoptosis. As described above, in the thymus and the spleen of SSI-1 (−/−) mice, apoptosis of lymphocytes with high expression of Bax becomes more frequent with aging. These results suggest that SSI-1 may repress the expression of Bax (either directly or indirectly). There is a report describing the involvement of p53 in the induction of Bax expression (55). However, the precise mechanism and the relationship between SSI-1 and Bax are still unclear.

A second possibility is that the apoptotic pathway induced by interferon regulatory factor-1 (IRF-1) (56,57) is enhanced by the lack of SSI-1. Deficiency of SSI-1 can cause the continuous transmission of cytokine signals, such as IFN-γ, followed by the continuous expression of IRF-1. This may lead to the continuous activation of caspase-1 and subsequent apoptosis of lymphocytes. Although the involvement of caspase-1 in apoptosis is suggested by in vitro data that the forced expression of caspase-1 promotes apoptosis, it is a controversial issue whether caspase-1 is the essential factors for apoptosis in vivo, because apoptosis is not inhibited and normally proceeds in the caspase-1 deficient mice.

Third, it is possible that the apoptosis pathway is promoted by the activation of STATs, as recently reported. There have not been any reports on definite relationships between STATs and apoptosis in vivo yet. However, in vitro studies, it has been reported that apoptosis-promoting factors, such as caspase-1, -3, -4, -7, -8, -10, Fas, Fas L, TNFR1-I, and bak, are induced by signals, like INF-γ, which activate STAT1 (58–61), and that apoptosis induced by TNF-α is remarkably inhibited in the STAT1-deletion mutant cells. It has been also reported that by the treatment with TNF-α, STAT1 without any modifications, such as dimerization or phospholyration, is able to control expression of pro-apoptotic genes via the formation of a transcriptionally active complex with a still unidentified transcription partner in the nucleus (62,63). Recently, the combination between TNFR-I and JAK1 is reported to lead to the activation of STAT1, 3, 5, and 6 (64). With consideration given to these reports, it may be hypothesized that apoptosis observed in the SSI-1 (−/−) mice is caused by the continuous activation of STATs. There is also another possibility that lymphocytes of SSI-1 (−/−) mice could be driven to apoptosis via some other pathways. In fact, it has been reported that cross linking of major histocompatibility (MHC-1) in human Jurkat cells causes the activation of both Tyk2 and STAT3 leading to apoptosis of cells, and that this apoptosis is disturbed by the forced expression of dominant negative counterpart of STAT3 (65).

Fourth, with lack of SSI-1, which is a negative regulator of cytokine signaling, signals of cytokine such as INF-γ and IL-4 are excessively transmitted, which may result in accelerated apoptosis of lymphocytes. In fact, according to reports on transgenic (Tg) mice with forced expression of INFγ or IL-4 in lymphocytes, both Tg mice were actually found to have a thymic and splenic atrophy, disappearance of thymic cortex and decrease in the numbers of thymocytes and splenocytes.

As described above, we suppose that there are four possibilities in the mechanism of enhanced apoptosis of lymphocytes observed in SSI-1 (−/−) mice. In regard to the JAK-STAT signal-transduction pathway, however, it is somehow difficult to believe that the continuous activated JAK-STAT signal is the only cause of the phenotype observed in SSI-1 (−/−) mice, because partially overlapped inhibitory actions of SSI-1 and SSI-3 against cytokine signals are proved in vitro, and SSI-3 may compensate for a negative regulation mechanism. Moreover, there are other mechanisms, to suppress the activity of STATs, such as soluble receptors, endocytosis of receptors, dephosphorylation by SHP-1, and PIAS. Therefore, SSI-1 may be a molecule that has a novel function to control some signal transduction pathways other than the JAK-STAT signal pathway essential for cell survival.

CONCLUSIONS

The present study has shown that inactivation of SSI-1, which is a negative-feedback molecule for JAK-STAT, is likely to induce expression of Bax in lymphocytic organs promoting apoptosis. It is still not clear, however, how the contradictory phenomena of cell proliferation and apoptosis in the JAK-STAT and Ras-Raf-MAP systems are controlled and the result of the present study may provide a key to the resolution of this contradiction.

As abnormal transduction of cytokine signals is one of the causes of diseases such as rheumatoid arthritis, not only the analysis of the positive control mechanisms of cytokines, but also of the negative control mechanism is important. It is expected that analysis of these mechanisms will lead to a further understanding of cancers as well as of chronic inflammations.

REFERENCES

1. JN Ihle. Cytokine receptor signaling. Nature 377:591–594, 1995.
2. JE Darnell Jr, IM Kerr, GR Stark. Jak-STAT pathways and transcriptional activation in response to IFNs and other extracellular signaling proteins. Science 264: 1415–1421, 1994.
3. T Kishimoto, S Akira, M Narazaki, T Taga. Interleukin-6 family of cytokines and gp130. Blood 86:1243–1254, 1995.
4. S Akira, T Taga, T Kishimoto. Interleukin-6 in biology and medicine. Adv Immunol 54:1–78, 1993.
5. M Kawano, A Kuramoto, T Hirano, T Kishimoto. Cytokines as autocrine growth factors in malignancies. Cancer Surv 8:905–919, 1989.
6. T Fukada, M Hibi, Y Yamanaka, M Takahashi-Tezuka, Y Fujitani, T Yamaguchi, K Nakajima, T Hirano. Two signals are necessary for cell proliferation induced by a cytokine receptor gp130: involvement of STAT3 in anti-apoptosis. Immunity 5:449–460, 1996.
7. T Takeshita, T Arita, M Higuchi, H Asao, K Endo, H Kuroda, N Tanaka, K Murata, N Ishii, K Sugamura. STAM, signal transducing adaptor molecule, is

associated with Janus kinases and involved in signaling for cell growth and c-myc induction. Immunity 6:449–457, 1997.
8. T Naka, M Narazaki, M Hirata, T Matsumoto, S Minamoto, A Aono, N Nishimoto, T Kajita, T Taga, K Yoshizaki, S Akira, T Kishimoto. Structure and function of a new STAT-induced STAT inhibitor. Nature 387:924–929, 1997.
9. R Starr, TA Willson, EM Viney, LJL Murray, JR Rayner, BJ Jenkins, TJ Gonda, WS Alexander, D Metcalf, NA Nicola, DJ Hilton. A family of cytokine-inducible inhibitors of signaling. Nature 387:917–921, 1997.
10. TA Endo, M Masuhara, M Yokouchi, R Suzuki, H Sakamoto, K Mitsui, A Matsumoto, S Tanimura, M Ohtsubo, H Misawa, T Miyazaki, N Leonor, T Taniguchi, T Fujita, Y Kanakura, S Komiya, A Yoshimura. A new protein containing an SH2 domain that inhibits JAK kinases. Nature 387:921–924, 1997.
11. ML Heaney, DW Golde. Soluble cytokine receptors. Blood 87:847–857, 1996.
12. M Narazaki, K Yasukawa, T Saito, Y Ohsugi, H Fukui, Y Koishihara, GD Yancopoulos, T Taga, T Kishimoto. Soluble forms of the interleukin-6 signal-transducing receptor component gp130 in human serum possessing a potential to inhibit signals through membrane-anchored gp130. Blood 82:1120–1126, 1993.
13. E Dittrich, S Rose-Jhon, C Gerhartz, J Mullberg, T Stoyan, K Yasukawa, PC Heinrich, I Graeve. Identification of region within the cytoplasmic domain of the interleukin-6 (IL-6) signal transducer gp130 important for ligand-induced endocytosis of the IL-6 receptor. J Biol Chem 269:19014–19020, 1994.
14. E Dittrich, CR Haft, L Muys, PC Heinrich, L Graeve. A di-leucine motif and an upstream serine in the interleukin-6 (IL-6) signal transducer gp130 mediate ligand-induced endocytosis and down-regulation of the IL-6 receptor. J Biol Chem 271:5487–5494, 1996.
15. U Klingmuller, U Lorenz, LC Cantley, BG Neel, HF Lodish. Specific recruitment of SH-PTP-1 to the erythropoietin receptor causes inactivation of JAK2 and termination of proliferative signals. Cell 80:729–738, 1995.
16. T Yi, AL Mui, G Krystal, JN Ihle. Hematopoietic cell phosphatase associates with the interleukin-3 (IL-3) receptor beta chain and down-regulates IL-3–induced tyrosine phosphorylation and mitogenesis. Mol Cell Biol 13:7577–7586, 1993.
17. H Jiao, K Berrada, W Yang, M Tabrizi, LC Platanias, T Yi. Direct association with and dephosphorylation of jak2 kinase by the SH2-domain-containing protein tyrosine phosphatase SHP-1. Mol Cell Biol 16:6985–6992, 1996.
18. LD Shultz, A Schweitzer, TV Rajan, T Vi, JN Ihle. Mutations at the murine motheaten locus are within the hematopoietic cell protein-tyrosine phosphatase (Hcph) gene. Cell 73:1445–1454, 1993.
19. CD Chung, J Liao, B Liu, X Rao, P Jay, P Berta, K Shuai. Specific inhibition of stat3 signal transduction by PIAS3. Science 278:1803–1805, 1997.
20. B Liu, J Liao, X Rao, SA Kushner, CD Chung, DD Chang, K Shuai. Inhibition of stat1-mediated gene activation by PIAS1. Proc Natl Acad Sci USA 95:10626–10631, 1998.
21. R Govers, P van Kerkhof, AL Schwartz, GJ Strous. Linkage of the ubiquitin-

conjugating system and the endocytic pathway in ligand-induced internalization of the growth hormone receptor. EMBO J 16:4851–4558, 1997.
22. BA Callus, B Mathey-Prevot. Interleukin-3-induced activation of the JAK/STAT pathway is prolonged by proteasome inhibitors. Blood 91:3182–3192, 1998.
23. TK Kim, T Maniatis. Regulation of interferon-gamma-activated STAT1 by the ubiquitin-proteasome pathway. Science 273:1717–1719, 1996.
24. TK Sengupta, EM Schmitt, LB Ivashkiv. Inhibition of cytokines and JAK-STAT activation by distinct signaling pathways. Proc Natl Acad Sci USA 93:9499–9504, 1996.
25. A Yoshimura, T Ohkubo, T Kiguchi, NA Jenkins, DJ Gilbert, NG Copeland, T Hara, A Miyajima. A novel cytokine-inducible gene CIS encodes an SH2-containing protein that binds to tyrosine-phosphorylated interleukin 3 and erythropoietin receptors. EMBO J 14:2816–2826, 1995.
26. S Minamoto, K Ikegame, K Ueno, M Narazaki, T Naka, H Yamamoto, T Matsumoto, H Saito, S Hosoe, T Kishimoto. Cloning and functional analysis of new members of STAT induced STAT inhibitor (SSI) family: SSI-2 and SSI-3. Biochem Biophys Res Commun 237:79–83, 1997.
27. M Masuhara, H Sakamoto, A Matsumoto, R Suzuki, H Yasukawa, K Mitsui, T Wakioka, S Tanimura, A Sasaki, H Misawa, M Yokouchi, M Ohtsubo, A Yoshimura. Cloning and characterization of novel CIS family genes. Biochem Biophys Res Commun 239:439–446, 1997.
28. DJ Hilton, RT Richardson, WS Alexander, EM Viney, TA Willson, NS Sprigg, R Starr, SE Nicholson, D Metcalf, NA Nicola. Twenty proteins containing a C-terminal SOCS box form five structural classes. Proc Natl Acad Sci USA 95:114–119, 1998.
29. A Matsumoto, M Masuhara, K Mitsui, M Yokouchi, M Ohtsubo, H Misawa, A Miyajima, A Yoshimura. CIS, a cytokine inducible SH2 protein, is a target of the JAK-STAT5 pathway and modulates STAT5 activation. Blood 89:3148–3154, 1997.
30. F Verdier, S Chrétien, O Muller, P Varlet, A Yoshimura, S Gisselbrecht, C Lacombe, P Mayeux. Proteasomes regulate erythropoietin receptor and signal transducer and activator of transcription 5 (STAT5) activation. Possible involvement of the ubiquitinated CIS protein. J Biol Chem 273:28185–28190, 1998.
31. K Yamakawa, R Morita, E Takahashi, T Hori, J Ishikawa, Y Nakamura. A detailed deletion mapping of the short arm of chromosome 3 in sporadic renal cell carcinoma. Cancer Res 51:4707–4711, 1992.
32. S Yokoyama, K Yamakawa, E Tsuchiya, M Murata, S Sakiyama, Y Nakamura. Deletion mapping on the short arm of chromosome 3 in squamous cell carcinoma and adenocarcinoma of the lung. Cancer Res 52:873–877, 1992.
33. M Minami, M Inoue, S Wei, K Takeda, M Matsumoto, T Kishimoto, S Akira. STAT3 activation is a critical step in gp130-mediated terminal differentiation and growth arrest of myeloid cell line. Proc Natl Acad Sci USA 93:3963–3966, 1996.
34. K Nakajima, Y Yamanaka, K Nakae, H Kojima, M Ichiba, N Kiuchi, T Kitaoka,

T Fukada, M Hibi, T Hirano. A central role for stat3 in IL-6–induced regulation of growth and differentiation in M1 leukemia cells. EMBO J 15:3651–3658, 1996.

35. M Narazaki, M Fujimoto, T Matsumoto, Y Morita, H Saito, T Kajita, K Yoshizaki, T Naka, T Kishimoto. Three distinct domains of SSI-1/SOCS-1/JAB protein are required for its suppression of interleukin 6 signaling. Proc Natl Acad Sci USA 95:13130–13134, 1998.
36. H Yasukawa, H Misawa, H Sakamoto, M Masuhara, A Sasaki, T Wakioka, S Ohtsuka, T Imaizumi, T Matsuda, JN Ihle, A Yoshimura. The JAK-binding protein JAK inhibits Janus tyrosine kinase activity through binding in the activation loop. EMBO J 18:1309–1320, 1999.
37. SE Nicholson, TA Willson, A Farley, R Starr, JG Zhang, M Baca, WS Alexander, D Metcalf, DJ Hilton, NA Nicola. Mutational analyses of the SOCS proteins suggest a dual domain requirement but distinct mechanisms for inhibition of LIF and IL-6 signal transduction. EMBO J 18:375–385, 1999.
38. T Kamura, S Sato, D Haque, L Liu, WG Kaelin Jr, RC Conaway, JW Conaway. The Elongin BC complex interacts with the conserved SOCS-box motif present in members of the SOCS, ras, WD-40 repeat, and ankyrin repeat families. Genes Dev 12:3872–3881, 1998.
39. JG Zhang, A Farley, SE Nicholson, TA Willson, LM Zugaro, RJ Simpson, RL Moritz, D Cary, R Richardson, G Hausmann, BJ Kile, SBH Kent, WS Alexander, D Metcalf, DJ Hilton, NA Nicola, M Baca. The conserved SOCS box motif in suppressors of cytokine signaling binds to elongins B and C and may couple bound proteins to proteasomal degradation. Proc Natl Acad Sci USA 96:2071–2076, 1999.
40. K Ohya, S Kajigaya, Y Yamashita, A Miyazato, K Hatake, Y Miura, U Ikeda, K Shimada, K Ozawa, H Mano. SOCS-1/JAB/SSI-1 can bind to and suppress Tec protein-tyrosine kinase. J Biol Chem 272:27178–27182, 1997.
41. BR Dey, SL Spence, P Nissley, RW Furlanetto. Interaction of human suppressor of cytokine signaling (SOCS)-2 with the insulin-like growth factor-1 receptor. J Biol Chem 273:24095–24101, 1998.
42. TE Adams, JA Hansen, R Starr, NA Nicola, DJ Hilton, N Billestrup. Growth hormone preferentially induces the rapid, transient expression of SOCS-3, a novel inhibitor of cytokine receptor signaling. J Biol Chem 273:1285–1287, 1998.
43. C Bjørbæk, JK Elmquist, JD Frantz, SE Shoelson, JS Flier. Identification of SOCS-3 as a potential mediator of central leptin resistance. Mol Cell 1:619–625, 1998.
44. CJ Auernhammer, V Chesnokova, C Bousquet, S Melmed. Pituitary corticotroph SOCS-3: Novel intracellular regulation of leukemia-inhibitory factor-mediated proopiomelanocortin gene expression and adrenocorticotropin secretion. Mol Endocrinol 12:954–961, 1998.
45. R Suzuki, H Sakamoto, H Yasukawa, M Masuhara, T Wakioka, A Sasaki, K Yuge, S Komiya, A Inoue, A Yoshimura. CIS3 and JAB have different regulatory

roles in interleukin-6 mediated differentiation and STAT3 activation in M1 leukemia cells. Oncogene 17:2271–2278, 1998.
46. K Matuoka, H Miki, K Takahashi, T Takenawa. A novel ligand for an SH3 domain of the adoptor protein Nck bears an SH2 domain and nuclear signaling motifs. Biochem Biophys Res Commun 239:488–492, 1997.
47. T Naka, T Matsumoto, M Narazaki, M Fujimoto, Y Morita, Y Ohsawa, H Saito, T Nagasawa, Y Uchiyama, T Kishimoto. Accelerated apoptosis of lymphocytes by augmented induction of Bax in SSI-1 (STAT-induced STAT inhibitor-1) deficient mice. Proc Natl Acad Sci USA 95:15577–15582, 1998.
48. R Starr, D Metcalf, AG Elefanty, M Brysha, TA Willson, NA Nicola, DJ Hilton, WS Alexander. Liver degeneration and lymphoid deficiencies in mice lacking suppressor of cytokine signaling-1. Proc Natl Acad Sci USA 95:14395–14399, 1998.
49. G Packham, EL White, CM Eischen, H Yang, E Parganas, JN Ihle. Selective regulation of Bcl-XL by a jak kinase-dependent pathway is bypassed in murine hematopoietic malignancies. Genes Dev 12:2475–2487, 1998.
50. M Adachi, T Torigoe, S Takayama, K Imai. BAG-1 and Bcl-2 in IL-2 signaling. Leuk Lymphoma 30:483–491, 1998.
51. FW Quelle, J Wang, J Feng, D Wang JL Cleveland JN Ihle, GP Zambetti. Cytokine rescue of p53-dependent apoptosis and cell cycle arrest is mediated by distinct Jak kinase signaling pathway. Genes Dev 12:1099–1107, 1998.
52. J Gomez, AC-Martine, A Gonzalez, A Garcia, A Rebollo. The Bcl-2 gene is differentially regulated by IL-2 and IL-4: role of the transcription factor NF-AT. Oncogene 17:1235–1243, 1998.
53. F Mor, IR Cohen. IL-2 rescues antigen-specific T cells from radiation or dexamethasone-induced apoptosis; correlation with induction of Bcl-2. J Immunol 156:515–522, 1996.
54. M Armant, G Delespesse, M Sarfati. IL-2 and IL-7 but not IL-12 protect natural killer cells from death by apoptosis and up-regulation bcl-2 expression. Immunology 85:331–337, 1995.
55. T Miyashita, JC Reed. Tumor suppression p53 is a direct transcriptional activator of human bax gene. Cell 80:293–299, 1995.
56. T Tamura, M Ishihara, MS Lamphier, N Tanaka, I Oishi, S Aizawa, T Matsuyama, TW Mak, S Taki, T Taniguchi. DNA damage-induced apoptosis and ICE gene induction in mitogenically activated T lymphocytes require IRF-1. Leukemia 11:439–440, 1997.
57. V Giandomenico, F Lanillotti, G Fiorucci, ZA Percario, R Rivabene, W Malorini, E Affabris, G Romeo. Retinoic acid and IFN inhibition of cell proliferation is associated with apoptosis in squamous cartinoma cell line. Cell Growth Differ 8:91–100, 1997.
58. T Tamura, M Ishihara, MS Lamphier, N Tanaka, I Oishi, S Aizawa, T Matsuyama, TW Mak, S Taki, T Taniguchi. An IRF-1-dependent pathway of DNA damage-induced apoptosis in mitogen-activated T lymphocytes. Nature 376:596–599, 1995.

59. NK Ossina, A Cannas, VC Powers, PA Fitzpatrick, JD Knight, JR Gilbert, EM Shekhtman, LD Tomei, SR Umansky, MC Kiefer. Interferon-gamma modulates a p53-indipendent apoptosis pathway and apoptosis-related gene expression. J Biol Chem 272:16351–16357, 1997.
60. X Xu, XY Fu, J Plate, AS Chong. IFN-gamma induces cell growth inhibition by Fas-mediated apoptosis: requirement of STAT1 protein for up-regulation of Fas and FasL expression. Cancer Res 58:2832–2837, 1998.
61. Y Chin, M Kitagawa, K Kuida, RA Flavell, XY Fu. Activation of the STAT signaling pathway can cause expression of caspase1 and apoptosis. Mol Cell Biol 17:5328–5337, 1997.
62. T Hoey. A new player in cell death. Science 278:1578–1579, 1997.
63. A Kumar, M Commane, TW Flickinger, CM Horvath, GR Stark. Defective TNF-alpha-induced apoptosis in STAT1-null cells due to low constituve level of caspase. Science 278:1630–1632, 1997.
64. D Guo, JD Dunbar, CH Yang, LM Pfeffer, DB Donner. Induction of Jak/STAT signaling by activation of the type1 TNF-receptor. J Immunol 160:2742–2750, 1998.
65. S Skov, M Nielsen, S Bregenholt, N Odum, MH Claesson. Activation of stat-3 is involved in the induction of apoptosis after ligation of major histocompatibility complex class 1 molecules on human Jurkay T cell. Blood 91:3566–3573, 1998.

10

Inhibitors of the Janus Kinase–Signal Transducers and Activators of Transcription (JAK/STAT) Signaling Pathway

Robyn Starr
The Walter and Eliza Hall Institute of Medical Research, Parkville, Victoria, Australia

INTRODUCTION

Cytokines are small, secreted polypeptides which regulate a diverse array of cellular functions including proliferation, differentiation, and survival. Cytokines exert their effects by binding to specific receptors at the cell surface. This signal is transmitted through the cytoplasm to the nucleus, where changes in gene transcription effect the cytokine-induced biological response. The Janus kinase–signal transducers and activators of transcription (JAK-STAT) pathway is one of the better characterized signal transduction pathways. Binding of ligand to its receptor induces dimerization of the receptor chains, bringing together two associated JAKs which are activated by transphosphorylation. Activated JAKs phosphorylate various substrates in the cell, including the receptor itself and STATs. Phosphorylated STATs then dimerize and migrate to the nucleus where they activate gene transcription. Initially described through studies of interferon (IFN) signal transduction, the JAK-STAT pathway is now known to be a general mechanism for connecting receptor activation to changes in gene transcription, and it is utilized by many different cytokine receptors.

In contrast to our detailed knowledge of the proteins that mediate cytokine signal transduction, little is understood about how these pathways are negatively regulated. Clearly, there is a need for tight regulation of signal transduction pathways, and the absence of appropriate regulation has been shown to result in uncontrolled proliferation, cellular transformation, or apoptosis. Significant advances have been made recently toward identifying molecules which switch off cytokine signal transduction pathways. This chapter focuses on these negative regulators, and discusses the possible therapeutic benefits of effectors which interfere with or enhance their function.

JAK Kinases

Unlike growth factor receptors, which have intrinsic tyrosine kinase activity, cytokine receptors rely on noncovalently associated kinases to couple ligand binding at the cell surface to tyrosine phosphorylation in the cytoplasm. JAKs are constitutively associated in an inactive state with the cytoplasmic domains of cytokine receptors. The JAK family consists of four members, TYK2, JAK1, JAK2, and JAK3, which range in molecular weight from 125 to 135 kD (1–3). Most are widely expressed, with the exception of JAK3, which is mainly expressed in myelocytic and lymphocytic lineages. Each cytokine receptor selectively associates with a distinct subset of JAK kinases (Table 1). Certain receptors, such as those for growth hormone or erythropoietin (EPO), bind a single JAK family member (JAK2), in contrast to gp130, which can bind TYK2, JAK1, or JAK2 (4). The importance of these molecules is evident by the presence of JAK homologs in diverse species, including carp (*JAK1*) and *Drosophila melanogaster* (*hopscotch*).

The general structure of JAK kinases is conserved, and can be arranged into seven JAK homology (JH) domains. Two kinase domains, JH1 and JH2, are located at the carboxyl-terminus of the protein. The JH1 domain conforms to all the consensus sequences associated with tyrosine kinases (5) and is likely to be the major catalytic domain. JAK activity is regulated by the phosphorylation state of two adjacent conserved tyrosine residues within the activation loop of the kinase (6). In particular, tyrosine phosphorylation of Y^{1007} in the JH1 domain is critical for JAK2 catalytic activity (7). The JH2 domain lacks several critical residues of the consensus and is most likely a pseudokinase domain. The JH2 domain appears to fulfill a regulatory function, although there is discordance in the literature as to whether this is a positive or negative role. Mutant TYK2 lacking this domain is no longer active in in vitro kinase assays (8). In contrast, an increase in kinase activity is seen in *hopscotch* and

Table 1 Activation of JAKs and STATs by Cytokines and Growth Factors

Ligand	JAK	STAT
IFN-α/β	1, TYK2	1, 2
IFN-γ	1, 2	1
IL-2	1, 3	3, 5
IL-4	1, 3	6
IL-7	1, 3	1, 5
IL-9	1, 3	1, 3, 5
IL-15	1, 3	3, 5
IL-13	2	6
IL-3	2	5
IL-5	2	1, 5
GM-CSF	2	5
IL-6	1, 2, TYK2	1, 3
LIF	1, 2, TYK2	1, 3
OSM	1, 2, TYK2	1, 3
CNTF	1, 2, TYK2	1, 3
G-CSF	1, 2	3
EPO	2	5
GH	2	1, 3, 5
PRL	2	1, 3, 5
IL-12	2, TYK2	3, 4
EGF	1	1, 3, 5
PDGF	1, 2, TYK2	1, 3
CSF-1	1, 2, TYK2	1, 3

JAK2 carrying analagous mutations in a conserved residue of the JH2 domain (9). The N-terminal region of JAKs (JH7–JH3) is essential for receptor recognition and association (10,11). It remains unclear whether this domain can be defined more specifically, since the whole of this region of JAK1 was required for its association with the IFN-γR1 chain, whereas only the JH7 and/or JH6 regions of JAK2 were necessary for appropriate interaction with the IFN-γR2 chain (11).

Since JAKs promote cellular proliferation and/or survival in response to certain cytokines, it is perhaps surprising that activating mutations of JAKs have not been commonly found in malignancies. This may be due to the regulatory mechanisms operating in the cell which function to limit JAK activity (see below). Tight control of JAK activity is, however, necessary for normal cellular function. For example, constitutive activation of JAKs can occur as a result of mutations within the cytokine receptors themselves, in which constitutive dimerization of the receptor chains leads to continuous JAK activation and cellular transformation (12,13).

The essential and nonredundant roles of JAK kinases in cytokine signaling have been confirmed by the analysis of JAK-deficient mice. JAK3, the most specific of the JAKs, interacts exclusively with a single cytokine receptor subunit, the γ_c chain shared by the receptors for interleukins IL-2, IL-4, IL-7, IL-9, and IL-15. It was therefore not surprising that mutations in JAK3 were found to cause severe combined immunodeficiency (SCID) in humans, similar to the X-linked form of the disease caused by mutations in γ_c (14,15). JAK3-deficient mice have been generated which are also immunodeficient (16–18).

In contrast, JAK1, JAK2, and TYK2 are ubiquitously expressed and associate with many different cytokine receptor subunits. In vitro studies produced conflicting data regarding the specificity of JAKs in signaling, with one study suggesting that the JAKs were fully interchangeable with certain cytokine receptor subunits (19) in contrast to another which showed that JAK1 was essential for IL-6 signal transduction (20). JAK1-deficient mice, however, manifested defective signaling in response to a specific subset of cytokines, including those cytokines utilizing class II cytokine receptors, the γ_c subunit, or the gp130 subunit for signaling, whereas JAK1−/− cells remained responsive to other cytokines such as granulocyte colony-stimulating factor (G-CSF) (21). JAK2 deficiency resulted in embryonic lethality due to the absence of definitive erythropoiesis. Cells from JAK2−/− mice did not respond to EPO, IL-3, thrombopoietin (TPO), or IFN-γ but responded normally to IL-6 and IFN-α/β, an almost complementary pattern of defects to those of JAK1−/− mice (22,23). Although these results suggest that JAKs contribute to some extent to the specificity of cytokine signaling, it is clear that this contribution is only partial. Fine tuning of the signal specificity is provided by the STAT family of transcription factors.

STAT TRANSCRIPTION FACTORS

STATs are a family of latent transcription factors which are located in an inactive state in the cytoplasm of resting cells (1,2,24,25). Stimulation of a cell with cytokine results in rapid phosphorylation and dimerization of STATs, and active STAT dimers are then translocated to the nucleus. Seven STAT proteins have been described in mammals, STAT1, 2, 3, 4, 5A, 5B, and 6, and a STAT homolog, D-STAT/*marelle*, has been identified in *Drosophila*

(26,27). Similar to JAKs, a specific set of STATs is activated in response to a given cytokine (see Table 1).

Several structural and functional domains are conserved between STAT family members. Of critical importance is the conserved *src*-homology-2 (SH2) domain, essential for STAT dimerization and for the interaction of STATs with activated receptors. SH2 domains are specific sequence modules common to a variety of cytoplasmic proteins which recognize short peptide sequences containing phosphotyrosine residues (28). Phosphorylation by JAKs of receptor tyrosine residues generates docking sites for a variety of SH2-containing signaling molecules, including STATs. STATs are then phosphorylated by JAKs on a critical tyrosine residue located immediately carboxyl-terminal to the SH2 domain (29,30). Phosphorylation of this residue is essential for STAT function, since mutation of this tyrosine blocks ligand-induced STAT dimerization, DNA binding activity, and gene activation (31). Phosphorylated STATs dissociate from the receptor and form either homodimers or heterodimers, which then migrate to the nucleus. Dimerization of STAT molecules is required for DNA-binding activity, and it is thought to occur through the interaction of an SH2 domain from one STAT partner with the phosphotyrosine of another (32).

Other functional domains within STATs include a transactivation domain at the C-terminal end which is partially regulated by serine phosphorylation (33). The domains or means by which the STATs are translocated to the nucleus are unknown. Since STATs do not possess a recognizable nuclear localization signal, nuclear translocation may possibly occur through binding to an intermediary protein which carries the necessary sequence motif. As discussed below, recent evidence suggests that the N-terminus of STATs is required for interaction with unidentified phosphatases which inactivate STATs. Although this chapter is restricted to the JAK-STAT pathway, it should be noted that STATs can be activated independently of JAKs by receptor tyrosine kinases and other receptors, suggesting that STATs may function as adaptors to link receptor activation to other signaling pathways (34–36).

The generation of STAT-deficient mice has been immensely valuable in dissecting the biological roles of each STAT. In contrast to in vitro studies, in which specificity of signaling was difficult to detect, a nonredundant biological role for each STAT has been seen in STAT-deficient mice (37–45). STAT1−/− mice were completely unresponsive to IFN-α/β or IFN-γ but displayed a normal response to other cytokines which activate STAT1 in vitro (37,38). Similar studies have shown that STAT4 is specific for IL-12–mediated biological responses (40), whereas STAT6 is essential for IL-4 and IL-13 functions (43–45). A specific function for STAT5 in vivo has also been confirmed by the analysis of STAT5-deficient mice. Consistent with its isolation as a mammary growth factor, mice lacking STAT5A are unable to lactate and do not develop normal breast tissue (41). Mice lacking STAT5B undergo normal lactation but display abnormalities resulting from a block in growth hormone signaling, including growth retardation (42). STAT3, on the other hand, plays an essential role during early development, since STAT3 deficiency leads to early embryonic lethality (39).

REGULATION OF THE JAK-STAT PATHWAY

Suppressors of Cytokine Signaling

A functional cloning strategy designed to isolate inhibitors of cytokine signal-transduction pathways identified a new family of negative regulators, the suppressors of cytokine signaling (SOCS) family of proteins (46). The prototype of this family, SOCS-1, was isolated by its ability to inhibit macrophage differentiation of M1 myeloid leukemic cells in response to IL-6. Database searches revealed the presence of at least seven homologs of SOCS-1 (SOCS-2–SOCS-7, CIS), all of which share a similar structure, but overall share limited primary sequence identity (47–49). One such homolog is cytokine-inducible SH2-containing protein (CIS), identified previously as an early response gene and inhibitor of IL-3 signaling (50).

The eight SOCS proteins share a common structure, characterized by a central SH2 domain flanked by an N-terminal region of variable length and sequence and a conserved C-terminal motif termed the SOCS box (Fig. 1) (46,47). The SOCS box is defined by a stretch of 40 amino acids common to these eight proteins. Database searches have subsequently identified an additional 12 proteins containing a C-terminal SOCS box, which can be classified into different families of proteins based on the structural domains located N-terminal to the SOCS box (47). Instead of the SH2 domain present in the SOCS proteins, WSB proteins contain WD-40 repeats, ASB proteins contain ankyrin repeats, SPRY domains are found in SSB proteins, and a class of small guanosine triphosphatases (GTPases) also contains a C-terminal SOCS box (47).

SOCS proteins are not present constitutively within a cell, but their expression is rapidly induced by cytokine both in vitro and in vivo (46,47). The STAT proteins themselves have been implicated in regulating the transcription of the SOCS genes, with SOCS-1 expression probably controlled by STAT3 (51), whereas CIS expression is regulated by STAT5 (52). Once synthesized,

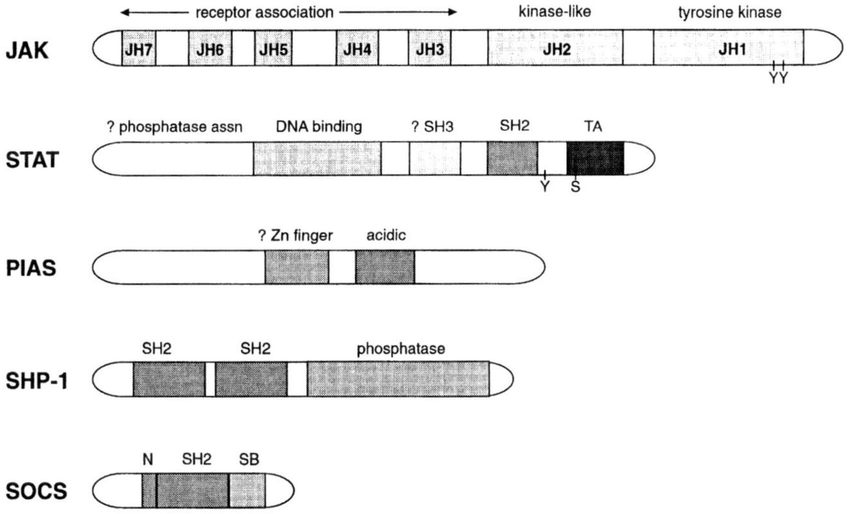

Figure 1 Domain structure of JAKs and STATs and proteins which regulate the JAK-STAT pathway. The putative function of each domain is indicated. Important tyrosine (Y) and serine (S) residues known to be phosphorylated are shown. TA, transactivation domain; N, N-terminal domain; SB, SOCS box.

the SOCS proteins act to suppress the signal transduction pathway that induced their expression, thereby completing a classic negative-feedback loop (Fig. 2).

The SOCS proteins suppress cytokine signaling in at least two distinct ways. Expression of SOCS-1 inhibited tyrosine phosphorylation of both the receptor component, gp130, and STAT3 in response to IL-6, suggesting that the primary site of action of SOCS-1 is upstream of both signaling molecules (46). Independent discoveries of SOCS-1, also referred to as JAK binding protein (JAB) and STAT-induced STAT inhibitor-1 (SSI-1) confirmed that SOCS-1 suppresses cytokine signaling by directly interacting with and inhibiting JAK kinase catalytic activity (51,53). In contrast, although SOCS-3 expression inhibited STAT3 activation in response to leukemia inhibitory factor (LIF), SOCS-3 was unable to inhibit JAK kinase activity in vitro, suggesting that SOCS-1 and SOCS-3 act in different ways to inhibit the JAK-STAT pathway (54). CIS, on the other hand, appears to inhibit signaling by directly binding to phosphorylated receptors, competing with signaling intermediates such as STAT5 for binding to the receptor (50,52).

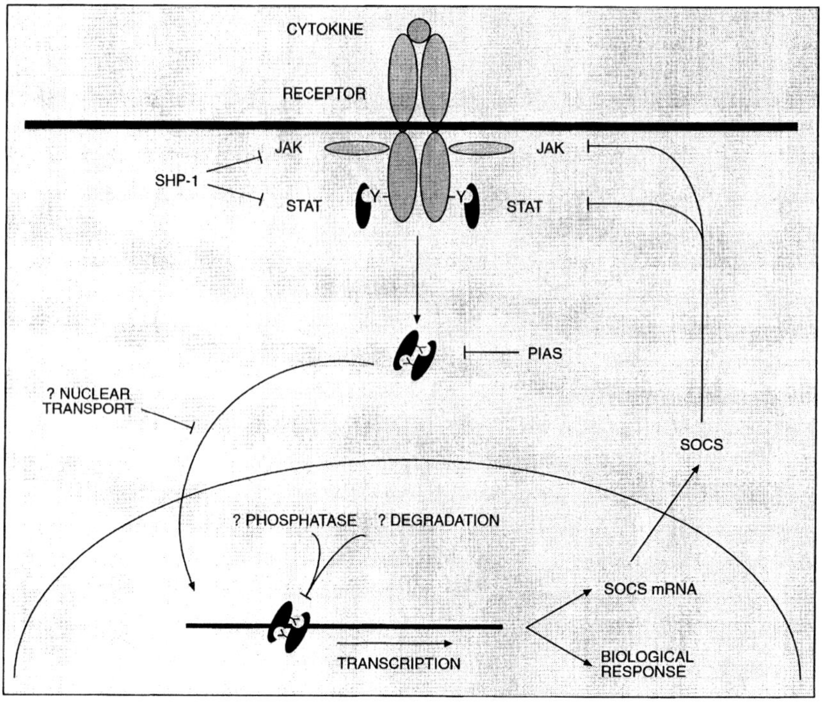

Figure 2 Negative regulation of the JAK-STAT pathway. Regulation of JAK activity by SHP-1 and SOCS is shown. Unlike SHP-1, which is present constitutively, SOCS mRNA is transcribed in response to cytokine stimulation. Regulation of STATs can occur at several levels. At the receptor, access to STAT-binding sites can be blocked either by dephosphorylation of tyrosines by SHP-1 or by SOCS competing with STATs for binding to phosphotyrosine residues. Activated STATs can be inhibited by PIAS by an unknown mechanism. In addition, transport of activated STAT dimers to the nucleus could be the target of regulation. Activated STAT dimers in the nucleus are probably also downregulated by degradation and dephosphorylation by unknown phosphatases.

SOCS Structure and Function

Initial structure-function analyses of SOCS-1 showed that the SH2 domain alone was insufficient for inhibiting cytokine signal transduction, indicating that at least one of the flanking domains was required for this function (53). Recent studies have further defined the functional domains of SOCS-1 and SOCS-3 (54,55). First, a functional SH2 domain was shown to be critical for

SOCS inhibition of signal transduction, since mutation of critical residues within the SH2 domain abolished this function (54,55). Given that overall the N-terminal domain is quite diverse within the SOCS family, it was surprising that deletion of this region of SOCS-1 abrogated protein function. More detailed analyses defined a region directly N-terminal of the SH2 domain (amino acids 50–78) critical for SOCS-1 function (54,55). Sequence alignment of this region of SOCS-1, SOCS-3, SOCS-4, and SOCS-5 revealed limited homology in this region among these four proteins (54,55).

Considering that both the N-terminal region and the SH2 domain of SOCS-1 are required for function, what is the specific role of each of these domains and how does SOCS-1 inhibit JAK activity? The SH2 domain of SOCS-1 has been shown to interact with the JH1 domain of JAK2, and this interaction is dependent on tyrosine phosphorylation of the JH1 domain (53). It is therefore likely that the SH2 domain interacts with JAK by binding to a phosphotyrosine residue within the JH1 domain. Coimmunoprecipitation studies showed that the SOCS-1 N-terminal domain was also able to bind to JAK2, albeit weakly, in the absence of the SH2 domain (54). Although the N-terminal and SH2 domains of SOCS-1 are required for recognition and binding to activated JAKs, current data suggest that it is the SOCS box which then acts to turn the signal off by targeting both the activated JAKs and the associated SOCS proteins for degradation (56).

The SOCS proteins would be expected to have a short half-life in the cell in order for a cell to respond to cytokine again after a previous cytokine-activation cycle. Indeed, early studies of CIS showed that this protein undergoes rapid turnover (50). More recently, CIS has been shown to be monoubiquitinated and subject to proteasomal degradation (57). Following the finding that the SOCS box motif binds to elongins B and C, which are thought to target proteins to proteasomal degradation (58,59), a model to explain the action of SOCS proteins has been proposed (56). In this model, SOCS proteins interact with activated signaling molecules such as JAKs via the N-terminus and SH2 domain. The SOCS box then brings elongins B and C into the complex, which are then thought either to interact directly with the proteasome or indirectly following ubiquitination (56). Since the SOCS box of other family members, including WSB-2 and ASB-2, also interacts with the elongin BC complex, it is likely that this motif serves to target all SOCS box–containing proteins and associated proteins for degradation (56). In this way, the SOCS box–containing proteins would function as adaptors to link activated proteins to the degradation machinery of the cell. Importantly, this model would result in degradation of the SOCS proteins themselves, which is required before a cell can respond to further cytokine stimulation. Kamura and colleagues have also demonstrated an interaction of elongins B and C with the SOCS box of several proteins, but in contrast to the model described above, they conclude that the interaction protects the SOCS box–containing proteins from degradation (60).

Little is known about specificity within the SOCS family of proteins. To date, most studies of SOCS function have depended on the use of in vitro overexpression systems. These studies have shown that expression of SOCS-1 suppresses signaling in response to a variety of cytokines, including LIF, oncostatin M (OSM), IFNγ, TPO, and growth hormone (GH) in addition to IL-6 (46,51,53,61–63). SOCS-3 expression also inhibits signaling in response to IL-6, LIF, GH, and IFN-γ (54,61,63,64) in addition to blocking leptin-induced signal transduction (65). CIS expression suppresses cell proliferation in response to IL-3 and EPO cell proliferation (50). However, SOCS-2 has yet to be shown to inhibit cytokine signal transduction pathways.

Recent reports have suggested that SOCS action may not be restricted to the JAK-STAT pathway. Although signaling through tyrosine kinase receptors does not appear to be regulated by SOCS (46,51,53), SOCS-1 has been shown to interact with and suppress the cytoplasmic tyrosine kinase, Tec (66). The association between SOCS-1 and Tec is phosphorylation independent, and it is mediated by the N-terminal region of SOCS-1 (66). Furthermore, the N-terminal domain of SOCS-7 appears to mediate an association between SOCS-7 and the adaptor molecule, Nck (67), and SOCS-2 reportedly interacts with the insulin-like growth factor I receptor (68). Further studies are needed to confirm that these associations occur in a physiological setting.

SOCS-1 Deficient Mice

Although SOCS-1 has been shown to suppress signaling in response to a broad spectrum of cytokines in vitro, the primary targets of SOCS-1 in vivo are unknown. The generation of SOCS-1 null mice has provided some insight into the physiological role of this negative regulator (69,70). SOCS-1 deficient mice were born in expected numbers and were indistinguishable from their normal littermates at birth. However, SOCS-1−/− mice became severely runted and died before weaning with hematopoietic infiltration of several organs and major abnormalities in the liver, including profound fatty degeneration of parenchymal cells which may be sufficient to account for the neonatal mortality of these mice (69).

A striking abnormality in SOCS-1−/− mice was severe lymphopenia evident in the bone marrow, spleen, and peripheral blood. The thymus was also reduced in size in SOCS-1−/− mice. FACS (fluorescence-activated cell sorter) analyses showed that the reduction in thymic cellularity did not reflect selective loss of a specific subset of thymocytes. In contrast, specific depletion of pre-B and mature B cells was evident in the bone marrow and spleen of SOCS-1 −/− mice. This defect did not represent an intrinsic inability of immature B cells to differentiate, since pre-B cells from SOCS-1 −/− mice were able to develop into Ig-positive cells in vitro with a similar efficiency to pre-B cells from littermates (69).

Since SOCS-1 acts to inhibit cytokine signaling, mice lacking this protein would be expected to exhibit inappropriate or prolonged responses to cytokine stimulation. The phenotype of the SOCS-1 −/− mice does not immediately indicate dysregulated responses to any particular cytokine. However, aspects of the disease seen in these mice, for instance, the specific depletion of B cells, suggest inappropriate signaling in response to an inhibitory cytokine. Indeed, overexpression of IFN-γ in vivo has been seen to cause very similar defects, including liver damage and B-cell loss (71,72). Alternatively, it has been proposed that SOCS-1 (SSI-1) functions to inhibit expression of the proapoptotic molecule Bax (70). Loss of this regulation in SOCS-1 −/− mice has been suggested to lead to increased apoptosis; possibly accounting for the depletion of lymphocytes (70). Further examination of the SOCS-1 −/− mice should help define which cytokine signal transduction pathways require regulation by SOCS-1 in vivo.

SHP-1

Tyrosine phosphorylation of signaling intermediates is a critical component of many signal transduction pathways. Since phosphorylation is a rapid and transient process, protein tyrosine phosphatases are likely candidates for molecules that regulate these signals. Indeed, phosphatase inhibitors have been shown to replace the action of cytokines in mitogenic responses to a certain extent (73). It is somewhat surprising, therefore, that few phosphatases have been identified that function as negative regulators of cytokine signaling (74). The best characterized of these is SHP-1.

SHP-1 is an SH2-containing protein tyrosine phosphatase expressed primarily in hematopoietic lineages. The critical importance of SHP-1 in regulating cytokine signaling is highlighted in motheaten mice, which lack functional SHP-1 expression due to mutations in the SHP-1 gene (75–77). Motheaten mice are hyperresponsive to hematopoietic growth factors, resulting in multiple hematopoietic abnormalities, including hyperproliferation and abnormal activation of granulocytes and macrophages. The broad spectrum of defects in the motheaten mouse reflects the fact that SHP-1 regulates a variety of growth factor– and cytokine-stimulated signaling pathways, including EPO, IL-3, IFN-α, CSF-1, and Steel factor (78–82).

SHP-1 has been most closely studied in the context of EPO signal transduction, and it appears that SHP-1 terminates EPO-induced proliferative responses by binding to the activated EPO receptor and dephosphorylating JAK2 (79). This interaction with JAK2 does not appear to require the SH2 domain (83). Additional substrates of SHP-1 have recently been identified, suggesting that the mechanism of SHP-1 function may be more complex (84,85). Furthermore, in other cytokine-receptor systems, SHP-1 inhibits signaling by direct dephosphorylation of the receptor itself (see Fig. 2) (86).

The importance of SHP-1 in regulating EPO signaling is indicated by a rare proliferative disorder of erythroid progenitor cells known as primary familial and congenital polycythemia (PFCP) (87). Most, but not all, patients with this disease harbor EPO receptor mutations, the majority of which result in a truncated cytoplasmic domain (88–90). This mutant receptor lacks a C-terminal negative regulatory domain of the receptor which is likely to contain the docking site for SHP-1 (91).

STAT Inhibitors

In addition to the mechanisms described above which turn off the signal upstream of STATs, activity of the STAT proteins themselves is tightly regulated. Activation of STATs in response to cytokine occurs transiently, persisting for at most a few hours after stimulation (92,93). STAT activity is likely to be regulated at several different levels; formation of active STAT dimers in the cytoplasm, transport of the dimers to the nucleus, and inactivation of active dimers by specific inhibitors, degradation, or dephosphorylation (see Fig. 2).

Protein Inhibitor of Activated STAT

The recent identification of the protein inhibitor of activated STAT (PIAS) family has introduced a novel mechanism of STAT regulation. PIAS1 was identified using a yeast two-hybrid screen for proteins that interact with STAT1 (94). EST database and library searches yielded an additional four family members which share greater than 50% homology (94,95). PIAS1, but not other PIAS proteins, specifically associates with STAT1 in response to ligand stimulation and inhibits the ability of STAT1 to bind DNA and activate transcription (94). Phosphorylation of STAT1 on the critical tyrosine residue Tyr-701 is required for the binding of PIAS1. Similarly, PIAS3 functions as a specific inhibitor of STAT3 (95), suggesting the possibility that each STAT protein may be regulated by a specific member of the PIAS family. PIASxα and PIASxβ differ only in their C-terminal regions, and they are likely differentially spliced products of a single gene. The ability of these family members, and that of PIASy, to inhibit STAT activity is unknown at present (94).

Several sequence domains are highly conserved within the PIAS family, including a putative zinc-binding motif and a highly acidic region. It is particularly intriguing that PIAS1 is essentially identical to Gu/RNA helicase II–binding protein (GBP), independently identified in a yeast two-hybrid screen as an interaction partner of Gu protein (RNA helicase II) (96). Gu/RNA helicase II is a nucleolar enzyme which has an adenosine triphosphate (ATP)–dependent RNA helicase activity exclusively in a 5′ to 3′ direction and a separate RNA folding activity (97). The function of GBP is unclear, but it may be involved in the regulation of Gu activity, since GBP appears to mediate proteolytic cleavage of the helicase (96). The apparent ability of a single protein to perform two such disparate functions remains to be reconciled.

How does PIAS inhibit STAT activity? Interaction of PIAS with STAT dimers may prevent DNA binding either by directly binding to the DNA-binding domain or by causing a conformational change so that the complex can no longer bind DNA. Alternatively, PIAS may bind to activated STAT monomers, preventing dimer formation, or may mediate the dissociation of active STAT dimers. It is interesting that PIAS3 inhibited the DNA-binding activity of STAT1-STAT3 heterodimers (95). Since STAT1 was not coimmunoprecipitated with PIAS3 (95), PIAS3 appears likely to bind to activated STAT3 monomers. It is presently unclear which cellular compartment the PIAS proteins are normally located, which may help distinguish between the possible mechanisms of action.

Phosphatases and Degradation

Activated STAT dimers translocate to the nucleus where they activate gene transcription. However, the presence of activated STATs in the nucleus is transient. The most likely mechanisms by which activated STATs are removed from the nucleus are either by dephosphorylation and recycling of STAT monomers to the cytoplasm or by degradation. Evidence in support of both models has been submitted in recent years. The use of phosphatase inhibitors, although not specifically showing that STATs are the targets of the putative phosphatases, has indicated a role for protein tyrosine phosphatases in suppressing responses to cytokine (98,99). Further evidence has shown that the half-life of phosphorylated STATs is rapid, but that the proteins themselves are stable for a much longer time (100,101). In contrast to STAT1, which appears to require the N-terminal region for its dephosphorylation, the C-terminal region of STAT5 may interact with phosphatases, since C-terminal truncated forms remain tyrosine phosphorylated for substantially longer periods than wild-type protein (102,103). Alternatively, proteolytic degradation may be involved in clearing the nucleus of activated STAT dimers. STAT activation is stabilized by treatment with proteasome inhibitors, but it is difficult to establish whether the proteasomes specifically target STATs or other intermediates in the pathway (100,104,105).

CONCLUSIONS

Until recently, our knowledge of how cells control their responses to cytokine has been limited. With the recent identification of several negative regulators, it has become increasingly clear that cells utilize a variety of mechanisms, targeted at different levels of the signal transduction pathway, to ensure that cytokine responses are controlled appropriately. The rate at which the signal is terminated would be due to the net effect of all of these regulatory mechanisms (see Fig. 2).

The timing and specificity of these regulators remains to be determined, as does the degree with which the inhibitors interact and communicate with each other. Unlike SOCS proteins, which are induced in response to cytokine, SHP-1 and PIAS proteins are present constitutively, and therefore may function in a more acute manner in response to cytokine. In addition, it is likely that the list of negative regulatory molecules is not yet complete. Dephosphorylation of the cytokine receptor, JAKs, and STATs must occur to completely turn off the signal. Aside from SHP-1, the phosphatases responsible for this regulation remain to be identified.

The fact that negative regulation targets a variety of different levels within the signal transduction pathway provides the opportunity to design small molecule effectors which may specifically modify the function of particular signaling intermediates. It is, however, too early to speculate on the effectiveness of targeting one particular signaling molecule in preference to another. In general terms, as our understanding of these molecules increases, it should be feasible to generate effectors which enhance the function of negative regulators to intervene in situations in which cytokine signaling needs to be controlled. It should be noted also that, in certain situations, it may be beneficial to block the action of negative regulators when prolonged signaling (e.g., a prolonged antiviral response) is a desirable outcome. In these circumstances, antagonists of negative regulators should prove invaluable.

ACKNOWLEDGMENTS

The original work described in this chapter was supported by the Anti-Cancer Council of Victoria, Melbourne, Australia, AMRAD Operations Pty. Ltd., Melbourne, Australia, The National Health and Medical Research Council, Canberra, Australia, The J.D. and L. Harris Trust, The National Institutes of Health, Bethesda, Maryland, and the Australian Federal Government Cooperative Research Centres Programme. The author is supported by a Postdoctoral Fellowship from the Australian Research Council.

REFERENCES

1. Pellegrini, S, Dusanter-Fourt, I. The structure, regulation and function of the Janus kinases (JAKs) and the signal transducers and activators of transcription (STATs). Eur J Biochem 248:615–633, 1997.
2. Heinrich, PC, Behrmann, I, Muller-Newen, G, Schaper, F, Graeve, L. Interleukin-6–type cytokine signalling through the gp130/Jak/STAT pathway. Biochem J 334:297–314, 1998.
3. Duhe, RJ, Farrar, WL, Structural and mechanistic aspects of Janus kinases: how the two-faced God wields a double-edged sword. J Interferon Cytokine Res 18: 1–15, 1998.
4. Lutticken, C, Wegenka, UM, Yuan, J, Buschmann, J, Schindler, C, Ziemiecki, A, Harpur, AG, Wilks, AF, Yasukawa, K, Taga, T, Kishimoto, T, Barbieri, G, Pelligrini, S, Sendtner, M, Heinrich, PC, Horn, F. Association of transcription factor APRF and protein kinase Jak1 with the interleukin-6 signal transducer gp130. Science 263:89–92, 1994.
5. Hanks, SK, Quinn, AM, Hunter, T. The protein kinase family: conserved features and deduced phylogeny of the catalytic domains. Science 241:42–52, 1988.
6. Gauzzi, MC, Velazquez, L, McKendry, R, Mogensen, KE, Fellous, M, Pelligrini, S. Interferon-alpha–dependent activation of Tyk2 requires phosphorylation of positive regulatory tyrosines by another kinase. J Biol Chem 271:20494–20500, 1996.
7. Feng, J, Witthuhn, BA, Matsuda, T, Kohlhuber, F, Kerr, IM, Ihle, JN. Activation of Jak2 catalytic activity requires phosphorylation of Y^{1007} in the kinase activation loop. Mol Cell Biol 17:2497–2501, 1997.
8. Velazquez, L, Mogensen, KE, Barbieri, G, Fellous, M, Uze, G, Pelligrini, S. Distinct domains of the protein tyrosine kinase tyk2 required for binding of interferon-alpha/beta and for signal transduction. J Biol Chem 270:3327–3334, 1995.
9. Luo, H, Rose, P, Barber, D, Hanratty, WP, Lee, S, Roberts, TM, D'Andrea,

AD, Dearolf, CR. Mutation in the Jak kinase JH2 domain hyperactivates Drosophila and mammalian Jak-Stat pathways. Mol Cell Biol 17:1562–1571, 1997.
10. Chen, M, Cheng, A, Chen, YQ, Hymel, A, Hanson, EP, Kimmel, L, Minami, Y, Taniguchi, T, Changelian, PS, O'Shea, JJ. The amino terminus of JAK3 is necessary and sufficient for binding to the common gamma chain and confers the ability to transmit interleukin 2–mediated signals. Proc Nat Acad Sci 94: 6910–6915, 1997.
11. Kohlhuber, F, Rogers, NC, Watling, D, Feng, J, Guschin, D, Briscoe, J, Witthuhn, BA, Kotenko, SV, Pestka, S, Stark, GR, Ihle, JN, Kerr, IM. A JAK1/JAK2 chimera can sustain alpha and gamma interferon responses. Mol Cell Biol 17:695–706, 1997.
12. Souyri, M, Vigon, I, Penciolelli, JF, Heard, JM, Tambourin, P, Wendling, F. A putative truncated cytokine receptor gene transduced by the myeloproliferative leukemia virus immortalizes hematopoietic progenitors. Cell 63:1137–1147, 1990.
13. Longmore, GD, Lodish, HF. An activating mutation in the murine erythropoietin receptor induces erythroleukemia in mice: a cytokine receptor superfamily oncogene. Cell 67:1089–1102, 1991.
14. Macchi, P, Villa, A, Gillani, S, Sacco, MG, Frattini, A, Porta, F, Ugazio, AG, Johnson, JA, Candotti, F, O'Shea, JJ, Vezzoni, P, Notarangelo, LD. Mutations of Jak-3 gene in patients with autosomal severe combined immune deficiency (SCID). Nature 377:65–68, 1995.
15. Russell, SM, Tayebi, N, Nakajima, H, Riedy, MC, Roberts, JL, Aman, MJ, Migone, TS, Noguchi, M, Markert, ML, Buckley, RH, O'Shea, JJ, Leonard, WJ. Mutation of Jak3 in a patient with SCID: essential role of Jak3 in lymphoid development. Science 270:797–800, 1995.
16. Thomis, DC, Gurniak, CB, Tivol, E, Sharpe, AH, Berg, LJ. Defects in B lymphocyte maturation and T lymphocyte activation in mice lacking Jak3. Science 270:794–797, 1995.
17. Nosaka, T, van Deursen, JM, Tripp, RA, Thierfelder, WE, Witthuhn, BA, McMickle, AP, Doherty, PC, Grosveld, GC, Ihle, JN. Defective lymphoid development in mice lacking Jak3. Science 270:800–802, 1995.
18. Park, SY, Saijo, K, Takahashi, T, Osawa, M, Arase, H, Hirayama, N, Miyake, K, Nakauchi, H, Shirasawa, T, Saito, T. Developmental defects of lymphoid cells in Jak3 kinase–deficient mice. Immunity 3:771–782, 1995.
19. Stahl, N, Boulton, TG, Farruggella, T, Ip, NY, Davis, S, Witthuhn, BA, Quelle, FW, Silvennoinen, O, Barbiere, G, Pelligrini, S, Ihle, JN, Yancopoulous, GD. Association and activation of Jak-Tyk kinases by CNTF-LIF-OSM-IL-6 beta receptor components. Science 263:92–95, 1994.
20. Guschin, D, Rogers, N, Briscoe, J, Witthuhn, BA, Watling, D, Horn, F, Pelligrini, S, Yasukawa, K, Heinrich, P, Stark, GR, Ihle, JN, Kerr, IM. A major role for the protein tyrosine kinase Jak1 in the Jak/STAT signal transduction pathway in response to interleukin-6. EMBO J 14:1421–1429, 1995.

21. Rodig, SJ, Meraz, MA, White, JM, Lampe, PA, Riley, JK, Arthur, CD, King, KL, Sheehan, KCF, Yin, L, Pennica, D, Johnson, EM Jr, Schreiber, RD. Disruption of the Jak1 gene demonstrates obligatory and nonredundant roles of the Jaks in cytokine-induced biologic responses. Cell 93:373–383, 1998.
22. Parganas, E, Wang, D, Stravopodis, D, Topham, DJ, Marine, J-C, Teglund, S, Vanin, EF, Bodner, S, Colamonici, OR, van Deursen, JM, Grosveld, G, Ihle, JN. Jak2 is essential for signalling through a variety of cytokine receptors. Cell 93:385–395, 1998.
23. Neubauer, H, Cumano, A, Muller, M, Wu, H, Huffstadt, U, Pfeffer, K. Jak2 deficiency defines an essential development checkpoint in definitive hematopoiesis. Cell 93:397–409, 1998.
24. Ihle, JN. STATs: Signal transducers and activators of transcription. Cell 84: 331–334, 1996.
25. Darnell, JE Jr. STATs and gene regulation. Science 277:1630–1635, 1997.
26. Yan, R, Small, S, Desplan, C, Dearolf, CR, Darnell, JE Jr. Identification of a Stat gene that functions in Drosophila development. Cell 84:421–430, 1996.
27. Hou, XS, Melnick, MB, Perrimon, N. Marelle acts downstream of the Drosophila HOP/JAK kinase and encodes a protein similar to the mammalian STATs. Cell 84:411–419, 1996.
28. Pawson, T. Protein modules and signalling networks. Nature 373:573–580, 1995.
29. Gouilleux, J, Wakao, H, Mundt, M, Groner, B. Prolactin induces phosphorylation of Tyr694 of Stat5 (MGF), a prerequisite for DNA binding and induction of transcription. EMBO J 13:4361–4369, 1994.
30. Improta, T, Schindler, C, Horvath, CM, Kerr, IM, Stark, GR, Darnell, JE Jr. Transcription factor ISGF-3 formation requires phosphorylated Stat91 protein, but Stat113 protein is phosphorylated independently of Stat91 protein. Proc Natl Acad Sci USA 91:4776–4780, 1994.
31. Shuai, K, Stark, GR, Kerr, IM, Darnell, JE Jr. A single phosphotyrosine residue of Stat91 required for gene activation by interferon-gamma. Science 261:1744–1746, 1993.
32. Shuai, K, Horvath, CM, Huang, LH, Qureshi, SA, Cowburn, D, Darnell, JE Jr. Interferon activation of the transcription factor Stat91 involves dimerization through SH2-phosphotyrosyl peptide interactions. Cell 76:821–828, 1994.
33. Wen, Z, Zhong, Z, Darnell, JE Jr. Maximal activation of transcription by Stat1 and Stat3 requires both tyrosine and serine phosphorylation. Cell 82:241–250, 1995.
34. Leaman, DW, Pisharody, S, Flickinger, TW, Commane, MA, Schlessinger, J, Kerr, IM, Levy, DE, Stark, GR. Roles of JAKs in activation of STATs and stimulation of c-fos gene expression by epidermal growth factor. Mol Cell Biol 16:369–375, 1996.
35. David, M, Wong, L, Flavell, R, Thompson, SA, Wells, A, Larner, AC, Johnson,

GR. STAT activation by epidermal growth factor (EGF) and amphiregulin. Requirement for the EGF receptor kinase but not for tyrosine phosphorylation sites or JAK1. J Biol Chem 271:9185–9188, 1996.
36. Pfeffer, LM, Mullersman, JE, Pfeffer, SR, Murti, A, Shi, W, Yang, CH. STAT3 as an adaptor to couple phosphatidylinositol 3-kinase to the IFNAR1 chain of the type I interferon receptor. Science 276:1418–1420, 1997.
37. Durbin, JE, Hackenmiller, R, Simon, MC, Levy, DE. Targeted disruption of the mouse Stat1 gene results in compromised innate immunity to viral disease. Cell 84:443–450, 1996.
38. Meraz, MA, White, JM, Sheehan, KCF, Bach, EA, Rodig, SJ, Dighe, AS, Kaplan, DH, Riley, JK, Greenlund, AC, Campbell, D, Carver-Moore, K, Dubois, RN, Clark, R, Aguet, M, Schreiber, RD. Targeted disruption of the Stat1 gene in mice reveals unexpected physiologic specificity in the JAK-STAT signaling pathway. Cell 84:431–442, 1996.
39. Takeda, K, Noguchi, K, Shi, W, Tanaka, T, Matsumoto, M, Yoshida, N, Kishimoto, T, Akira, S. Targeted disruption of the mouse Stat3 gene leads to early embryonic lethality. Proc Nat Acad Sci USA 94:3801–3804, 1997.
40. Kaplan, MH, Sun, Y-L, Hoey, T, Grusby, MJ. Impaired IL-12 responses and enhanced development of Th2 cells in Stat4-deficient mice. Nature 382:174–177, 1996.
41. Liu, X, Robinson, GW, Wagner, K-U, Garrett, L, Wynshaw-Boris, A, Henninghausen, L. Stat5a is mandatory for adult mammary gland development and lactogenesis. Genes Dev 11:179–186, 1997.
42. Udy, GB, Towers, RP, Snell, RG, Wilkins, RJ, Park, S-H, Ram, PA, Waxman, DJ, Davey, HW. Requirement of STAT5b for sexual dimorphism of body growth rates and liver gene expression. Proc Natl Acad Sci USA 94:7239–7244, 1997.
43. Takeda, K, Tanaka, T, Shi, W, Matsumoto, M, Minami, M, Kashiwamura, S, Nakanishi, L, Yoshida, N, Kishimoto, T, Akira, S. Essential role of Stat6 in IL-4 signalling. Nature 380:627–630, 1996.
44. Shimoda, K, Van Deursen, J, Sangster, MY, Sarawar, SR, Carson RT, Tripp, RA, Chu, C, Quelle, FW, Nosaka, T, Vignali, DAA, Doherty, PC, Grosveld, G, Paul, WE, Ihle, JN. Lack of IL-4–induced Th2 response and IgE class switching in mice with disrupted Stat6 gene. Nature 380:630–633, 1996.
45. Kaplan, MH, Schindler, U, Smiley, ST, Grusby, MJ. Stat6 is required for mediating responses to IL-4 and for the development of Th2 cells. Immunity 4:313–319, 1996.
46. Starr, R, Willson, TA, Viney, EM, Murray, LJ, Rayner, JR, Jenkins, BJ, Gonda, TJ, Alexander, WS, Metcalf, D, Nicola, NA, Hilton, DJ. A family of cytokine-inducible inhibitors of signalling. Nature 387:917–921, 1997.
47. Hilton, DJ, Richardson, RT, Alexander, WS, Viney, EM, Willson, TA, Sprigg, NS, Starr, R, Nicholson, SE, Metcalf, D, Nicola, NA. Twenty proteins containing a C-terminal SOCS box form five structural classes. Proc Natl Acad Sci USA 95:114–119, 1998.

48. Masuhara, M, Sakamoto, H, Matsumoto, A, Suzuki, R, Yasukawa, H, Mitsui, K, Wakioka, T, Tanimura, S, Sasaki, A, Misawa, H, Yokouchi, M, Ohtsubo, M, Yoshimura, A. Cloning and characterization of novel CIS family genes. Biochem Biophys Res Commun 239:439–446, 1997.
49. Minamoto, S, Ikegame, K, Ueno, K, Narazaki, M, Naka, T, Yamamoto, H, Matsumoto, T, Saito, H, Hosoe, S, Kishimoto, T. Cloning and functional analysis of new members of STAT induced STAT inhibitor (SSI) family: SSI-2 and SSI-3. Biochem Biophys Res Commun 237:79–83, 1997.
50. Yoshimura, A, Ohkubo, T, Kiguchi, T, Jenkins, NA, Gilbert, DJ, Copeland, NG, Hara, T, Miyajima, A. A novel cytokine-inducible gene CIS encodes an SH2-containing protein that binds to tyrosine-phosphorylated interleukin 3 and erythropoietin receptors. EMBO J 14:2816–2826, 1995.
51. Naka, T, Narazaki, M, Hirata, M, Matsumoto, T, Minamoto, S, Aono, A, Nishimoto, N, Kajita, T, Taga, T, Yoshizaki, K, Akira, S, Kishimoto, T. Structure and function of a new STAT-induced STAT inhibitor. Nature 387:924–929, 1997.
52. Matsumoto, A, Masuhara, M, Mitsui, K, Yokouchi, M, Ohtsubo, M, Misawa, H, Miyajima, A, Yoshimura, A. CIS, a cytokine inducible SH2 protein, is a target of the JAK-STAT5 pathway and modulates STAT5 activation. Blood 89:3148–3154, 1997.
53. Endo, TA, Masuhara, M, Yokouchi, M, Suzuki, R, Sakamoto, H, Mitsui, K, Matsumoto, A, Tanimura, S, Ohtsubo, M, Misawa, H, Miyazaki, T, Leonor, N, Taniguchi, T, Fujita, T, Kanakura, Y, Komiya, S, Yoshimura, A. A new protein containing an SH2 domain that inhibits JAK kinases. Nature 387:921–924, 1997.
54. Nicholson, SE, Willson, TA, Farley, A, Starr, R, Zhang, J-G, Baca, M, Alexander, WS, Metcalf, D, Hilton, DJ, Nicola, NA. Mutational analyses of the SOCS proteins suggest a dual domain requirement but distinct mechanisms for inhibition of LIF and IL-6 signal transduction. EMBO J 18:375–385, 1999.
55. Narazaki M, Fujimoto, M, Matsumoto, T, Morita, Y, Saito, H, Kajita, T, Yoshizaki, K, Naka, T, Kishimoto, T. Three distinct domains of SSI-1/SOCS-1/JAB protein are required for its suppression of interleukin-6 signaling. Proc Natl Acad Sci 95:13130–13134, 1998.
56. Zhang, J-G, Farley, A, Nicholson, SE, Willson, TA, Zugaro, LM, Simpson, RJ, Moritz, RL, Cary, D, Richardson, R, Hausmann, G, Kile, BJ, Kent, SBH, Alexander, WS, Metcalf, D, Hilton, DJ, Nicola, NA, Baca, M. The conserved SOCS box motif in suppressors of cytokine signaling binds to elongins B and C and may couple bound proteins to proteasomal degradation. Proc Natl Acad Sci USA 96:1999.
57. Verdier, F, Chretien, S, Muller, O, Varlet, P, Yoshimura, A, Gisselbrecht, S, Lacombe, C, Mayeux, P. Proteasomes regulate erythropoietin receptor and signal transducer and activator of transcription 5 (STAT5) activation. J Biol Chem 273:28185–28190, 1998.
58. Conaway, JW, Kamura, T, Conaway, RC. The Elongin BC complex and the

von Hippel-Lindau tumor suppressor protein. Biochim Biophys Acta 1377: M49–M54, 1998.
59. Kaelin, WGJ, Maher, ER. The VHL tumour-suppressor gene paradigm. Trends Genet 14:423–426, 1998.
60. Kamura, T, Sato, S, Haque, D, Liu, L, Kaelin, WG Jr, Conaway, RC, Conaway, JW. The elongin BC complex interacts with the conserved SOCS-box motif present in members of the SOCS, ras, WD-40 repeat and ankyrin repeat families. Genes Dev 12:3872–3881, 1998.
61. Adams, TE, Hansen, JA, Starr, R, Nicola, NA, Hilton, DJ, Billestrup, N. Growth hormone preferentially induces the rapid, transient expression of SOCS-3, a novel inhibitor of cytokine receptor signaling. J Biol Chem 273: 1285–1287, 1998.
62. Sakamoto, H, Yasukawa, H, Masuhara, M, Tanimura, S, Sasaki, A, Yuge, K, Ohtsubo, M, Ohtsuka, A, Fujita, T, Ohta, T, Furukawa, Y, Iwase, S, Yamada, H, Yoshimura, A. A Janus kinase inhibitor, JAB, is an interferon-gamma-inducible gene and confers resistance to interferons. Blood 92:1668–1676, 1998.
63. Song, MM, Shuai, K. The suppressor of cytokine signaling (SOCS) 1 and SOCS3 but not SOCS2 proteins inhibit interferon-mediated antiviral and antiproliferative activities. J Biol Chem 273:35056–35062, 1998.
64. Auernhammer, CJ, Chesnokova, V, Bousquet, C, Melmed, S. Pituitary corticotroph SOCS-3—novel intracellular regulation of leukemia-inhibitory factor–mediated proopiomelanocortin gene expression and adrenocorticotropin secretion. Mol Endocrinol 12:954–961, 1998.
65. Bjorbaek, C, Elmquist, JK, Frantz, JD, Shoelson, SE, Flier, JS. Identification of SOCS-3 as a potential mediator of central leptin resistance. Mol Cell 1:619–625, 1998.
66. Ohya, K, Kajigaya, S, Yamashita, Y, Miyazato, A, Hatake, K, Miura, Y, Ikeda, U, Shimada, K, Ozawa, K, Mano, H. SOCS-1/JAB/SSI-1 can bind to and suppress Tec protein–tyrosine kinase. J Biol Chem 272:27178–27182, 1997.
67. Matuoka, K, Miki, H, Takahashi, K, Takenawa, T. A novel ligand for an SH3 domain of the adaptor protein Nck bears an SH2 domain and nuclear signalling motifs. Biochem Biophys Res Commun 239:488–492, 1997.
68. Dey, BR, Spence, SL, Nissley, P, Furlanetto, RW. Interaction of human suppressor of cytokine signalling (SOCS)-2 with the insulin-like growth factor-I receptor. J Biol Chem 273:24095–24101, 1998.
69. Starr, R, Metcalf, D, Elefanty, AG, Brysha, M, Willson, TA, Nicola, NA, Hilton, DJ, Alexander, WA. Liver degeneration and lymphoid deficiencies in mice lacking suppressor of cytokine signaling-1. Proc Natl Acad Sci USA 95:14395–14399, 1998.
70. Naka, R, Matsumoto, T, Narazaki, M, Fujimoto, M, Morita, Y, Ohsawa, Y, Saito, H, Nagasawa, T, Uchiyama, Y, Kishimoto, T. Accelerated apoptosis of lymphocytes by augmented induction of Bax in SSI-1 (STAT-induced STAT inhibitor-1) deficient mice. Proc Natl Acad Sci USA 95:15577–15582, 1998.

71. Toyonaga, T, Hino, O, Sugai, S, Wakasugi, A, Abe, K, Shichiri, M, Yamamura, K. Chronic active hepatitis in transgenic mice expressing interferon-γ in the liver. Proc Natl Acad Sci USA 91:614–618, 1994.
72. Young, HA, Klinman, DM, Reynolds, DA, Grzegorzewski, KJ, Nii, A, Ward, JM, Winkler-Pickett, RT, Ortaldo, JR, Kenny, JJ, Komschlies, KL. Bone marrow and thymus expression of interferon-γ results in severe B-cell lineage reduction, T-cell lineage alterations and hematopoietic progenitor deficiencies. Blood 89:583–595, 1997.
73. Tojo, A, Kasuga M, Urabe, A, Takaku, F. Vanadate can replace interleukin 3 for transient growth of factor-dependent cells. Exp Cell Res 171:16–23, 1987.
74. Frearson, JA, Alexander DR. The role of phosphotyrosine phosphatases in haematopoietic cell signal transduction. BioEssays 19:417–427, 1997.
75. Shultz, LD, Schweitzer, PA, Rajan, TV, Yi, T, Ihle, JN, Matthews, RJ, Thomas, ML, Beier, DR. Mutations at the murine motheaten locus are within the hematopoietic cell protein-tyrosine phosphatase (Hcph) gene. Cell 73:1445–1454, 1993.
76. Tsui, HW, Siminovitch, KA, de Souza, L, Tsui, FW. Motheaten and viable motheaten mice have mutations in the haematopoietic cell phosphatase gene. Nat Genet 4:124–129, 1993.
77. Shultz, LD, Rajan, TV, Greiner, DL. Severe defects in immunity and hematopoiesis caused by SHP-1 protein–tyrosine-phosphatase deficiency. Trends Biotechnol 15:302–307, 1997.
78. Yi, T, Mui, AL, Krystal, G, Ihle, JN. Hematopoietic cell phosphatase associates with the interleukin-3 (IL-3) receptor beta chain and down-regulates IL-3–induced tyrosine phosphorylation and mitogenesis. Mol Cell Biol 13:7577–7586, 1993.
79. Klingmuller, U, Lorenz, U, Cantley, LC, Neel, BG, Lodish, HF. Specific recruitment of SH-PTP1 to the erythropoietin receptor causes inactivation of JAK2 and termination of proliferative signals. Cell 80:729–738, 1995.
80. David, M, Chen, HE, Goelz, S, Larner, AC, Neel, BG. Differential regulation of the alpha/beta interferon–stimulated Jak/Stat pathway by the SH2 domain–containing tyrosine phosphatase SHPTP1. Mol Cell Biol 15:7050–7058, 1995.
81. Chen, HE, Chang, S, Trub, T, Neel, BG. Regulation of colony-stimulating factor 1 receptor signaling by the SH2 domain-containing tyrosine phosphatase SHPTP1. Mol Cell Biol 16:3685–3697, 1996.
82. Paulson, RF, Vesely, S, Siminovitch, KA, Bernstein, A. Signalling by the W/Kit receptor tyrosine kinase is negatively regulated in vivo by the protein tyrosine phosphatase Shp1. Nat Genet 13:309–315, 1996.
83. Jiao, H, Berrada, K, Yang, W, Tabrizi, M, Platanias, LC, Yi, T. Direct association with and dephosphorylation of Jak2 kinase by the SH2-domain containing protein tyrosine phosphatase SHP-1. Mol Cell Biol 16:6985–6992, 1996.
84. Timms, JF, Carlberg, K, Gu, H, Chen, H, Kamatkar, S, Nadler, MJ, Rohrschneider, LR, Neel, BG. Identification of major binding proteins and substrates

for the SH2-containing protein tyrosine phosphatase SHP-1 in macrophages. Mol Cell Biol 18:3838–3850, 1998.

85. Yang, W, Tabrizi, M, Berrada, K, Yi, T. SHP-1 phosphatase C-terminus interacts with novel substrates p32/p30 during erythropoietin and interleukin-3 mitogenic responses. Blood 91:3746–3755, 1998.
86. Keilhack, H, Tenev, T, Nyakatura, E, Godovac-Zimmerman, J, Nielsen, L, Seedorf, K, Bodmer, FD. Phosphotyrosine 1173 mediates binding of the protein-tyrosine phosphatase SHP-1 to the epidermal growth factor receptor and attenuation of receptor signaling. J Biol Chem 273:24839–24846, 1998.
87. Prchal, JT, Sokol, L. ''Benign erythrocytosis'' and other familial and congenital polycythemias. Eur J Haematol 57:263–268, 1996.
88. de la Chapelle, A, Traskelin, AL, Juvonen, E. Truncated erythropoietin receptor causes dominantly inherited benign erythrocytosis. Proc Natl Acad Sci USA 90:4495–4499, 1993.
89. Sokol, L, Luhovy, M, Guan, Y, Prchal, JF, Semenza, GL, Prchal, JT. Primary familial polycythemia: a frameshift mutation in the erythropoietin receptor gene and increased sensitivity of erythroid progenitors to erythropoietin. Blood 86: 15–22, 1995.
90. Kralovics, R, Indrak, R, Stopka, T, Berman, BW, Prchal, JF, Prchal, JT. Two new EPO receptor mutations: truncated EPO receptors are most frequently associated with primary familial and congenital polycythemias. Blood 90:2057–2061, 1997.
91. Yi, T, Zhang, J, Miura, O, Ihle, JN. Hematopoietic cell phosphatase associates with erythropoietin (Epo) receptor after Epo-induced receptor tyrosine phosphorylation: identification of potential binding sites. Blood 85:87–95, 1995.
92. Shuai, K, Schindler, C, Prezioso, VR, Darnell, JE Jr. Activation of transcription by IFN-γ: tyrosine phosphorylation of a 91kDa DNA binding protein. Science 259:1808–1812, 1992.
93. Pallard, C, Gouilleux, F, Benit, L, Cocault, L, Souyri, M, Levy, D, Groner, B, Gisselbrecht, S, Dunsanter-Fourt, I. Thrombopoietin activates a STAT5-like factor in hematopoietic cells. EMBO J 14:2847–2856, 1995.
94. Liu, B, Liao, J, Rao, X, Kushner, SA, Chung, CD, Chang, DD, Shuai, K. Inhibition of Stat1-mediated gene activation by PIAS1. Proc Natl Acad Sci USA 95: 10626–10631, 1998.
95. Chung, CD, Liao, J, Liu, B, Rao, X, Jay, P, Berta, P, Shuai, K. Specific inhibition of Stat3 signal transduction by PIAS3. Science 278:1803–1805, 1997.
96. Valdez, BC, Henning, D, Perlaky, L, Busch, RK, Busch, H. Cloning and characterization of Gu/RH-II binding protein. Biochem Biophys Res Commun 234: 335–340, 1997.
97. Flores-Rozas, H, Hurwitz, J. Characterization of a new RNA helicase from nuclear extracts of HeLa cells which translocates in the 5′ to 3′ direction. J Biol Chem 268:21372–21383, 1993.
98. David, M, Grimley, PM, Finbloom DS, Larner AC. A nuclear tyrosine phospha-

tase downregulates interferon-induced gene expression. Mol Cell Biol 13: 7515–7521, 1993.

99. Haque, SJ, Flati, V, Deb, A, Williams, BRG. Roles of protein-tyrosine phosphatases in Stat1α-mediated cell signalling. J Biol Chem 270:25709–25714, 1995.
100. Haspel, RL, Salditt-Georgieff, M, Darnell, JE Jr. The rapid inactivation of nuclear tyrosine phosphorylated Stat1 depends upon a protein tyrosine phosphatase. EMBO 15:6262–6268, 1996.
101. Lee, C-K, Bluyssen, HAR, Levy, DE. Regulation of interferon-α responsiveness by the duration of Janus kinase activity. J Biol Chem 272:21872–21877, 1997.
102. Shuai, K, Liao, J, Song, MM. Enhancement of antiproliferative activity of gamma interferon by the specific inhibition of tyrosine dephosphorylation of Stat1. Mol Cell Biol 16:4932–4941, 1996.
103. Wang, D, Stravopodis, D, Teglund, S, Kitazawa, J, Ihle, JN. Naturally occurring dominant negative variants of Stat5. Mol Cell Biol 16:6141–6148, 1996.
104. Kim, TK, Maniatis, T. Regulation of interferon-γ–activated STAT1 by the ubiquitin-proteasome pathway. Science 273:1717–1719, 1996.
105. Yu, C-L, Burakoff, SJ. Involvement of proteasomes in regulating Jak/STAT pathways upon interleukin-2 stimulation. J Biol Chem 272:14017–14020, 1997.

11

Inhibition of Vascular Endothelial Growth Factor (VEGF) and Stem-Cell Factor (SCF) Receptor Kinases as Therapeutic Targets for the Treatment of Human Diseases

Kenneth E. Lipson, Li Sun, Congxin Liang, and Gerald McMahon
SUGEN, Inc., South San Francisco, California

INTRODUCTION

The receptors for cytokines and growth factors are required to propagate intracellular signals from the outside to the inside of the cell. In a general sense, receptor-mediated signaling can lead to a wide variety of cellular changes. Receptors that contain intrinsic tyrosine kinase catalytic activity (RTK) have been and continue to be associated with human diseases (Plowman et al., 1994). Such receptors undergo conformational changes that activate the kinase catalytic domain on interaction with their cytokine ligands. This catalytic activation leads to adenosine triphosphate (ATP)–dependent phosphorylation of tyrosine residues located on the receptor and nearby protein substrates leading to a cascade of protein association, downstream catalytic activities, and intracellular movement of proteins. The result of the activation of this cascade leads to changes in cell metabolism, movement, mitosis, adhesion, differentiation, morphology, and survival. In human diseases, increased RTK activity has been shown to result from (1) aberrant protein expression of the ligand or RTK or both; (2) activating mutations of RTK genes leading to constitutive (ligand-independent) receptor forms; and (3) RTK gene amplification resulting in overexpression of the protein. Such changes are often cell type specific and may be localized to those cells involved in the pathophysiology associated with the diseased tissue. In this regard, RTK have been implicated as therapeutic targets for inhibition of human cancers, cardiovascular diseases, growth and metabolic disorders, and inflammatory syndromes.

In this chapter, we have focused our attention on RTK that bind stem cell factor (SCF) and vascular endothelial growth factor (VEGF) cytokines in order to provide recent evidence to support an association of increased activities for these RTK with human disease. Table 1 summarizes the cell type distribution and diseases associated with these receptors. To reduce the number of references, we have cited previous literature reviews that focus on the biology of these RTK forms. Our objective in this chapter is to provide a rationale to use small synthetic inhibitors to block the catalytic function of these RTK in order to ameliorate clinical symptoms and improve patient outcome. This objective derives, in part, from the recent discovery of synthetic inhibitors of RTK that allow for pharmacological regulation of RTK activities, including antitumor efficacy in rodent models of cancer when these compounds were administered by the parenteral or oral route (McMahon et al., 1998). Moreover, some of these compounds have entered clinical evaluation for the treatment of human cancers and metastasis and have exhibited good tolerability (Strawn and Shawver, 1998).

SCF RECEPTOR *c-kit*

For recent reviews on c-*kit*, SCF, and their role in normal cellular function and pathology, see the following recent publications and the references therein: Tsujimura, 1996; Broudy, 1997; Loveland and Schlatt, 1997; Pignon, 1997; Vliagoftis et al., 1997; Lyman and Jacobsen, 1998. The receptor tyrosine kinase c-*kit* was originally identified as the cellular homolog of the viral oncogene v-*kit* of the Hardy-Zuckerman 4 feline sarcoma virus (Besmer et al., 1996; Yarden et al., 1987; Qiu et al., 1988). It is closely related to receptors for colony-stimulating factor-1 (CSF-1) and platelet-derived growth factor (PDGF) with respect to protein structure (five immunoglobulin-like motifs in the extracellular domain and a cytoplasmic "split" kinase domain) (Fig. 1) and in the primary amino acid sequence of the kinase domain (see the above reviews and references therein). Mapping experiments have identified c-*kit* as the protein expressed by the white-spotting (W) genetic locus on mouse chromosome 5 (Chabot et al., 1988; Geissler et al., 1988). Mice with mutations

Table 1 SCF and VEGF Receptors, Cell Types, and Human Diseases

RTK	Cell types[a]	Implicated diseases	
SCF receptor	Hematopoietic stem cells	Cancer	acute myelogenous leukemia, chronic myelogenous leukemia
	Melanocytes		small cell lung cancer
	Germ cells		seminoma
	Interstitial cells of Cajal		gastrointestinal stromal tumors
	Neural cells		central nervous system tumors
	Mast cells	Mast cell diseases	mastocytosis
			asthma and allergy
VEGF receptors	Vascular endothelial cells	Angiogenesis	tumor growth and metastasis
	Lymphatic endothelial cells		rheumatoid arthritis
	Monocytes and macrophages		psoriasis
	Smooth muscle cells		ocular diseases
		Cancer	acute myelogenous leukemia
			Kaposi's sarcoma
		Inflammation	inflammatory bowel disease
			HIV-associated dementia
			sarcoidosis
			rheumatoid arthritis
			allergy and asthma
		Cardiovascular	atherosclerosis
			restenosis

[a] The normal cell types in which the receptor has been observed to be expressed and mediate a biological function.

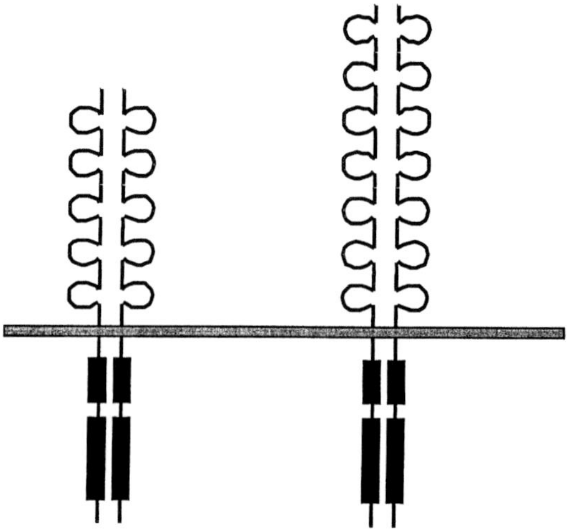

CSF-1r/c-*fms* **VEGFr1/*Flt*1**
SCFr/c-*kit* **VEGFr2/*Flt*2/*Flk*1/KDR**
***Flt*3/*Flk*2** **VEGFr3/*Flt*4**

Figure 1 Structural motifs characteristic of SCF and VEGF receptors. The receptor for SCF is related to the RTK for PDGF and CSF-1 containing five immunoglobulin motifs in the extracellular domain and a cytoplasmic kinase containing a long nonkinase region insert (''split kinase''). The receptors for VEGF ligands contain seven immunoglobulin domains in the extracellular domain.

at the W locus (W mice) have characteristic white spots on their abdomen resulting from lack of skin pigmentation. W mice exhibit other abnormalities, including reproductive difficulties associated with abnormal germ cell development, diminished numbers of tissue mast cells, and macrocytic anemia resulting from defective hematopoiesis. These observations provide genetic evidence for the critical role of c-*kit* in the development of melanocytes and mast, germ, and hematopoietic cells.

Mice exhibiting the Steel (Sl) mutation have a very similar phenotype as those with W mutations. However, the Sl locus was found to map to mouse chromosome 10, indicating that the defect resulted from mutation of a protein other than c-*kit* and suggesting that this locus may encode the ligand for this receptor. Cloning of the cDNA for the ligand of c-*kit* (Huang et al., 1990; Martin et al., 1990; Williams et al., 1990) enabled the subsequent confirmation of this hypothesis. The protein encoded by the Sl locus has been called *kit* ligand (KL), stem cell factor (SCF), or mast cell growth factor (MGF) based on its biological properties used to identify it (see reviews). For simplicity, we will use SCF to designate the ligand for the c-*kit* RTK. SCF is synthesized as a transmembrane protein with a molecular weight of 220 or 248 D, depending on alternative splicing of the mRNA to encode exon 6. The larger protein can be proteolytically cleaved to form a soluble, glycosylated protein which noncovalently dimerizes. Both the soluble and membrane-bound forms of SCF can bind to and activate c-*kit*. However, there have been reports indicating that the soluble and membrane-associated forms may induce somewhat different biological responses (Miyazawa et al., 1995; Williams 1997). Under normal circumstances, SCF typically activates c-*kit* by a paracrine mechanism. For example, in the skin, SCF is predominantly expressed by fibroblasts, keratinocytes, and endothelial cells, which modulate the activity of melanocytes and mast cells expressing c-*kit*. In bone, marrow stromal cells express SCF and regulate hematopoiesis of c-*kit*–expressing stem cells (Fig. 2). In the gastrointestinal tract, intestinal epithelial cells express SCF and affect the intersti-

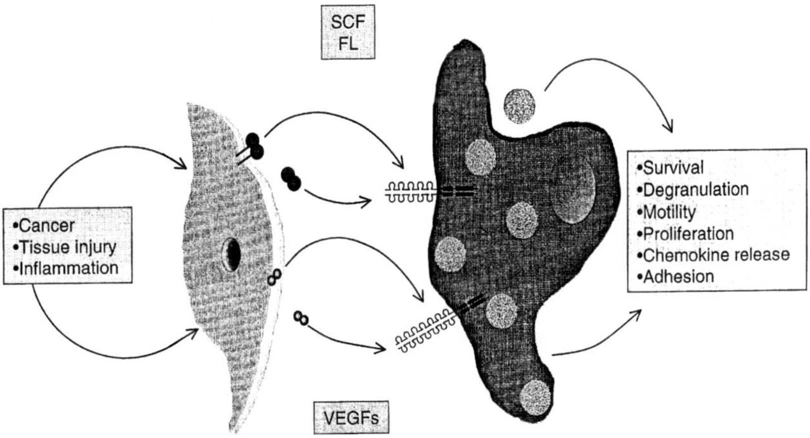

Figure 2 Paracrine mechanisms by which SCF, FL, and VEGF activate their RTK. SCF is produced as an active transmembrane protein capable of receptor activation, which can be proteolytically cleaved to form a soluble dimer that retains biological activity for activation of c-*kit*. VEGF and related ligands are synthesized as soluble proteins that dimerize and activate various VEGF receptors. Some splice variants contain motifs that enable them to interact with heparin sulfate proteoglycans on the surface of cells or in the extracelluar matrix. Some splice variants can also bind to the accessory receptor, neuropilin-1.

tial cells of Cajal and intraepithelial lymphocytes. In the testis, Sertoli cells and granulosa cells express SCF, which regulates spermatogenesis by interaction with c-*kit* on germ cells.

In mice, a variety of mutations have been identified that inactivate c-*kit* RTK activity leading to the W phenotype. In addition, several mutations of c-*kit* have also been identified that lead to the constitutive activation of c-*kit* kinase and development of cell-specific pathologies. In addition to mutations, c-*kit* can be incorrectly activated by aberrant expression or processing of either the receptor (e.g., overexpression) or ligand (e.g., formation of an autocrine loop or excessive soluble SCF). A wide variety of diseases have been associated with aberrant activation of the c-*kit* RTK.

Malignancies

Aberrant expression and/or activation of c-*kit* has been implicated in a variety of tumors. Although many tumor cells express SCF, which may indirectly stimulate tumor cell growth by inducing cytokine release from mast cells or other cells that normally respond to SCF, the discussion in this chapter will be limited to c-*kit* activation that has been directly implicated in tumor cell growth. In this regard, the strongest evidence for a contribution of c-*kit* to neoplastic pathology is associated with leukemias and mast cell tumors, small cell lung cancer, testicular cancer, and some cancers of the gastrointestinal tract and central nervous system (see below). In addition, c-*kit* has been implicated in playing a role in carcinogenesis of the female genital tract (Inoue et al., 1994), sarcomas of neuroectodermal origin (Ricotti et al., 1998), and Schwann cell neoplasia associated with neurofibromatosis (Ryan et al., 1994). It is of interest that c-*kit* has not been found to be associated with melanoma, given that activation of c-*kit* is essential for normal melanocyte development, and mice with mutations of either c-*kit* or SCF have abnormal pigmentation. Although SCF appears to induce proliferation of normal melanocytes, activation of c-*kit* appears to inhibit the proliferation of, and induce apoptosis in melanoma cells (Zakut et al., 1993; Huang et al., 1996). This may explain why c-*kit* is not expressed in the majority of melanoma cells or tumors (Lassam and Bickford, 1992; Natali et al., 1992b; Zakut et al., 1993; Ohashi et al., 1996).

Leukemias

SCF binding to the c-*kit* RTK protects hematopoietic stem and progenitor cells from apoptosis (Lee et al., 1997), thereby contributing to colony formation and hematopoiesis. Expression of c-*kit* is frequently observed in acute myelocytic leukemia (AML), but is less common in acute lymphocytic leukemia (ALL) (for reviews, see Sperling et al., 1997, and Escribano et al., 1998). Although c-*kit* is expressed in the majority of AML cells, its expression does not appear to be prognostic of disease progression (Sperling et al., 1997). However, SCF protected AML cells from apoptosis induced by chemotherapeutic agents (Hassan and Zander, 1996), suggesting that inhibition of c-*kit* would enhance the efficacy of these agents. The clonal growth of cells from patients with myelodysplastic syndrome (Sawada et al., 1996) or chronic myelogenous leukemia (CML) (Sawai et al., 1996) was found to be significantly enhanced by SCF in combination with other cytokines. CML is characterized by expansion of Philadelphia chromosome–positive cells of the marrow (Verfaillie, 1998), which appears to primarily result from inhibition of apoptotic death (Jones, 1997). The product of the Philadelphia chromosome, $p210^{BCR\text{-}ABL}$, has been reported to mediate inhibition of apoptosis (Bedi et al., 1995). Since $p210^{BCR\text{-}ABL}$ and the c-*kit* RTK both inhibit apoptosis and $p62^{dok}$ has been suggested as a substrate (Carpino et al., 1997), it is possible that clonal expansion mediated by these kinases occurs through a common signaling pathway. However, c-*kit* has also been reported to interact directly with $p210^{BCR\text{-}ABL}$ (Hallek et al., 1996), which suggests that c-*kit* may have a more causative role in CML pathology.

Lung Cancers

SCF is expressed in lung epithelial cells (Pietsch et al., 1998) and lung cancer cells (Hibi et al., 1991; Turner et al., 1992). In contrast, c-*kit* is not expressed in normal lung tissues (Hida et al., 1994; Pietsch et al., 1998), but it becomes aberrantly expressed in a subset of non–small cell lung cancers (Pietsch et al., 1998) and in the majority of small cell lung cancers (SCLC) (Hibi et al., 1991, Sekido et al. 1991). In fact, more than 70% of SCLC appear to express both SCF and c-*kit*, suggesting that this autocrine loop may contribute to the malignant features of SCLC (Hibi et al., 1991; Turner et al., 1992; Rygaard et al., 1993). Although it has been reported that addition of SCF to SCLC cells exerts, at best, a modest stimulation of cell growth (Papadimitriou et al., 1995; Shui et al., 1995) and no significant enhancement of survival after irradiation (Shui et al., 1995), there is evidence that the SCF/c-*kit* autocrine loop that is often observed in SCLC cells may contribute to growth factor–independent proliferation and protection of the tumor cells from apoptosis. When an autocrine loop was introduced into a SCLC cell line, it conferred a significant growth advantage in serum-free medium and enabled the cells to grow well in the presence of other growth factors (Krystal et al., 1996). Conversely, introduction of a dominant-negative c-*kit* receptor into a SCLC autocrine cell line significantly inhibited the ability of the cells to grow in serum-free medium (Krystal et al., 1996). The growth of SCLC cells has also been shown to be inhibited by c-*kit* antisense RNA transcripts introduced with an adenoviral vector (Yamanish et al., 1996), recombinant SCF-exotoxin (Nishida et al., 1997), or the synthetic tyrosine kinase inhibitor AG1296 (Krystal et al., 1997). In the latter case, the inhibition of c-*kit* RTK induced apoptosis in SCLC cells, suggesting that inhibition of c-*kit* kinase may have anticancer utility for SCLC (Krystal et al., 1997). Since SCLC is characterized by an aggressive primary disease that often relapses after chemotherapy (see Ihde et al., 1997 for a review), treatment of relapsed SCLC with an agent that targets c-*kit* may provide an additional means to treat recurrent disease in these patients.

Gastrointestinal Cancers

Normal colorectal mucosa does not express c-*kit* (Bellone et al., 1997). However, c-*kit* is frequently expressed in colorectal carcinoma (Bellone et al., 1997), and autocrine loops of SCF and c-*kit* have been observed in several colon carcinoma cell lines (Toyota et al., 1993; Lahm et al., 1995; Bellone et al., 1997). Furthermore, disruption of the autocrine loop by the use of neutralizing antibodies (Lahm et al., 1995) and downregulation of c-*kit* and/or SCF significantly inhibits cell proliferation (Lahm et al., 1995; Bellone et al., 1997). SCF/c-*kit* autocrine loops have been observed in gastric carcinoma cell lines (Turner et al., 1992; Hassan et al., 1998), and constitutive c-*kit* activation also appears to be important for gastrointestinal stromal tumors (GIST). GIST are the most common mesenchymal tumors of the digestive system. More than 90% of GIST express c-*kit*, which is consistent with the putative origin of these tumor cells from interstitial cells of Cajal (ICC) (Hirota et al., 1998). ICC are thought to regulate contraction of the gastrointestinal tract, and patients lacking c-*kit* in their ICC exhibited a myopathic form of chronic idiopathic intestinal pseudo-obstruction (Isozaki et al., 1997). The c-*kit* expressed in GIST from several different patients was observed to have mutations in the intracellular juxtamembrane domain leading to constitutive activation of this RTK (Hirota et al., 1998).

Testicular Cancers

Male germ cell tumors have been histologically categorized into seminomas, which retain germ cell characteristics and nonseminomas which can display characteristics of embryonal differentiation. Both seminomas and nonseminomas are thought to initiate from a preinvasive stage designated carcinoma in situ (CIS) (Murty and Chaganti, 1998). Both c-*kit* and SCF are essential for normal gonadal development during embryogenesis (Loveland and Schlatt,

1997). Loss of either the receptor or the ligand resulted in animals devoid of germ cells. In postnatal testes, c-*kit* has been found to be expressed in Leydig cells and spermatogonia, whereas SCF was expressed in Sertoli cells (Loveland and Schlatt, 1997). Testicular tumors develop from Leydig cells with high frequency in transgenic mice expressing human papilloma virus 16 (HPV16) E6 and E7 oncogenes (Kondoh et al. 1991, 1994). These tumors express both c-*kit* and SCF, suggesting that an autocrine loop may contribute to the tumorigenesis (Kondoh et al., 1995) associated with cellular loss of functional p53 and the retinoblastoma gene product by association with E6 and E7 (Dyson et al., 1989 Scheffner et al., 1990 Werness et al., 1990). The observation that defective signaling mutants of SCF (Kondoh et al., 1995) or c-*kit* (Li et al., 1996) inhibited formation of testicular tumors in mice expressing HPV16 E6 and E7 indicates that c-*kit* activation is pivotal to tumorigenesis in these animals. Expression of c-*kit* on germ cell tumors has been examined by several groups with similar results. The receptor is expressed by the majority of carcinomas in situ and seminomas, but c-*kit* is expressed in only a minority of nonseminomas (Strohmeyer et al., 1991; Rajpert-de Meyts and Skakkebaek, 1994; Izquierdo et al., 1995; Strohmeyer et al., 1995; Bokemeyer et al., 1996; Sandlow et al., 1996). There is only one report of coexpression of c-*kit* and SCF (Bokemeyer et al., 1996) and no reports of activation mutations of c-*kit* in testicular tumors.

Central Nervous System Cancers

SCF and c-*kit* are expressed throughout the CNS of developing rodents, and the pattern of expression suggests a role in growth, migration, and differentiation of neuroectodermal cells. Expression of both receptor and ligand have also been reported in the adult brain (Hamel and Westphal, 1997). Expression of c-*kit* has also been observed in normal human brain tissue (Tada et al., 1994). Glioblastoma and astrocytoma, which define the majority of intracranial tumors, arise from neoplastic transformation of astrocytes (Levin et al., 1997). Expression of c-*kit* has been observed in glioblastoma cell lines and tissues (Berdel et al., 1992; Tada et al., 1994; Stanulla et al., 1995). However, exogenous addition of SCF to glioblastoma cell lines appears to be mitogenic in only a minority of cases (Berdel et al., 1992; Stanulla et al., 1995), and antibodies that block the interaction of SCF with c-*kit* did not inhibit the proliferation of cells with an autocrine loop (Stanulla et al., 1995). The association of c-*kit* with astrocytoma pathology is less clear. Reports of expression of c-*kit* in normal astrocytes have been made (Natali, et al., 1992a; Tada et al., 1994), whereas others report it is not expressed (Kristt et al., 1993). In the latter case, high levels of c-*kit* expression in high-grade tumors were observed

(Kristt et al., 1993), whereas the former groups were unable to detect any expression in astrocytomas. In addition, contradictory reports of c-*kit* and SCF expression in neuroblastomas also exist. One study found that neuroblastoma cell lines often express SCF but rarely express c-*kit*. In primary tumors, c-*kit* was only detected in about 8% of neuroblastomas, whereas SCF was found in only 18% of tumors (Beck et al., 1995). In contrast, other studies (Cohen et al., 1994) have reported that all 14 neuroblastoma cell lines examined contained c-*kit*/SCF autocrine loops, and expression of both the receptor and ligand were observed in 45% of tumor samples examined. In two cell lines, anti–c-*kit* antibodies inhibited cell proliferation, suggesting that the SCF/c-*kit* autocrine loop contributed to growth (Cohen et al., 1994).

Mast Cell Diseases

Mastocytosis As mentioned above, SCF (also known as mast cell growth factor) stimulation of c-*kit* has been shown to be essential for the growth and development of mast cells (Galli and Hammel, 1994; Kitamura et al., 1995). Mice with mutations of c-*kit* that attenuate its signaling activity have exhibited significantly fewer mast cells in their skin (Tsujimura et al., 1996). Therefore, it is not surprising that excessive activation of c-*kit* might be associated with diseases resulting from an over abundance of mast cells. *Mastocytosis* is the term used to describe a heterogeneous series of disorders characterized by excessive mast cell proliferation (Metcalfe, 1991; Valent, 1996; Golkar and Bernhard, 1997). Mastocytosis is limited to the skin in the majority of patients, but it can involve other organs in 15–20% of patients (Valent, 1996; Golkar and Bernhard, 1997). Even among patients with systemic mastocytosis, the disease can range from having a relatively benign prognosis to aggressive mastocytosis and mast cell leukemia (Valent, 1996; Golkar and Bernhard, 1997). c-*kit* Has been observed on malignant mast cells from canine mast cell tumors (London et al., 1996), as well as on mast cells from patients with aggressive systemic mastocytosis (Baghestanian et al., 1996; Castells, et al., 1996). Elevated c-*kit* expression was reported on mast cells from patients with aggressive mastocytosis but not on mast cells from patients with indolent mastocytosis (Nagata et al., 1998), suggesting that overexpression may contribute to the pathology associated with more aggressive forms of the disease in some patients.

SCF has been shown to be expressed on stromal cells as a membrane-bound protein, and its expression can be induced by fibrogenic growth factors such as PDGF (Hiragun et al., 1998). It has also been shown to be expressed on keratinocytes as a membrane-bound protein in normal skin. However, in the skin of patients with mastocytosis, an increased amount of soluble SCF has been observed (Longley et al., 1993). Mast cell chymase has been reported to cleave membrane-associated SCF to a soluble and biologically active form. This mast cell–mediated process could serve to generate a feedback loop to enhance mast cell proliferation and function (Longley et al., 1997), and may be important for the etiology of mastocytosis. Transgenic mice overexpressing a form of SCF that could not be proteolytically released from keratinocytes did not develop mastocytosis, whereas similar animals expressing normal SCF in keratinocytes exhibited a phenotype resembling human cutaneous mastocytosis (Kunisada et al., 1998). This observation suggested that formation of large amounts of soluble SCF can contribute to the pathology associated with mastocytosis in some patients. Several different mutations of the c-*kit* RTK that resulted in constitutive kinase activity have been found in human and rodent mast cell tumor cell lines (Furitsu et al., 1993; Tsujumura et al., 1994, 1995, 1996). In addition, activating mutations of the c-*kit* gene have been observed in peripheral mononuclear cells isolated from patients with mastocytosis and associated hematological disorders (Nagata et al., 1998) and in mast cells from a patient with urticaria pigmentosa and aggressive mastocytosis (Longley et al., 1996). These reports indicate that, in some patients, activating mutations of the c-*kit* RTK may be responsible for the pathogenesis of the disease. SCF activation of c-*kit* has been shown to prevent mast cell apoptosis, which may be critical for maintaining cutaneous mast cell homeostasis (Iemura et al., 1994; Mekori and Metcalfe, 1994, 1995; Yee et al., 1994). Inhibition of mast cell apoptosis could lead to the mast cell accumulation associated with mastocytosis. Thus, observation of c-*kit* activation resulting from overexpression of the receptor, excessive formation of soluble SCF, or mutations of the c-*kit* gene that constitutively activate its kinase provides a rationale that inhibition of the kinase activity of c-*kit* may decrease the number of mast cells, and this aspect may provide benefit for patients with mastocytosis.

Asthma and Allergy Mast cells and eosinophils represent key cells in parasitic infection, allergy, inflammation, and asthma (Thomas and Warner, 1996; Costa et al., 1997; Holgate, 1997; Metcalfe et al., 1997; Naclerio and Solomon, 1997). SCF has been shown to be essential for mast cell development, survival, and growth (Kitamura et al., 1995; Metcalfe et al., 1997). In addition, SCF cooperates with the eosinophil-specific regulator, interleukin-5 (IL-5) to increase the development of eosinophilic progenitors (Metcalf, 1998). SCF has also been shown to induce mast cells to secrete factors (Okayama et al., 1997, 1998) that promote the survival of eosinophils (Kay et al., 1997), which may contribute to chronic, eosinophil-mediated inflammation (Okayama et al., 1997, 1998). In this regard, SCF directly and indirectly regulates activation of both mast cells and eosinophils. SCF induces mediator release from mast cells, as well as priming these cells for IgE-induced degranulation (Columbo et al., 1992) and sensitizing their responsiveness to eosinophil-derived granular major basic protein (Furuta et al., 1998). Among the factors released by activated mast cells are IL-5, granulocyte-macrophage colony-stimulating factor (GM-CSF), and tumor necrosis factor-α (TNF-α), which influence eosinophilic protein secretion (Okayama et al., 1997, 1998). In addition to inducing histamine release from mast cells (Luckacs et al., 1996; Hogaboam et al., 1998), SCF promotes the mast cell production of the eosinophilic chemotactic factor eotaxin (Hogaboam et al., 1998) and eosinophilic infiltration (Luckacs et al., 1996). SCF also directly influences the adhesion of both mast cells (Dastych and Metcalfe, 1994; Kinashi and Springer, 1994) and eosinophils (Yuan et al., 1997), which in turn regulates tissue infiltration. Thus, SCF can influence the primary cells involved in allergy and asthma through multiple mechanisms. Currently, corticosteroids are the most effective treatment for chronic rhinitis and inflammation associated with allergy (Meltzer, 1997; Naclerio and Solomon, 1997). These agents work through multiple mechanisms, including reduction of circulating and infiltrating mast cells and eosinophils and diminished survival of eosinophils associated with inhibition of cytokine production (Meltzer, 1997). Steroids have also been reported to inhibit the expression of SCF by fibroblasts and resident connective tissue cells, which leads to diminished mast cell survival (Finotto et al., 1997). Because of the mutual regulation of mast cell and eosinophilic function, and the role that SCF can play in this regulation, inhibition of c-*kit* kinase may provide a means to treat allergy-associated chronic rhinitis, inflammation, and asthma.

VEGF RECEPTORS

There are numerous excellent review articles describing VEGF, which is also known as vascular permeability factor (VPF), and related proteins, their receptors, and the biological responses they mediate (Ferrara, 1995; Shibuya, 1995; Klagsbrun and D'Amore, 1996; Neufeld et al., 1996; Terman and Dougher-Vermazen, 1996; Breier et al., 1997; Joukov et al., 1997; Plate and Warnke, 1997; Shawver et al., 1997). In this chapter, we will focus on biological activities of VEGF receptors and their putative role in human disease.

Receptors

At present, there are three known members of the VEGFr family of receptor tyrosine kinases. VEGFr1/*Flt1* (*fms*-like tyrosine kinase 1) was originally cloned by low-stringency hybridization using v-*ros* as a DNA probe (Shibuya et al., 1990). The cDNA encoded a receptor with cysteine motifs in the extracellular domain and a cytoplasmic tyrosine kinase containing a long insert region, which resembled motifs observed in the CSF-1 receptor, c-*fms* (Shibuya et al., 1990). *Flt1* was subsequently demonstrated to be a receptor for VEGF/vascular permeability factor (VPF) ligand (de Vries et al., 1992). Although the VEGFr family of receptor kinases have structural (see Fig. 1) and sequence (Fig. 3) similarities to the c-*fms*/PDGF-r/c-*kit* superfamily, the VEGFr family receptors contain seven immunoglobulin-like motifs in the extracellular domain rather than five of SCF receptor.

VEGFr2/*Flt*2/*Flk*1/KDR was cloned independently by two different groups. Using degenerate oligonucleotide primers to generate a hybridization probe, Terman et al., 1991 isolated a cDNA from a human endothelial cell library named KDR (kinase insert domain–containing receptor). The murine homolog of KDR was cloned simultaneously from mouse cell populations enriched in stem and progenitor cells, and designated *Flk*1 (fetal liver kinase 1) (Matthews et al., 1991). KDR/*Flk*1 was subsequently demonstrated to be a receptor for VEGF by cellular expression, ligand binding, and cellular response (Terman et al., 1992; Quinn et al., 1993). It is common in the literature to refer to the mouse homolog of VEGFr2 as *Flk*1 and to the human homologue as KDR. The third VEGFr family member, VEGFr3/*Flt*4, was also cloned by a degenerate polymerase chain reaction (PCR) amplification/hybridization strategy (Pajusola et al., 1992; Finnerty et al., 1993; Galland et al.,

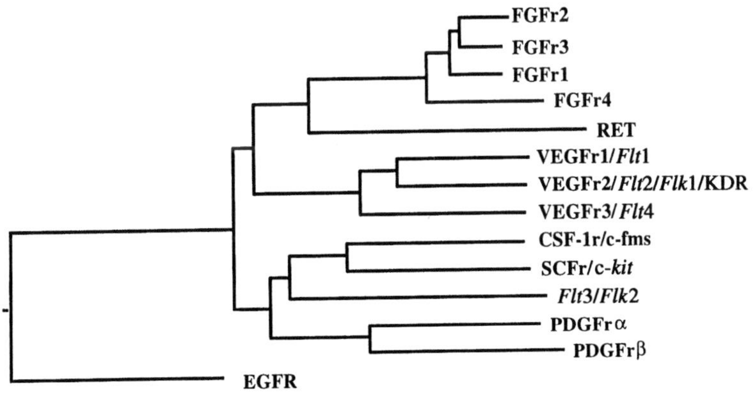

Figure 3 A comparison of the catalytic core of receptor tyrosine kinases containing kinase insert domains. The relatedness of various "split kinases" catalytic domains to each other is diagrammed in this dendrogram.

1993). Two different *Flt*4 mRNA transcripts were observed (Pajusola et al., 1992; Galland et al., 1993) and subsequently shown to result from alternative splicing and to encode proteins with different carboxyl-termini (Pajusola et al., 1993; Borg et al., 1995).

The third *fms*-like tyrosine kinase receptor, *Flt*3/*Flk*2, is not a VEGF receptor family member (for reviews on *Flt*3/*Flk*2, its ligand, and its biological function, see Lyman, 1995, 1998; Drexler, 1996; Namikawa et al., 1996; Lyman and Jacobsen, 1998). Structurally, it is more related to c-*kit*, c-*fms*, and the PDGF receptors, since it contains five immunoglobulin-like domains in the extracellular domain. *Flt*3 ligand (FL) is also more similar to SCF than to the VEGFr ligands, since it is synthesized as a biologically active transmembrane protein that can be proteolytically cleaved to form a biologically active soluble protein. As with SCF, FL also has several splice variants that either cannot be proteolytically cleaved or are constitutively secreted. *Flt*3/*Flk*2 is predominantly expressed on hematopoietic stem cells, and FL activation of the receptor synergizes with other cytokines to regulate hematopoiesis. It has been implicated as possibly having a role in leukemia, but there is little evidence that it contributes to the pathology of other diseases. FL has been considered to have therapeutic potential to treat hematopoietic disorders, whereas the *flt*3/*flk*2 RTK has not generally been considered as a target for inhibition.

In addition to the three tyrosine kinase receptors for VEGF-related ligands, another receptor lacking a kinase domain has been identified that can bind to some splice variants of the ligands. This receptor is named neuropilin-1 (Migdal et al., 1998; Soker et al., 1998), which was previously identified as the receptor for the collapsin/semaphorin family that mediates neural cell guidance (Fujisawa et al., 1995; Kawakami et al., 1996). Although neuropilin-1 expression provides a ''low''-affinity ligand-binding site and augments the affinity of and biological response to some VEGF-related ligands, there is presently no evidence that ligand binding to neuropilin-1 leads to signal transduction (Migdal et al., 1998; Soker et al., 1998).

Ligands

There are at least five different ligands related to VEGF/VPF, several of which are known to be differentially spliced. VEGF/VPF was originally isolated based on principles related to the two major biological responses following receptor activation, including increased vascular permeability (Senger et al., 1983) and induction of the proliferation of endothelial cells (Ferrara, 1989; Plouet et al., 1989). The subsequent cloning of VPF (Keck et al., 1989) and VEGF (Leung et al., 1989; Tischer et al., 1989) revealed that the same protein

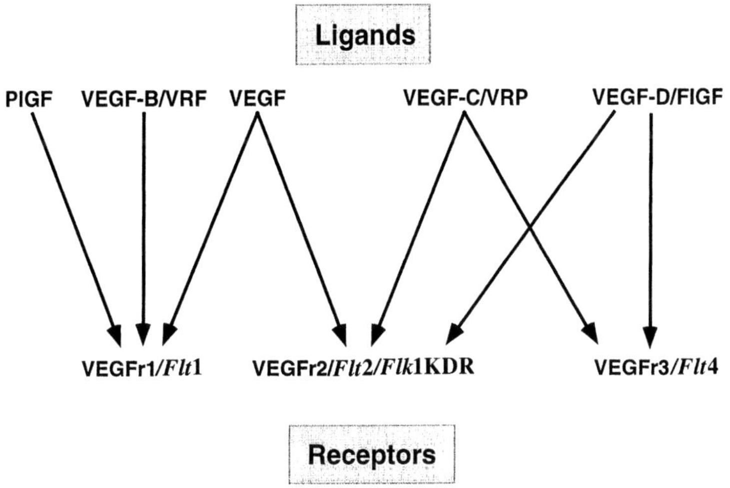

Figure 4 Interaction of VEGF ligands with their receptors. The schematic denotes the potential of VEGF and VEGF-related ligands to bind to VEGFr1, VEGFr2, or VEGFr3.

induced both biological responses. Subsequently, four other proteins related to VEGF were identified: placental growth factor (PlGF) (Maglione et al., 1991), VEGF-B/VRF (VEGF-related factor) (Grimmond et al., 1996; Olofsson et al., 1996), VEGF-C/VRP (VEGF-related protein) (Joukov et al., 1996; Lee et al., 1996), and VEGF-D/FIGF (c-fos–induced growth factor) (Orlandini et al., 1996; Yamada et al., 1997). VEGFr ligands are structurally related to PDGF, containing eight conserved cysteines and forming dimers (see reviews). Splice variants have been reported for most of the ligand family members which produce proteins of various sizes with different abilities to interact with heparan sulfate proteoglycans (HSPG) and neuropilin-1. Generally, the shortest splice variants bind exclusively to the VEGF receptors, whereas the longer splice variants also contain motifs that enable binding to HSPG or neuropilin-1 (see reviews and above references). The VEGFr ligands exhibit a complex pattern of binding (Fig. 4). Homodimers of PlGF and VEGF-B bind exclusively to VEGFr1/*Flt*1, whereas all of the other ligands have been reported to bind to at least two of the VEGFr family members. VEGF/VPF binds to both VEGFr1/*Flt1* and VEGFr2/*Flk*1/KDR. VEGF-C and VEGF-D both bind to VEGFr2/*Flk*1/KDR as well as to VEGFr3/*Flt*4. The ability of these ligands to heterodimerize (Cao et al., 1996; Olofsson et al., 1996) suggested that the pattern of ligand binding and cellular activation may be even more complex. Thus, heterodimerization, coupled with differential splicing which regulates ligand interaction with auxiliary receptors and extracellular matrix, may serve as an exquisite mechanism to finely regulate cellular responses to VEGFr activation.

Biology

VEGF receptors are predominantly localized to endothelial cells and associated with regulating endothelial cell function, vasculogenesis during development, and angiogenesis. VEGFr1/*Flt*1 and VEGFr2/*Flk*1/KDR are both expressed in vascular endothelial cells (see reviews), whereas the expression of VEGFr3/*Flt*4 is more restricted, and is found primarily in endothelial cells of the lymphatic system in adult animals (Kaipainen et al., 1995). As implied its name, VEGF/VPF is a potent, endothelium-specific mitogen as well as a potent factor regulating vascular permeability. As VEGF, it is an important regulator of both normal and pathological angiogenesis. As VPF, it may play a key role in renal function, make a significant contribution to edema associated with inflammation, and provide an extravascular fibrin matrix to support the growth of tumor cells. Because there are numerous articles reviewing the role of angiogenesis in different diseases, and the contribution of VEGF receptors to angiogenesis and vascular permeability (see above reviews and Colville-Nash and Scott, 1992; Battegay, 1995; Dvorak et al., 1995; Folkman, 1995; Risau and Flamme, 1995; Claffey and Robinson, 1996; Feldman et al., 1996; Isner, 1996; Amoroso et al., 1997; Carmeliet and Collen, 1997; Miller, 1997; Paques et al., 1997; Paleolog, 1997; Molema et al., 1998; Shawver et al., 1997), this chapter will focus on VEGFr regulation of cells other than endothelial cells.

Although VEGF receptors are often described as specific to endothelial cells, they have been reported to be expressed and exhibit activity in other types of cells. For example, VEGFr1/*Flt*1 and VEGFr2/*Flk*1/KDR as well as VEGF/VPF are expressed in retinal pigment epithelial cells in culture and promote their autocrine growth (Guerrin et al., 1995). Uterine smooth muscle cells have also been reported to express both receptors and (in culture) to proliferate in response to VEGF/VPF (Brown et al., 1997). Vascular smooth muscle cells express VEGFr1/*Flt1*, which appeared to regulate motility (Klagsbrun and D'Amore, 1996) and sensitized a proliferative response to FGF2/bFGF (Couper et al., 1997). Consequently, we have focused this chapter on diseases and cell types that are not directly associated with the angiogenic process.

Inflammation

The observation that VEGF/VPF induces monocyte activation and chemotaxis (Clauss et al., 1990) in response to binding to a receptor (Shen et al., 1993) suggested that VEGFr ligands might contribute to the recruitment and activation of monocytes and macrophages in some inflammatory syndromes. Monocytes express VEGFr1 but not other VEGF receptors (Barleon et al., 1996; Clauss et al., 1996). Monocyte activation by lipopolysaccharide significantly increased the expression of *Flt*1 (Barleon et al., 1996). Since peritoneal macrophages have been shown to express VEGF/VPF (Berse et al., 1992), the presence of both ligand and receptor may create a positive-feedback loop to augment macrophage-mediated effects. In support of this, macrophages from mice containing a homozygous deletion of the kinase domain of VEGFr1/*Flt1* did not migrate in response to VEGF/VPF or PlGF (Hiratsuka et al., 1998), demonstrating that the kinase activity of VEGFr1/*Flt*1 may be essential for monocyte and macrophage responses following ligand binding. These observations and others suggest that *Flt*1 kinase inhibitors may ameliorate macrophage-mediated inflammatory events.

Inflammatory Bowel Disease

Macrophages can play a central role in inflammation, since they respond to and induce cytokine-mediated cellular responses (Hauser, 1996). Activation of macrophages is a key feature of many chronic inflammatory syndromes such as Crohn's disease and ulcerative colitis, which are both chronic inflammatory bowel diseases (Braegger and McDonald, 1994). It is of interest that patients with Crohn's disease and ulcerative colitis have been shown to possess significantly elevated serum levels of VEGF/VPF (Griga et al., 1998), suggesting a role of VEGF/VPF in chronic inflammatory bowel disease.

Acquired Immunodeficiency Syndrome

Macrophages also appear to play a role in HIV-associated dementia, which is the most severe neurological complication of AIDS (Achim and Wiley, 1996). Interestingly, the Tat protein encoded by human immunodeficiency virus (HIV-1) has been demonstrated to induce chemotaxis of human monocytes through binding of Tat to VEGFr1/*Flt*1 and inducing receptor phosphorylation (Mitola et al., 1997). In this case, inhibition of VEGFr1/*Flt*1 kinase activity may block macrophage recruitment leading to an amelioration of the macrophage-associated symptoms.

Sarcoidosis

Pulmonary sarcoidosis is a granulomatous disorder of unknown etiology. However, key events in granuloma formation include activation and migration of mononuclear phagocytes. Lung tissue biopsies from patients with pulmonary sarcoidosis have demonstrated elevated levels of VEGF/VPF and overexpression of VEGFr1/*Flt*1 in alveolar macrophages as well as in epitheloid and multinuclear giant cells (Tolnay et al., 1998). These observations suggest that inhibition of VEGFr1/*Flt*1 kinase may help to prevent granuloma formation associated with this disease.

Rheumatoid Arthritis

Rheumatoid arthritis (RA) is characterized by chronic joint inflammation and infiltration of activated T cells and macrophages (see Feldmann et al., 1996 for a review). VEGF/VPF has been reported to be elevated in synovial fluid of RA patients (Fava et al., 1994; Koch et al., 1994), and has been implicated as a primary factor contributing to angiogenesis in the diseased joint (see reviews). However, since TNF-α has been implicated as a central cytokine contributing to the pathology associated with RA (Feldmann et al., 1996; Paleolog, 1997), and since activated macrophages are a source of TNF-α (Beutler and Cerami, 1988), VEGF/VPF may play a role in macrophage recruitment associated with RA. It is possible that VEGFr1/*Flt*1 kinase inhibitors may be effective for the treatment of RA by preventing macrophage infiltration, thereby reducing one source of TNF-α and other proinflammatory cytokines.

Allergy and Asthma

Although all of the above examples of inflammatory diseases have mononuclear phagocytes as central cellular mediators, mast cells may also contribute to some inflammatory syndromes such as asthma and allergy. As mentioned above VEGF/VPF has been shown to induce mast cell migration at picomolar concentrations (Gruber et al., 1995). Eosinophils, which also have important roles in asthma and allergy, have been shown to express VEGF/VPF, which is upregulated by IL-5 and GM-CSF (Horiuchi and Weller, 1997). This observation suggests that eosinophils may help to recruit mast cells by releasing VEGF/VPF.

Cardiovascular Diseases

Atherosclerosis is a cardiovascular disease associated with significant involvement of leukocytes. One of the early events in development of an atherosclerotic plaque is the adherence of monocytes to activated endothelial cells. Subsequently, monocytes and smooth muscle cells migrate into the intima followed by lipid uptake to form foam cells (Libby et al., 1996; Murakami and Yamada, 1996; Rosenfeld, 1996; Watanabe et al., 1997). Arterial restenosis is a tissue injury response characterized in part by the migration of vascular smooth muscle cells into the intima followed by proliferation until they occlude the artery. Arterial restenosis occurs after balloon angioplasty in 30–50% of patients (Landzberg et al., 1997). Elevated levels of VEGF/VPF have been observed in the majority of atherosclerotic plaques and restenotic lesions (Couffinhal et al., 1997). VEGF/VPF appears to be expressed by both macrophages and smooth muscle cells (Ramos et al., 1998), and as described above, both of these types of cells express VEGFr1/*Flt*1 and respond to ligand binding. Oxidized low-density lipoprotein (LDL), which may play a key role in regulating cellular events in atherogenesis (Rosenfeld, 1996), has been shown to augment secretion of VEGF/VPF by macrophages (Ramos et al., 1998). These observations suggest that VEGFr1/*Flt*1 could contribute to atherosclerotic plaque development and arterial restenosis. In the latter case, VEGF/VPF may enhance both smooth muscle cell migration (see above) and the proliferative response of smooth muscle cells to FGF2/bFGF (Couper et al., 1997).

Cancer and Tumor Angiogenesis

VEGF receptors obviously play a critical role in cancer because of their active regulation of angiogenesis and vascular permeability (see above reviews). There is also some evidence that they may have a more direct role in some cancers by regulating the proliferation or survival of the tumor cells. For example, VEGFr1/*Flt*1 has been reported to have the potential to transform cells when appropriately activated (Maru et al., 1998). In endothelial cells overexpressing VEGFr1/*Flt*1, PlGF binding has been shown to stimulate cell proliferation (Landgren et al., 1998). The specific ligands for VEGFr1/*Flt*1 (PlGF and VEGFB/VRF) also induce the secretion of urokinase plasminogen activator (uPA) from endothelial cells (Landgren, 1998; Olofsson et al., 1998). Activation of the uPA is considered to be important for the invasive growth and metastasis of tumor cells (Conese and Blasi, 1995; Schmitt et al., 1995). Thus, these observations suggest that aberrant expression and/or activation of VEGF receptors could contribute to neoplasia of cell types other than endothelial cells. At present, there is some evidence for a direct effect of VEGF on tumor cells derived from leukemias and Kaposi's sarcoma.

Leukemia

VEGFr2/*Flt2*/*Flk*1/KDR was reported to be expressed in hematopoietic stem cells, megakaryocytes, and platelets, and VEGF/VPF was shown to protect stem cells from apoptosis induced by gamma irradiation (Katoh et al., 1995). VEGF/VPF was also shown to synergize with some cytokines to suppress the proliferation of myeloid progenitor cells, suggesting that it may contribute to the regulation of hematopoiesis (Broxmeyer et al., 1995; 1996). VEGFr2/*Flt2*/*Flk*1/KDR was also observed on some leukemic cell lines, and it was shown to protect at least one of these cell lines from irradiation-induced apoptosis (Katoh et al., 1995). In clinical samples, 60–70% of AML cell lines were shown to express VEGF/VPF, which may contribute to paracrine growth regulation by inducing endothelial cells to secrete GM-CSF (Fiedler et al., 1997). VEGFr1/*Flt*1 was also expressed by 50–60% of AML cells from patients, and 20–25% of them expressed VEGFr2/*Flt2*/*Flk*1/KDR (Fiedler et al., 1997). The observation of both ligand and receptor suggests the possibility that an autocrine loop may contribute to AML pathology; perhaps by protecting these cells from apoptosis.

Kaposi's Sarcoma

Kaposi's sarcoma (KS) is the most prevalent form of cancer associated with AIDS. Proliferation of KS spindle cells is accompanied by proliferation of endothelial cells, angiogenesis, and enhanced vascular permeability as a result of VEGF/VPF and FGF2/bFGF secretion from the KS spindle cells (Samaniego et al., 1998). KS spindle cells also express VEGFr1/*Flt*1 and VEGFr2/*Flt2*/*Flk*1/KDR (Masood et al., 1997; Samaniego et al. 1998), suggesting that they may have a functional autocrine loop. Antisense oligonucleotides appeared specifically to inhibit the proliferation of KS cells in culture as well as in vivo suggesting that VEGF/VPF may be an autocrine growth factor for these cells (Masood et al., 1997). However, this observation has not yet been confirmed, and at least one group has reported that the autocrine loop may not be functional (Samaniego et al., 1998). Thus, further studies are needed to determine if VEGF receptors contribute to the pathology of the KS cell.

SYNTHETIC TYROSINE KINASE INHIBITORS

In the past two decades, numerous small molecule kinase inhibitors with diverse chemical structures have emerged from compound collections of synthetic or natural origin. Early synthetic tyrosine kinase inhibitors, such as tyrphostins (Levitzki and Gazit, 1995), were based upon chemical structural features of erbstatin. The focus of this latter work to generate substrate inhibitors was logical, since various protein kinases were known to phosphorylate specific protein substrates.

The initial discovery of dianilinophthalimides (Trinks et al., 1994) and 4-phenylamino-quinazolines (Barker and Davies, 1992; Fry et al., 1994) as highly potent and selective inhibitors of tyrosine kinases created the possibility for developing highly selective inhibitors. Subsequently, numerous chemical entities have emerged as potent and selective protein kinase inhibitors (McMahon et al., 1998; Sun and McMahon, 2000). These chemical scaffolds have included quinazolines, pyrido[*d*]- and pyrimido[*d*]-pyrimidines, dianilinophthalimides, pyrazolo[*d*]pyrimidines, pyrrolo[*d*]pyrimidines, phenylaminopyrimidines, 1-oxo-3-aryl-1*H*-indene-2-carboxylic acid derivatives, substituted indolin-2-ones, 5-aminopyrazoles, pyrrolopyridines, quinoxalines, and pyridinylimidazoles (McMahon et al., 1998). The molecular basis underlying the potency and selectivity of these inhibitors has been elucidated by several recent cocrystal structures of synthetic inhibitors and various protein kinases (Zheng et al., 1993; Zhang et al., 1994; Schulze-Gahmen et al., 1995; Mohammadi et al., 1997; Sicheri et al., 1997; Xu et al., 1997). These studies have indicated that it is possible to develop inhibitors with altered specificity profiles utilizing common core structures (chemical scaffolds) that mimic, in part, the adenine moiety of adenosine triphosphate (ATP).

VEGFr TYROSINE KINASE INHIBITORS

3-Arylidenyl indolin-2-ones have been described as inhibitors of tyrosine kinases, including VEGFr1 (*Flk*1), FGF-R, PDGFr, ABL, EGF-R, and Her-2 (Sun et al., 1998, 1999, 2000). In this case, inhibitory activity was measured using cell-based kinase assays following ligand-dependent receptor activation and autophosphorylation (Sun et al., 1998, 1999, 2000; Tang et al., 1998). In these studies, compounds possessing different arylidenyl moieties at the C-3 position of the indolin-2-one core scaffold were used to explore potency and specificity requirements for this class of compounds. 3-[(Substituted pyrrol-2-yl)methylidenyl]indolin-2-ones exhibited high potency and selectivity against the VEGF receptor tyrosine kinase. SU-5416 (compound 1 in Fig. 5) was found to inhibit VEGF-dependent phosphorylation of the *Flk*1 receptor in *Flk*1-overexpressing NIH 3T3 cells with an IC_{50} of 1 μM and was inactive toward fibroblast growth factor (FGF)–, insulin-like growth factor[1] (IGF[1])–, and epidermal growth factor (EGF)–R tyrosine kinases. It showed high potency and selectivity against VEGF-stimulated mitogenesis in human umbilical vein endothelial cells (IC_{50} = 40 nM) and was inactive against FGF-stimulated endothelial cell proliferation (Fong et al., 1999). Animal studies have

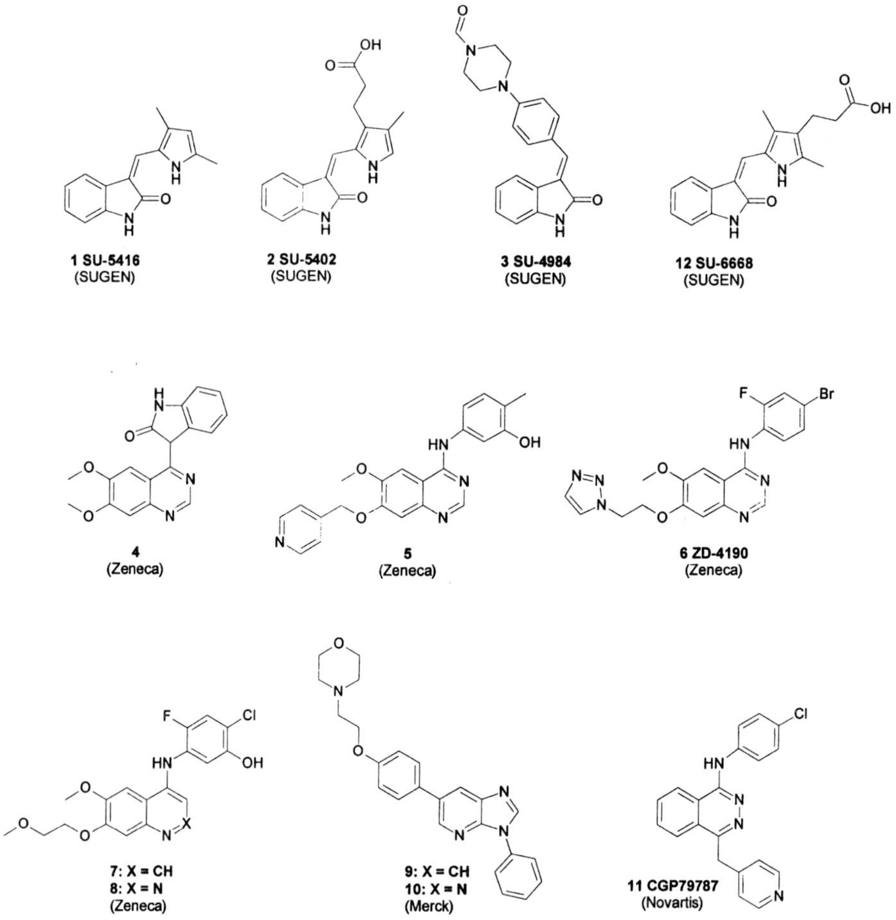

Figure 5 Prototype synthetic inhibitors of VEGFr tyrosine kinases.

revealed that SU-5416 has a broad antitumor spectrum and its in vivo antitumor activity correlated with its antiangiogenesis mechanism. In addition, SU-5416 inhibits tumor vascularization (Fong et al., 1999). SU-5416 is currently in clinical evaluation for the inhibition of tumor angiogenesis associated with the growth and spread of human cancers.

X-ray crystallographic studies of the catalytic core of the FGF-R with two 3-substituted indolin-2-ones, SU-5402 (2) and SU-4984 (3) (see Fig. 5) revealed differences in receptor structure that may influence inhibitory potency and selectivity of different compounds (Mohammadi et al., 1997). The oxindole ring was found to occupy the relatively conserved adenine-binding site forming a bidentate hydrogen bond donor-acceptor system with the peptide backbone atoms of Glu562 and Ala564 that define the base of the adenine-binding pocket. The arylidenyl substituents of SU-5402 and SU-4984 occupied a hydrophobic pocket at the entrance to the ATP-binding site, which was not occupied by ATP and was distinct from the sugar-binding region or characteristic hydrophobic region deep in the ATP-binding site. Although both SU-5402 and SU-4984 exhibited good protein tyrosine kinase (PTK) selectivity, SU-5402 was substantially more selective for inhibition of FGF-R kinase activity compared to SU-4984. The SU-5402 cocrystallographic studies indicated a conformational displacement of the nucleotide-binding loop and the creation of a hydrogen bond interaction between the 3′-carboxyethyl group of SU-5402 and the side chain of Asn568. This type of effect was not observed with the SU-4984 cocrystallographic study and may help to explain differences in kinase specificity for these two compounds.

Based on the crystallographic studies of the catalytic domain of FGFr1 with indolinones (Mohammadi et al., 1997; Sun et al., 2000; Laird et al., 2000), several classes of indolinones have emerged as inhibitors of various split kinases. Among them, SU6668 (Fig. 5, compound 12) was identified as an inhibitor of VEGFr2/Flk-1/KDR, FGFr1, and PDGFr tyrosine kinases (Laird et al., 2000). SU6668 was found to lack inhibitory activity towards the MET or EGFr kinases. In contrast to SU5416, SU6668 inhibited both VEGF- and FGF-dependent proliferation of HUVEC (human umbilical vein endothelial cells) which was consistent with its increased inhibitory activity towards FGFr. SU6668 was found to induce tumor growth stasis or regression when administered by the oral route in mice bearing a wide variety of tumor xenografts (Laird et al., 2000). The anti-angiogenic properties of SU6668 were shown using intravital multi-fluorescence videomicroscopy in the dorsal skin chamber model, and the lack of blood vessels in models of colon metastasis (Shaheen et al., 1999). SU6668 is currently in early stage clinical evaluation for the treatment of human cancers.

4-Phenylamino-quinazolines were first described as potent and selective inhibitors of the EGF-R tyrosine kinase (Barker et al., 1992) and subsequently have been the subject of extensive studies. The modification of this core has led to the generation of inhibitors with altered protein kinase specificity. Two classes of quinazolines have been described as VEGFr inhibitors (see Fig. 5) (Thomas et al., 1997a, 1997b). Quinazolines containing oxindole substitutions at the C-4 position (e.g., compound 4 in Fig. 5) were described as VEGFr and FGF-R inhibitors (Thomas et al., 1997a, 1997b). Compound 5 represents the second series of quinazolines as VEGFr inhibitor (Lohmann et al., 1997). ZD-

4190 (compound 6 in Fig. 5) is an analog of compound 5 and was found to inhibit *Flt*[1] and *Flk*1 tyrosine kinase activities with IC_{50} values of 0.7 and 0.03 μM, respectively. ZD-4190 did not inhibit FGF-R tyrosine kinase activity (IC_{50} 100 μM), although it inhibited both VEGF- and FGF-dependent human umbilical vein endothelial cell (HUVEC) proliferation with IC_{50} values of 0.05 and 1.5 μM, respectively (Hennequin et al., 1999; Ogilvie et al., 1999).

Recently, ZD6474 (the structure of ZD6474 has not yet been disclosed) has emerged as an inhibitor of VEGFr with preferred phamacokinetic properties (Wedge et al., 2000). In this regard, ZD6474 was found to be 1400-fold more soluble than ZD4190 in phosphate buffer at pH 7.4. The pharmacokinetic properties of ZD6474 are compatible with a once-daily oral dosing regimen. When administered in this manner, ZD6474 exhibited broad anti-tumor efficacy. ZD6474 has been shown to exhibit a nanomolar inhibition of KDR with submicromolar potencies against EGFr, VEGFr1/Flt-1, and FGFr1. In cell-based assays, ZD6474 inhibited VEGF-stimulated HUVEC proliferation with submicromolar inhibitory activity toward EGF- and bFGF-stimulated HUVEC proliferation. In addition to inhibiting a broad spectrum of human tumor xenografts, ZD6474 has been shown to induce regression of some established tumors. ZD6474 is currently in phase I clinical evaluation.

Replacing the quinazoline core with either a quinoline or cinnoline core, 4-phenylaminoquinolines and 4-phenylaminocinnolines were also defined as inhibitors of VEGFr. These quinolines and cinnolines showed similar kinase profiles to the corresponding quinazolines (Thomas et al., 1998; Hennequin et al., 1999). Compound 7 (see Fig. 5) as found to inhibit *Flt*[1] and Flk1 with IC_{50} values of 3 and less than 2 nM, respectively, in contrast to its weaker inhibitory activity against FGF-R (IC_{50} = 1400 nM) (Hennequin et al., 1999). The corresponding cinnoline analog (compound 8 in Fig. 5) inhibited *Flt*[1] and *Flk*1 with IC_{50} values of 50 and 1.4 nM, respectively, and its inhibitory activity toward FGF-R was much lower ($IC_{50} > 33$ μM) (Hennequin et al., 1999).

Recently, Merck has developed a series of di-aryl substituted benzoimidazoles and imidazo[4,5-b]pyridines (compounds 9 and 10 in Fig. 5) as inhibitors of VEGFr tyrosine kinase (Bilodeau et al., 1999). These compounds were found to inhibit VEGF-stimulated mitogenesis of HUVEC with IC_{50} values ranging from 150 to 650 nM (Bilodeau et al., 1999).

Finally, a series of 1,4-disubstituted phthalazines (such as compound 11 in Fig. 5) have been described as VEGFr tyrosine kinase inhibitors (Traxler 1998, Bold et al., 1998). CGP 79787, also known as PTK787 or ZK22584, is an analog of phthalazines (Fig. 5, compound 11) and was found to inhibit VEGFr2/KDR and VEGFr1/Flt-1 with IC_{50} values of 37 and 77 nM, respectively (Bold et al., 2000). In addition, CGP 79787 also inhibits PDGFr and c-Kit at submicromolar level (Bold et al., 2000). In contrast, CGP 79787 had

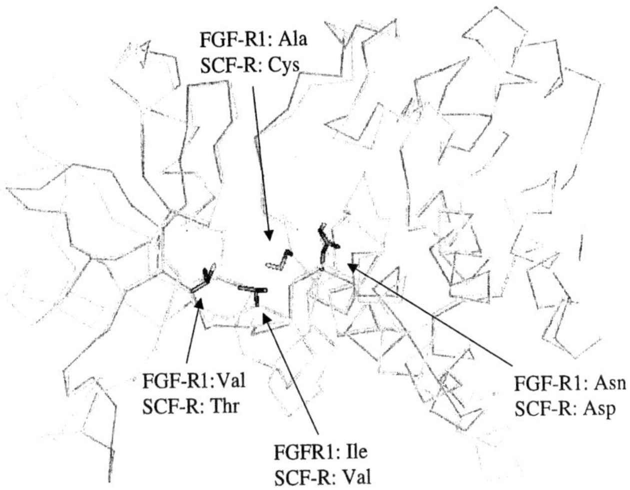

Figure 6 Superimposition of a SCF-R model (in black) over the crystalline structure of FGF-R1 (in gray). Key nonconserved residues in the ATP-binding site are highlighted. The sequence homology between SCF-R and FGF-R1 is 44%, whereas the RMS deviation of the C ∝ traces of the SCF-R model and FGF-R1 crystal structure is 0.63 Å.

no inhibitory effect on other receptor and non-receptor tyrosine kinases including EGFr, FGFr, Tek, c-Src, v-Abl. In cell based assay, CGP 79787 was found to inhibit VEGF-driven cellular receptor autophosphorylation in CHO cells transfected with the KDR receptor and VEGF-stimulated HUVEC proliferation at submicromolar concentrations (Traxler 1998). In vivo studies revealed that selected compounds from this series exhibited dose-dependent inhibition of VEGF- and PDGF-induced angiogenesis in mice (Traxler, 1998; Wood et al., 2000). CGP 79787 is undergoing clinical evaluation for the treatment of solid tumors.

SCF RECEPTOR INHIBITORS

To date, few inhibitors have been identified that potently inhibit the catalytic function of the SCF receptor. Krystal et al. (1997) have shown that AG1296 (quinoxaline chemical scaffold) was able to block SCF-R tyrosine phosphorylation leading to apoptosis of tumor cells. Compounds corresponding to other chemical scaffolds may inhibit the catalytic activity of the SCF-R (Bold et al.,

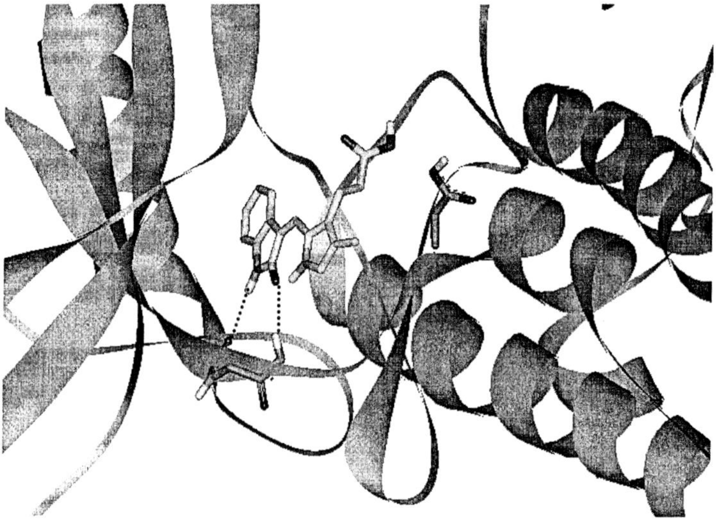

Figure 7 SU-5402 docked to the ATP-binding site of a SCF-R model. The two hydrogen bonds between the indolinone core and the receptor backbone at the hinge region are shown by dotted lines.

2000; Wood et al., 2000). In support of this, we have constructed molecular models and have provided design principles to generate such inhibitors. Thus, a homology model for the catalytic domain of SCF-R was generated using the Modeler program (MSI, 1996). The ''open form'' of FGF-R1 cocrystallized with SU-4984 (PDB code: 1agw) was used as reference and the sequence alignment was based on that of Hanks and Quinn (1991) with slight modifications. Since the sequence homology between FGF-R1 and SCF-R is very high (44%), the SCF-R model and FGF-R1 crystal structures have very similar overall folding (Fig. 6). The key nonconserved residues in the ATP-binding site are highlighted in Figure 6. When SU-5402 is docked into the SCF-R model, the indolinone core fits nicely into the adenine-binding site, forming two hydrogen bonds with the receptor peptide backbone at the hinge region (Fig. 7). Thus, potent SCF-R inhibitors can be derived from this class of compounds with proper substitutions.

CONCLUSIONS

In conclusion, kinase inhibitors for SCF (c-*kit*) and VEGF receptors may have many clinical uses for the treatment of human disease. Kinase inhibitors of c-*kit* may be beneficial for the treatment of a variety of cancers as well as mast cell diseases. VEGFr kinase inhibitors are currently in clinical development for the inhibition of tumor angiogenesis to treat cancer and metastatic disease. In addition, they may find utility for the treatment of inflammatory diseases associated with monocyte or macrophage infiltration and cardiovascular syndromes.

ACKNOWLEDGMENTS

We would like to thank Holly Gregorio and Martha Velarde for their assistance in the preparation of text and graphics, respectively.

REFERENCES

Achim, CL, Wiley, CA. Inflammation in AIDS and the role of the macrophage in brain pathology. Curr Opin Neurol 9:221–225, 1996.

Ahmed, A, Dunk, C, Kniss, D, Wilkes, M. Role of VEGF receptor 1 (Flt-1) in mediating calcium-dependent nitric-oxide release and limiting DNA synthesis in human trophoblast cells. Lab Invest 76:779–791, 1997.

Amoroso, A, Del Porto, F, Di Monaco, C, Manfredini, P, Afeltra A. Vascular Endothelial growth factor: a key mediator of neoangiogenesis. Eur Rev Med Pharmacol Sci 1:17–25, 1997.

Ankoma, SV, Matli, M, Chang, KB, Lalazar, A, Donner, DB, Wong, L, Warren, RS, Friedman, SL. Coordinated induction of VEGF receptors in mesenchymal cell types during rat hepatic wound healing. Oncogene 17:115–121, 1998.

Baghestanian, HC, Bankl, C, Sillaber, WJ, Beil, TH, Radazkiewicz, W, Füreder, J, Preiser, M, Vesely, G, Schernthaner, Lechner, K, Valent, P. A case of malignant mastocystosis with circulating mast cell precursors: biologic and phenotypic characterization of the malignant clone. Leukemia 10:159–166, 1996.

Barleon, B, Sozzani, S, Zhou, D, Weich, HA, Montovani, A, Marme, D. Migration of human monocytes in response to vascular endothelial growth factor (VEGF) is mediated via the VEGF receptor fit-1. Blood 87:3336–3343, 1996.

Battegay, EJ. Angiogenesis: mechanistic insights, neovascular diseases, and therapeutic prospects. J Mol Med 73:333–346, 1995.

Beck, D, Gross, N, Beretta, B, Brognara, C, Peruisseau, G. Expression of stem cell factor and its receptor by human neuroblastoma cells and tumors. Blood 86: 3132–3138, 1995.

Bedi, A, Barber, JP, Bedi, GC, El-Deiry, WS, Sidransy, D, Vala, MS, Akhtar, AJ, Hilton, J, Jones, R. BCR-ABL–mediated inhibition of apoptosis with delay of G2/M transition after DNA damage: a mechanism of resistance to multiple anticancer agents. Blood 86:1148–1158, 1995.

Bellone, G, Silvestri, S, Artusio, E, Tibaudi, D, Turletti, A, Geuna, M, Giachino C, Valente, G, Emanuelli, G, Rodeck, U. Growth stimulation of colorectal carcinoma cells via the c-kit receptor is inhibited by TGF-b1. J Cell Physiol 172:1–11, 1997.

Berdel, WE, deVos, S, Mauer, J, Oberberg, D, von Marschall, Z, Schroeder, JK, Li, J, Ludwig, WD, Kreuser, ED, Thiel, E, Herrman, F. Recombinant human stem cell factor stimulates growth of a human glioblastoma cell line expressing c-kit protooncogene. Cancer Res 52:3498–3502, 1992.

Berse, B, Brown, LF, Van der Water, L, Dvorak, HF, Senger, DR. Vascular permeability factor (vascular endothelial growth factor) gene is expressed differentially in normal tissues, macrophages, and tumors. Mol Biol Cell 3:211–220, 1992.

Besmer, P, Murphy, JE, George, PC, Qui, FH, Bergold, PJ, Lederman, L, Snyder, HWJ, Brodeur, D, Zuckerman, EE, WD, a. H. A new acute transforming feline retrovirus and relationship of its oncogene v-kit with the protein kinase gene family. Nature 320:415–421, 1986.

Beutler, B, Ceram, A. Cachectin (tumor necrosis factor): a macrophage hormone governing cellular metabolism and inflammatory response. Endocrinol Rev 9:57–66, 1988.

Bokemeyer, C, Kuczyk, MA, Dunn, T, Serth, J, Hartmann, K, Jonasson, J, Pietsch, T, Jonas, U, Schmoll, HJ. Expression of stem-cell Factor and its receptor c-kit protein in normal testicular and malignant germ-cell tumors. J Cancer Res Clin Oncol 122:301–306, 1996.

Bold, G, Altmann, K-H, Frei, J, Lang, M, Manley, PW, Traxler, P, Wietfeld, B, Brüggen, J, Buchdunger, E, Cozens, R, Ferrari, S, Furet, P, Hofmann, F, Martiny-Baron, G, Mestan, J, Rösel, J, Sills, M, Stover, D, Acemoglu, F, Boss, E, Emmenegger, R, Lässer, L, Masso, E, Roth, R, Schlachter, C, Vetterli, W, Wyss, D, Wood, JM. New anilinophthalazines as potent and orally well absorbed inhibitors of the VEGF receptor tyrosine kinases useful as antagonists of tumor-driven angiogenesis. J Med Chem 43:2310–2323, 2000.

Borg, JP, Delapeyiere, O, Noguchi, T, Rottapel, R, Dubreuil, P, Birnbaum, D. Biochemical characterization of two isoforms of FLT4, a VEGF receptor–related tyrosine kinase oncogene. 10:973–984, 1995.

Braegger, CP, MacDonald, TT. Immune mechanisms in chronic inflammatory bowel disease. Ann Allergy 72:135–141, 1994.

Breier, G, Damert, A, Plate, KH, Risau, W. Angiogenesis in embryos and ischemic diseases. Thromb Hemos 78:678–683, 1997.

Broudy, VC. Stem cell factor and hematopoiesis. Blood 90:1345–1364, 1997.

Brown, LF, Detmar, M, Tognazzi, K, Abu Jawdek, G. Iruela Arispe, ML. Uterine smooth muscle cells express functional receptors (fit-1 and KDR) for vascular permeability factor/vascular endothelial growth factor. Lab Invest 76:245–255, 1997.

Broxmeyer, HE, Cooper, S, Li, ZH, Lu, L, Song, HY, Kwon, BS, Warren, RS, Donner, DB. Myeloid progenitor cell regulatory effects of vascular endothelial cell growth factor. Int J Hematol 62:203–215, 1995.

Broxmeyer, HE, Cooper, S, Li, ZH, Lu, L, Sarris, A, Wang, MH, Chang, MS, Donner, DB, Leonard, EJ. Macrophage-stimulated protein, a ligand for the RON receptor protein tyrosine kinase, suppresses myeloid progenitor cell proliferation and synergizes with vascular endothelial cell growth factor and members of the chemokine family. Am Hematol 73:1–9, 1996.

Cao, Y, Chen, H, Zhou, CL, Chiang, MK, Anand-Aptes, B, Weatherbee, JA, Wang, Y, Fang, F, Flanagan, JG, Tsang, ML. Heterodimers of placenta growth factor/vascular endothelial growth factor: endothelial activity, tumor cell expression, and high affinity binding to Flk-1/KDR J Biol Chem 6:3154–3162, 1996.

Carmeliet, P, Collen, D. Molecular analysis of blood vessel formation and disease. Am Physiol Soc 273:H2091–H2104, 1997.

Carpino, N, Wisniewski, D, Strife, A, Marshak, D, Kobayashi, R, Stillman, B. Clarkson, B, $p62^{dok}$: A constitutively tyrosine-phosphorylated, GAP-associated protein in chronic myelogenous leukemia progenitor cells. Cell 88:197–204, 1997.

Castells, MC, Friend, DS, Bunnell, CA, Austen, KF. The presence of membrane-bound stem cell factor on highly immature nonmetachromatic mast cells in the peripheral blood of a patient with aggressive systemic mastocytosis J Allerergy Clin Immunol 98:831–840, 1996.

Chabot, B, Stephenson, DA, Chapman, VM, Besmer, P, Berstein, A. The proto-oncogene c-kit encoding a transmembrane tyrosine kinase receptor maps to the mouse W locus. Nature 335:88–89, 1988.

Claffey, KP, Robinson, GS. Regulation of VEGF/VPF expression in tumor cells: consequences for tumor growth and metastasis. Cancer Metast Rev 15:165–176, 1996.

Clauss, M, Gerlach, M, Gerlach, H, Brett, J, Wang, F, Familetti, PC, Pan YC, Olander, JV, Connolly, DT, Stern, D. Vascular permeability factor: a tumor-derived polypeptide that induces endothelial cell and monocyte procoagulant activity, and promotes monocyte migration. J Exp Med 172:1535–1545, 1990.

Clauss, M, Weich, H, Breier, G, Knies, U, Rockl, W, Watenberger, J, Risau, W. The vascular endothelial growth factor receptor Flt-1 mediates biological activities: implications for a functional role of placenta growth factor in monocyte activation and chemotaxis. J Biol Chem 271:17629–17634, 1996.

Cohen, PS, Chan, JP, Lipkunskaya, M, Biedler, JL, Seeger, RC. Expression of stem cell factor and c-kit in human neuroblastoma. The Children's Cancer Group. Blood 84:3465–3472, 1994.

Columbo, M, Horowitz, EM, Botana, LM, MacGlachan, DW, Jr, Bochner, BS, Gillis, S, Zsebo, KM, Galli, SJ, Liechtenstein, LM. The human recombinant c-kit receptor ligand, rhSCF, induces mediator release from human cutaneous mast cells and enhances IgE-dependent mediator release from both skin mast cells and peripheral blood basophils. J Immunol 149:599–602, 1992.

Colville-Nash, PR, Scott, DL. Angiogenesis and rheumatoid arthritis: pathogenic and therapeutic implications. J Rheum Dis 51:919–925, 1992.

Conese, M, Blasi, F. The urokinase/urokinase-receptor system and cancer invasion. Baillieres Clin Hematol 8:365–389, 1995.

Costa, JJ, Weller, PF, Galli, SJ. The cells of the allergic response: mast cells, basophils, and eosinophils. JAMA 278:1815–1822, 1997.

Couffinhal, T, Kearney, M, Witzenbichler, B, Chen, D, Murohara, T, Losordo, DW, Symes, J, Isner, JM. Vascular endothelial growth factor/vascular permeability factor (VEGF/VPF) in normal and atherosclerotic human arteries. Am J Pathol 150:1673–1685, 1997.

Couper, LL, Bryant, SR, Eldrup Jorgenson, J, Brendenberg, CE, Lindner, V. Vascular endothelial growth factor increases the mitogenic response to fibroblast growth factor-2 in vascular smooth muscle cells in vivo via expression on FMS-like tyrosine kinase-1. Cancer Res 81:932–939, 1997.

Dastych, J, Metcalfe, DD. Stem cell factor induces mast cell adhesion to fibronectin J Immunol 152:213–219, 1994.

de Vries, C, Escobedo, JA, Ueno, H, Houck, K, Ferrara, N, Williams, LT. The fms-like tyrosine kinase, a receptor for vascular endothelial growth factor. Science 255:989–991, 1992.

Drexler, HG. Expression of the FLT3 receptor and response to FLT3 ligand by leukemic cells. Leukemia 10:588–599, 1996.

Dvorak, HF, Detmar, M, Claffey, KP, Nagy, JA, Van De Water, V, Senger, DR. Vascular permeability factor/vascular endothelial growth factor: an important mediator of angiogenesis in malignancy and inflammation. Int Arch Allergy Immunol 107:233–235, 1995.

Dyson, N, Howley, PM, Münger, K, Harlow, E. The human papilloma virus-16 E7 oncoprotein is able to bind to the retinoblastoma gene product. Science 243: 934–937, 1989.

Escribano, L, Oqueteau, M, Almeida, J, Orfao, A, San Miguel, J. Expression of the c-kit (CD117) molecule in normal and malignant hematapoiesis. Leuk, Lymphoma 30:459–466, 1998.

Fava, RA, Olsen, NJ, Spencer-Green, G, Yeo, K-T, Yeo, T-K, Berse, B, Jackman, RW, Senger, DR, Dvorak, HF, Brown, LF. Vascular permeability factor/endothelial growth factor (VPF/VEGF): accumulation and expression in human synovial fluids and rheumatoid synovial tissue. J Exp Med 180:341–346, 1994.

Feldmann, M, Brennan, FM, Maini, RN. Role of cytokines in rheumatoid arthritis. Annu Rev Immunol 14:397–440, 1996.

Ferrara, N. The role of vascular endothelial growth factor in pathological angiogenesis. Breast Cancer Res Treat 36:127–137, 1995.

Ferrara, N, Henzel, WJ. Pituitary follicular cells secrete a novel heparin-binding growth factor specific for vascular endothelial cells. Biochem Biophys Res Commun 161:851–858, 1989.

Fiedler, W, Graeven, U, Ergun, S, Verago, S, Kilic, N, Stockschlader, M, Hossfeld, D. Vascular endothelial growth factor, a possible paracrine growth factor in human acute myeloid leukemia. Blood 89:1870–1875, 1997.

Finnerty, H, Kelleher, K, Morris, GE, Bean, K, Merberg, DM, Kriz, R, Morris, JC, Sookdeo, H, Turner, KJ, Wood, CR. Molecular cloning of murine FLT and FLT4 Oncogene 8:2293–2298, 1993.

Finotto, S, Mekori, YA, Metcalfe, DD. Glucocorticoids decrease tissue mast cell number by reducing the production of the c-kit ligand, stem cell factor, by resident cells (in vivo evidence in murine systems). J Clin Invest 99:1721–1728, 1997.

Folkman, J. Angiogenesis in cancer, vascular, rheumatoid and other disease. Nat Med 1:27–31, 1995.

Fong, TAT, Shawver, LK, Sun, L, Tang, C, App, H, Powell, TJ, Young, HK, Schreck, R, Wang, X, Risau, W, Ullrich, A, Hirth, KP, McMahon, G. SU5416 is a potent and selective inhibotor of the vascular endothelial growth factor receptor (Flk-1/KRD) that inhibits tyrosine kinase catalysis, tumor vascularization, and growth of mutiple tumor types. Cancer Res 59:99–106, 1999.

Fry, DW, Kraker, AJ, McMichael, A, Ambroso, LA, Nelson, JM, Leopold, WR, Conners, RW, Bridges, AJ. A specific inhibitor of the epidermal growth factor receptor tyrosine kinase. Science 265:1093–1095, 1994.

Fujisawa, H, Takagi, S, Hirata, T. Growth-associated expression of a membrane protein, neuropilin in Xenopus optic nerve fibers. Dev Neurosci 17:343–349, 1995.

Furitsu, T, Tsujimura, T, Tono, T, Ikeda, H, Kitayama, H, Koshimizu, U, Sugahara, H, Butterfield, JH, Ashman, LK, Kanayama, Y, Matsuzawa, Y, Kitamura, Y, Kanakura, Y. Identification of mutations in the coding sequence of the protooncogene c-kit in a human mast cell leukemia cell line causing ligand-independent activation of c-kit product. J Clin Invest 92:1736–1744, 1993.

Furuta, GT, Ackerman, SJ, Lu, L, Williams, RE, Wershil, BK. Stem cell factor influences mast cell mediator release in response to eosinophil-derived granule major basic protein. Blood 92:1055–1061, 1998.

Galland, F, Karamysheva, A, Pebusque, MJ, Borg, JP, Rottapel, R, Dubreuil, P, Rosnet, O, Birnbaum, D. The FLT4 gene encodes a transmembrane tyrosine kinase related to the vascular endothelial growth factor receptor. Oncogene 8:1233–1240, 1993.

Galli, SJ, Hammel, I. Mast cell and basophil development. Curr Opin Hematol 1:33–39, 1994.

Geissler EN, Ryan, MA, Housman, DE. The dominant-white spotting (W) locus of the mouse encodes the c-kit proto-oncogene. Cell 55:185–192, 1988.

Golkar L, Bernhard JD. Mastocystosis. Lancet 349:1379–1385, 1997.

Griga, T, Tromm, A, Spranger, J, May, B. Increased serum levels of vascular endothelial growth factor in patients with inflammatory bowel disease. Scand J Gastroenterol 33:504–508, 1998.

Grimmond, S, Lagercrantz, J, Drinkwater, C, Silins, G, Townson, S, Pollock, P, Gotley, D, Carson, E Rakar, S, Nordenskjold, M, Ward, L, Hayward, N, Weber, G. Cloning and characterization of a novel human gene related to vascular endothelial growth factor. Genome Res 6:124–131, 1996.

Gruber, BL, Marchese, MJ, Kew, R. Angiogenic factors stimulate mast-cell migration. Blood 86:2488–2493, 1995.

Guerrin, M, Moukadiri, H, Chollet, P, Moro, F, Dutt, K, Malecaze, F, Plouet, J. Vasculatropin/vascular endothelial growth factor is an autocrine growth factor for human retinal pigment epithelial cells cultured in vitro. J Cell Physiol 164: 385–394, 1995.

Hallek, M, Danhauser-Reidl, S, Herbst, R, Warmuth, M, Winkler, A, Kolb, H-J, Drucker, B, Griffin, JD, Emmerich, B, Ullrich, A. Interaction of the receptor tyrosine kinase p145 E-kit with the p $210^{Bcr/Ab}$I kinase in myeloid cells Br J Haematol 94:5–16, 1996.

Hamel, W, Westphal, M. The road less travelled: c-kit and stem cell factor J Neuro Oncol 35:327–333, 1997.

Hanks, S, Quinn, AM. Protein kinase catalytic domain sequence database: Identification of conserved features of primary structure and classification of family members Methods Enzymol 200:38–62, 1991.

Hassan, HT, Zander, A. Stem cell factor as a survival and growth factor in human normal and malignant hematopoiesis Acta Hematol 95:257–262, 1996.

Hassan, S, Kinoshita, Y, Kawanami, C, Kishi, K, Matsushima, Y, Ohashi, A, Funasaka, Y, Okada, A, Maekawa, T, Wang, H-Y Chiba, T. Expression of protooncogene c-kit and its ligand stem cell factor (SCF) in gastric carcinoma cell lines Dig Dis Science 43:8–14, 1998.

Hauser, CJ. Regional macrophage activation after injury and the compartmentalization of inflammation in trauma. New Horiz 4:235–251, 1996.

Hennequin, LF, Thomas, AP, Johnstone, C, Pie, P, Stokes, ESE, Ogilvie, DJ, Dukes, M, Wedge, SR. ZD4190: the design and synthesis of a novel, orally active VEGF receptor tyrosine kinase inhibitor. 90th Annual Meeting of the American Association for Cancer Research, Philadelphia, Abstracts 457, April 10–14, 1999.

Hibi, K, Takahashi, T, Sekido, Y, Ueda, R, Hida, T, Ariyoshi, Y, Takagi, H, Takahashi, T. Coexpression of the stem cell factor and the c-kit genes in small-cell lung cancer. Oncogene 6:2291–2296, 1991.

Hida, T, Ueda, R, Sekido, Y, Hibi, K, Matsuda, R, Ariyoshi, Y, Sugiura, T, Takahashi, T, Takahashi, T. Ectopic expression of c-kit in small-cell lung cancer. Int J Cancer 8(suppl):108–109, 1994.

Hiragun, T, Morita, E, Tanaka, T, Kameyoshi, Y, Yamamoto, S. A fibrogenic cytokine, platelet-derived growth factor (PDGF) enhances mast cell growth indirectly via a SCF- and fibroblast-dependent pathway. J Invest Dermatol 111:213–217, 1998.

Hiratsuka, S, Minowa, O, Kuno, J, Noda, T, Shibuya, M. Flt-1 lacking the tyrosine kinase domain is sufficient for normal development and angiogenesis in mice. Proc Natl Acad Sci USA 95:9349–9354, 1998.

Hirota, S, Isozaki, K, Moriyama, Y, Hashimoto, K, Nishida, T, Ishiguru, S, Kawano, K, Hanada, M, Kurata, A, Takeda, M, Tunio, MG, Matsuzawa, Y, Kanakura, Y, Shinomura, Y, Kitamura, Y. Gain-of-Function mutations of c-kit in human gastrointestinal stromal tumors. Science 279:577–580, 1998.

Hogaboam, C, Kunkel, SL, Streiter, RM, Taub, DD, Lincoln, P, Standiford, TJ Lukacs, NW. Novel role of transmembrane SCF fpr mast cell activation and eotaxin production J Immunol 160:6166–6171, 1998.

Holgate, ST. Asthma: a dynamic disease of inflammation and repair. CIBA Found Symp, 206:5–28, 1997.

Horiuchi, T, Weller, PF. Expression of vascular endothelial growth factor by human eosinophilic upregulation by granulocyte macrophage colony-stimulating factor and interleukin-5. Am J Respir Cell Mol Biol 17:70–77, 1997.

Huang, E, Nocka, K, Beier, DR, Chu, TY, Buck, J, Lahm, HW, Wellner, D, Leder, P, Besmer, P. The hematopoietic growth factor KL is encoded by the SI locus and is the ligand of the c-kit receptor, the gene product of the W locus. Cell 63:225–233, 1990.

Huang, S, Luca, M, Gutman, M, McConkey, DJ, Langley, KE, Lyman, SD, Bar-Eli, M. Enforced c-kit expression renders highly metastatic human melanoma cells suceptible to stem cell factor-induced apoptosis and inhibits their tumorigenic and metastatic potential. Oncogene 13:2339–2347, 1996.

Iemura, A, Tsai, M, Ando, A, Wershil, BK, Galli, SJ. The c-kit ligand, stem cell factor, promotes mast cell survival by suppressing apoptosis Am J Pathol 144:321–328, 1994.

Ihde, DC, Pass, HI, Glatstein, EJ, De Vita, VT, Hellman, S, Rosenberg, SA. Small cell lung cancer. In: Cancer, Principles & Practice of Oncology. 1997, pp 911–949.

Inoue, M, Kyo, S, Fujita, T, Enomoto, T, Kondoh, G. Coexpression of the c-kit receptor and the stem cell factor in gynecological tumors. Cancer Res 54:3049–3053, 1994.

Isner, JM. The role of angiogenic cytokines in cardiovascular disease. Clin Immunol Immunopathol 80:882–891, 1996.

Isozaki, K, Hirota, S, Miyagawa, JI, Taniguchi, MS, Hinomura, Y, Matsuzawa, Y. Deficiency of c-kit$^+$ cells in patients with a myopathic form of chronic idiopathic intestinal pseudo-obstruction Am J Gastroenterol 9:332–334, 1997.

Izquierdo, MA, Van Der Walk, P, Van Ark-Otte, J, Rubio, G, Germa-Lluchs, JR, Ueda, R, Scheiper, RJ, Takahashi, T, Giaccone, G. Differential Expression of the c-kit proto-oncogene in germ cell tumors. J Pathol 177:253–258, 1995.

Jones, RJ. Biology and treatment of chronic myeloid leukemia. Curr Opin Oncol 9: 3–7, 1997.

Joukov, V, Pajusola, K, Kaipeinen, A, Chilov, D, Lahtinen, I, Kukk, E, Saksela, O, Kalkkinen, N, Alitalo, K. A novel vascular endothelial growth factor, VEGF-C is a ligand for the FLT4 (VEGFR-3) and KDR (VEGF-2) receptor tyrosine kinases EMBO J 15:290–298, 1996.

Joukov, V, Kaipainen, A, Jeltsch, M, Pajusola, K, Olofsson, B, Kumar, V, Eriksson, U, Alitalo, K. Vascular endothelial growth factors VEGF-B and VEGF-C. J Cell Pysiol 173:211–215, 1997.

Kaipainen, A, Korhonen, J, Mustonen, T, Van Hinsbergh, VWM, Fang, GH, Dumont, D, Breitman, M, Alitalo, K. Expression of the fms-Like tyrosine kinase 4 gene becomes restricted to lymphatic endothelium during development. Proc Natl Acad Sci USA 92:3566–3570, 1995.

Katoh, O, Tauchi, H, Kawaishi, K, Kimura, A, Satow, Y. Expression of the vascular endothelial growth factor (VEGF) receptor gene, KDR, in hematopoietic cells and inhibitory effect of VEGF on apoptotic cell death caused by ionizing radiation. Cancer Res 55:5687–5692, 1995.

Kawakami, A, Kitsukawa, T, Takagi, S, Fujisawa, H. Developmentally regulated expression of a cell surface protein, neuropilin, in the mouse nervous system. J Neorobiol 29:1–17, 1996.

Kay, AB, Barata, L, Meng, Q, Durham, SR, Ying, S. Eosinophils and eosinophil-associated cytokines in allergic inflammation. Int Arch Allergy Immunol 113:196–199, 1997.

Keck, PJ, Hauser, SD, Krivi, G, Sanzo, K, Warren, T, Feder, J, Connolly, DT. Vascular permeability factor, an endothelial cell mitogen related to PDGF. Science 246: 1309–1312, 1989.

Kinashi, T, Springer, TA. Steel factor and the c-kit regulate cell-matrix adhesion. Blood 83:1033–1038, 1994.

Kitamura, Y, Tsujimura, T, Jippo, T, Kasugai, T, Kanakura, Y. Regulation of development, survival and neoplastic growth of mast cells through the c-kit receptor Int Arch Allergy Immunol 107:54–56, 1995.

Klagsbrun, M, D'Amore, P. Vascular endothelial growth factor and its receptors. Cytokine Growth Factor Rev 7:259–270, 1996.

Koch, AE, Harlow, LA, Haines, GK, Amento, EP, Unemori, EN, Wong, WL, Pope, RM, Ferrara, N. Vascular endothelial growth factor. J Immunol 152:41–49, 1994.

Kondoh, G, Murata, Y, Aozasa, K, Yutsudo, M, Hakura, A. Very high incidence of germ cell tumorigenesis (seminomagenesis) in human papilomavirus type 16 transgenic mice. J Virol 65:3335–3339, 1991.

Kondoh, G, Nishimune, Y, Nishizawa, Y, Hayasaka, N, Matsumoto, K, Hakura, A. Establishment and further charactarization of a line of transgenic mice showing testicular tumorigenesis at 100% incidence. J Urol 152:2151–2154, 1994.

Kondoh, G, Hayasaka, N, Li, Q, Nishimune, Y, Hakura, A. As in vivo model for receptor tyrosine kinase autocrine/paracrine activation: auto-stimulated KIT receptor acts as a tumor pormoting factor in papilomavirus-induced tumorigenesis. Oncogene 10:341–347, 1995.

Kristt, DA, Reedy, E, Yarden, Y. Receptor tyrosine kinase expression in astrocystic lesions: similar features in gliosis and glioma. Neurosurgery 33:106–115, 1993.

Krystal, GW, Hines, SJ, Organ, CP. Autocrine growth of small cell lung cancer mediated by coexpression of c-kit and stem cell factor. Cancer Res 56:370–376, 1996.

Krystal, GW, Carlson, P, Litz, J. Induction of apoptosis and inhibition of small cell lung cancer growth by the quinoxaline tyrphostins. Cancer Res 57:2203–2208, 1997.

Kunisada, T, La, S-Z, Yoshida, H, Nishikawa, S, Nishikawa, S, Mizoguchi, M, Hayashi, S, Tyrrell, L, Williams, DA, Wang, X, Longley, BJ. Murine cutaneous mastocytosis and epidermal melanocytosis induced by keratinocyte expression of transgenic stem cell factor. J Exp Med 187:1565–1573, 1998.

Laird, AD, Vajkoczy, P, Shawver, LK, Thurnher, A, Liang, C, Mohammadi, M, Schlessinger, J, Ullrich, A, Hubbard, SR, Blake, RA, Fong, TAT, Strawn, LM, Sun, L, Tang, C, Hawtin, R, Tang, F, Shenoy, N, Hirth, KP, McMahon, G, Cherrington, JM. SU6668 is a potent anti-angiogenic and anti-tumor agent which induces regression of established tumors. Cancer Res, in press, 2000.

Lahm, H, Amstad, P, Yilmaz, A, Borbenyi, Z, Wyniger, J, Fischer, JR, Suardet, L, Givel, JC, Odartchenko, N. Interleukin 4 down-regulates expression of c-kit and autocrine stem cell factor in human colorectal carcinoma cells. Cell Growth Differ 6:1111–1118, 1995.

Landgren, E, Schiller, P, Cao, Y. Claesson-Welsh, L. Placenta growth factor stimulates MAP kinase and mitogenicity but not phospholipase C-γ and migration of endothelial cells expressing Flt 1. Oncogene 16:359–367, 1998.

Landzberg, BR, Frishman, WH, Lerrick, K. Pathophysiology and pharmacological approaches for prevention of coronary artery restenosis following coronary artery balloon Prog Cardiovasc Dis 39:361–398, 1997.

Lassam, N, Bickford, S. Loss of c-kit expression in cultured melanoma cells. Oncogene 7:51–56, 1992.

Lee, J, Gray, A, Yuan, J, Luoh, SM, Avraham, H, Wood, WI. Vascular endothelial growth factor-related protein: a ligand and specific activator of the tyrosine kinase receptor Flt4. Proc Natl Acad Sci USA 93:1988–1992, 1996.

Lee, JW, Gersuk, GM, Keiner, PA, Beckham, C, Ledbetter, JA, Deeg, HJ. HLA-DR–triggered inhibition of hemopoiesis involves Fas/Fas ligand interactions and is prevented by c-kit ligand. J Immunol 159:3211–3219, 1997.

Leung, DW, Cachianes, G, Kuang, WJ, Goeddel, DV, Ferrara, N. Vascular endothelial growth factor is a secreted angiogenic mitogen. Science 246:1306–1309, 1989.

Levin, VA, Leibel, SA, Gutin, PH. Neoplasms of the central nervous system. In Cancer. Principles & Practice of Oncology. 5th Edition. Lippincott-Raven Publishers, Philadelphia, DeVita Jr, VT, Hellman, S, Rosenberg, SA, Eds. 1997. pp 2022–2082.

Levitsky, A, Gazit, A. Tyrosine kinase inhibition: an approach to drug development Science 267:1782–1788, 1995.

Li, Q, Kondoh, G, Inafuku, S, Nishimune, Y, Hakura, A. Abrogation of c-kit Steel factor–dependent tumorigenesis by kinase defective mutants of the c-kit receptor: c-kit Kinase defective mutants as candidate tools for cancer gene therapy. Cancer Res 56:4343–4346, 1996.

Libby, P, Geng, YJ, Aikawa, M, Schoenbeck, U, Mach, F, Clinton, SK, Sukhova, GK, Lee, RT. Macrophages and atherosclerotic plaque stability. Curr Opin Lipidol 7:330–335, 1996.

London, CA, Kisseberth, WC, Galli, SJ, Geissler, EN, Helfand, SC. Expression of stem cell factor receptor (c-kit) by the malignant mast cells from spontaneous canine mast cell Tumours. J Compar Pathol 115:399–414, 1996.

Longley, BJ, Jr., Morganroth, GS, Tyrrell, L, Ding, TG, Anderson, DM, Williams, DE, Halaban, R. Altered metabolism of mast-cell growth factor (c-kit ligand) in cutaneous mastocytosis. N Engl J Med 328:1302–1307, 1993.

Longley, JB, Tyrell, L, Lu, SZ, Ma, YS, Langley, K, Ding, TG, Duffy, T, Jacobs, P, Tang, L, Modlin, I. Somatic c-kit activating mutation in urticaria pigmentosa and aggressive mastocytosis: establishment of clonality in a human mast cell neoplasm. Nat Genet 12:312–314, 1996.

Longley, BJ, Tyrrell, L, Ma, Y, Williams, DA, Halaban, R, Langley, K, Lu, HS, Schechter, NM. Chymase cleavage of stem cell Factor/Yields a bioactive, soluble product Proc Natl Acad Sci USA 94:9017–9021, 1997.

Loveland, KL, Schlatt, S. Stem cell factor and c-kit in the mammalian testis: lessons originating from Mother Nature's gene knockouts. J Endocrinol 153:337–344, 1997.

Luckacs, NW, Streiter, RM, Lincoln, PM, Brownell, E, Pullen, DM, Schock, HJ, Chensue, Taub, DT, Kunkel, S. Stem cell factor (c-kit ligand) influences eosinophil recruitment and histamine levels in allergic airway inflammation. J Immunol 156:3945–3951, 1996.

Lyman, SD. Biology of the flt3 ligand and receptor. Int J Hematol 62:63–73, 1995.

Lyman, SD. Biologic effects and potential clinical applications of Flt3 ligand. Curr Opin Hematol 5:192–196, 1998.

Lyman, SD, Jacobsen, SEW. c-kit Ligand and Flt3 ligand: stem/progenitor cell factors with overlapping yet distinct activities. Blood 91:1101–1134, 1998.

Maglione, D, Guerriero, V, Vigliett, G,Delli-Bovi, P Persico, MG. Isolation of a human placenta cDNA coding for a protein related to the vascular permeability factor. Proc Natl Acad Sci USA 88:9267–9271, 1991.

Martin, FH, Suggs, SV, Langley, KE, Lu, HS, Ting, J, Okino, KH, Morris, CF, McNiece, IK, Jacobsen, FW, Mendiaz, EA, Birkett, NC, Smith, KA, Johnson, MJ,Parker, VP,Flores, JC, Patel AC, Fisher, EF, Erjavec, HO, Herrera, CJ, Wypych, J, Sachdev, RK, Pope, JA, Leslie, I, Wen, D, Lin, C, Cupples, RL Zsebo, KM. Primary structure and functional expression of rat and human stem cell factor DNAs. Cell 63:203–211, 1990.

Maru, Y, Yamaguchi, S, Shibuya, M. Flt-1, a receptor for vascular endothelial growth factor, has transforming and morphogenic potentials. Oncogene 16:2585–2595, 1998.

Masood, R, Cai, J, Zheng, T, Smith, DL, Naidu, Y, Gill, PS. Vascular endothelial growth factor/vascular permeability factor is an autocrine growth factor for AIDS–Kaposi sarcoma. Proc Natl Acad Sci USA 94:979–984, 1997.

Matthews, W, Jordan, CT, Gavin, M, Jenkins, NA, Copeland, NG, Lemischka, IR. A receptor tyrosine kinase cDNA isolated from a population of Enriched primitive hematapoietic Proc Natl Acad Sci USA 88:9026–9030, 1991.

McMahon, G, Sun, L, Liang, C, Tang, C. Protein kinase inhibitors: structural determinants for target specificity. Curr Opin Drug Disc Dev 1:131–146, 1998.

Mekori, YA, Metcalfe, DD. Transforming growth factor-beta prevents stem cell factor–mediated rescue of mast cells from apoptosis after IL-3 deprivation. J Immunol 153:2194–2203, 1994.

Mekori, YA, Oh, C, Metcalfe, DD. The role of c-kit and its ligand in stem cell apoptosis. Int Arch Allergy Immunol 107:137–138, 1995.

Meltzer, EO. The pharmacological basis for the treatment of perennial allergic rhinitis and non-allergic rhinitis with topical corticosteroids. Allererg 52:33–40, 1997.

Metcalf, D. Lineage commitment in the progeny of murine hematapoietic progenitor cells: influence of thrombopoietin and interleukin 5 Proc Natl Acad Sci USA 95:6408–6412, 1998.

Metcalfe, DD. Classification and diagnosis of mastocytosis: current status. J Invest Dermatol 93:2S–4S, 1991.

Metcalfe, DD, Baram, D, Mekori, YA. Mast cells. Physiol Rev 77:1033–1079, 1997.

Migdal, M, Huppertz, B, Tessler, S, Comforti, A, Shibuya, M, Reich, R, Baumann, H, Neufeld, G. Neuropilin-1 is a placenta growth factor-2 receptor. J Biol Chem 273:22272–22278, 1998.

Miller, JW. Vascular endothelial growth factor and ocular neovascularization. Am J Pathol 151:13–23, 1997.

Mitola, S, Sozzani, S, Luini, W, Primo, L, Borsatti, A, Weich, H, Bussolino, F. Tat-human immunodeficiency virus-1 induces human monocyte chemotaxis by activation of vascular endothelial growth factor receptor-1 Blood 90:1365–1372, 1997.

Miyazawa, K, Williams, DA, Gotoh, A, Nishimaki, J, Broxmeyer, HE, Toyama, K. Membrane-bound Steel factor induces more persistent tyrosine kinase activation and longer life span of c-kit gene-encoded protein than its soluble form. Blood 85:641–649, 1995.

Mohammadi, M, McMahon, G, Sun, L, Tang, PC, Hirth, P, Yeh, PK, Hubbard, SR, Schlessinger, J. Structures of the tyrosine kinase domain of fibroblast growth factor receptor in complex with inhibitors Science 267:955–960, 1997.

Mohammadi, M, Froum, S, Hamby, JM, Schroeder, MC, Panek, RL, Lu, GH, Eliseenkova, AV, Green, D, Schlessinger, J, Hubbard, SR. Crystal structure of an angiogenesis inhibitor bound to the FGF receptor tyrosine kinase domain. EMBO 20: 5896–5904, 1998.

Molema, G, Meijer, DKF, de Liej, LFMH. Tumor vasculature targeted therapies: getting the players organized. Biochem Pharmacol 55:1939–1945, 1998.

MSA, Modular User Guide. San Diego: Modular Simulations, 1996.

Murakami, T, Yamada, N. Modification of macrophage function and effects on atherosclerosis. Curr Opin Lipidol 7:320–323, 1998.

Murty, VVVS, Chaganti, RSK. A genetic perspective of male germ cell tumors. Semin Oncol 25:133–144, 1998.

Naclerio, R, Solomon, W. Rhinitis and inhalant allergens. JAMA 278:1842–1848, 1997.

Nagata, H, Worobec, AS, Oh, CK, Chowdhury, BA, Tannenbaum, S, Suzuki, Y, Metcalfe, DD. Identification of a point mutation in the catalytic domain of the protooncogene c-kit in the peripheral blood mononuclear cells of patients who have mastocytosis with an associated hematologic disorder. Proc Natl Acad Sci USA 92:10560–10564, 1995.

Nagata, H, Worobec, AS, Semere, T, Metcalfe, DD. Elevated expression of the protooncogene c-kit in patients with mastocytosis. Leukemia 12:175–181, 1998.

Namikawa, R, Muench, MO, Roncarolo, M-G. Regulatory roles of the ligand for Flk2/Flt3 tyrosine kinase receptor on human hematopoiesis. Stem Cells 14:388–395, 1996.

Natali, PG, Nicotra, MR, Sures, I, Santoro, E, Bigotti, A, Ullrich, A. Expression of c-kit receptor in normal and transformed human nonlymphoid tissues. Cancer Res 52:6139–6143, 1992a.

Natali, PG, Nicotra, MR, Winkler, AB, Cavaliere, R, Bigotti, A, Ullrich, A. Progression of human cutaneous melanoma is associated with loss of expression of c-kit proto-oncogene receptor. Int J Cancer 52:197–201, 1992b.

Neufeld, G, Cohen, T, Gitay-Goren, H, Poltorak, Z, Tessler, S, Sharon, R, Gengrinovitch, S, Levi, B-Z. Similarities and differences between the vascular endothelial growth factor (VEGF) splice variants. Cancer Metast Rev 15:153–158, 1996.

Nishida, K, Seto, M, Takahashi, T, Oshima, Y, Asano, S, Tojo, A, Ueda, R. In vitro effects of a recombinant toxin, mSCF-PE40, targeting c-kit receptors ectopically expressed in small cell lung cancers. Cancer Lett 113:153–158, 1997.

Ohashi, A, Funasaka, Y, Ueda, M, Ichihashi, M. c-kit receptor expression in cutaneous malignant melanoma and benign melanocytic naevi. Melanoma Res 6:25–30, 1996.

Okayama, Y, Kobayashi, H, Ashman, LK, Holgate, ST, Church, MK, Mori, M. Activation of eosinophils with cytokines produced by lung mast cells. Int Arch Aller Immunol 114:75–77, 1997.

Okayama, Y, Kobayashi, H, Ashman, LK, Dobashi, K, Nakazawa, T, Holgate, ST, Church, MK, Mori, M. Human lung mast cells are enriched in the capacity to produce granulocyte-macrophage colony-stimulating factor in response to IgE-dependent stimulation. Eur J Immunol 28:708–715, 1998.

Ogilvie, DJ, Wedge, SR, Dukes, M, Kendrew, J, Curwen, JO, Thomas, AP, Hennequin, LF, Pie, P, Stokes, ESE, Johnstone, C, Wadsworth, P, Richmond, GHP, Curry, B. ZD4190: an orally administered inhibitor of VEGF signaling with pan-xenograft anti-tumor activity. 90th Annual Meeting of the American Association for Cancer Research, Philadelphia, Abstract 457, April 10–14, 1999.

Olofsson, B, Pajusola, K, Kaipainen, A, von Euler, G, Joukoy, V, Saksela, O, Pettersson, RF, Alitalo, K, Eriksson, U. Vascular endothelial growth factor B, a novel growth factor for endothelial cells. Proc Natl Acad Sci USA 93:2576–2581, 1996.

Olofsson, B, Koppelainen, E, Pepper, MS, Mandriota, SJ, Aase, K, Kumar, V, Gunji, Y, Jeltsch, MM, Shibuya, M, Alitalo, K, Eriksson, U. Vascular endothelial growth factor B (VEGF-B) binds to VEGF receptor-1 and regulates plasminogen activator activity in endothelial Cells. Proc Natl Acad Sci USA 95:11709–11714, 1998.

Orlandini, M, Marocini, L, Ferruzzi, R, Oliviero, S. Identification of a c-fos–induced gene that is related to the platelet-derived growth factor/vascular endothelial growth factor family. Proc Natl Acad Sci USA 93:11675–11680, 1996.

Pajusola, K, Aprelikova, O, Korhonen, J, Kaipenen, A, Pertovaara, L, Alitalo, R, Alitalo, K. FLT4 receptor tyrosine kinase contains seven immunoglobulin-like loops and is expressed in multiple human tissues and cell lines. Cancer Res 52: 5738–5743, 1992.

Pajusola, K, Aprelikova, O, Armstrong, E, Morris, S, Alitalo, K. Two Human FLT4 receptor tyrosine kinase isoforms with distinct corboxy terminal tails are produced by alternative processing of primary transcripts. Oncogene 8:2931–2937, 1993.

Paleolog, E. Target effector role of vascular endothelium in the inflammatory response: insights from the clinical trial of anti-TNFα antibody in rheumatoid arthritis. J Clin Pathol Mol Pathol 50:225–233, 1997.

Papadimitriou, CA, Topp, MS, Serve, H, Oelmann, E, Koenigsmann, M, Maurer, J, Oberberg, B, Reufi, B, Thiel, E, Berdel, WE. Recombinant human stem cell factor does exert minor stimulation of growth in small cell lung cancer and melanoma cell lines. Eur J Cancer 31A:2371–2378, 1995.

Paques, M, Massin, P, Gaudric, A. Growth factors and diabetic retinopathy. Diabetes Metab 23:125–130, 1997.

Pietsch, T, Nicotra, MR, Fraioli, R, Wolf, HK, Montiolese, M, Natali, PG. Expression of the c-kit receptor and its ligand SCF in non–small-cell lung carcinomas. Int J Cancer 75:171–175, 1998.

Pignon, JM. C-kit mutations and mast cell disorders. Hematol Cell Ther 39:114–116, 1997.

Plate, KH, Warnke, PC. Vascular endothelial growth factor. J Neurol Oncol 35:365–372, 1997.

Plouët, J, Schilling, J, Gospodarowicz, D. Isolation and characterization of a newly identified endothelial cell mitogen produced by AtT-20 cells. EMBO J 8:3801–3806.

Plowman, G, Ullrich, A, Shawver, LK. Receptor tyrosine kinases as targets for drug intervention. DN&P 7:334–339, 1994.

Qiu, FH, Ray, P, Brown, K, Barker, PE, Jhanwar, S, Ruddle, FH, Besmer, P. Primary structure of c-kit relationship with the CSF-1/PDGF receptor kinase family—oncogenic activation of v-kit involves deletion of extracelluar domain and C terminus. EMBO J 7:1003–1011, 1988.

Quinn, TP, Peters, KG, DeVries, C, Ferrara, N, Williams, LT. Fetal liver kinase 1 is a receptor for vascular endothelial growth factor and is selectivly expressed in vascular endothelium. Proc Natl Acad Sci USA 90:7533–7577, 1993.

Rajpert-de Meyts, E, Skakkebaek, NE. Expression of the c-kit protein product in carcinoma-in-situ and invasive testicular germ cell tumors. Int J Androl 17:85–92, 1994.

Ramos, MA, Kuzuya, M, Esaki, T, Miura, S, Satake, S, Asai, T, Kanda, S, Hayashi, T, Iguchi, A. Induction of macrophage VEGF in response to oxidized LDL and VEGF accumulation in human atherosclerotic lesions. Arterioscler Thromb Vasc Biol 18:1188–1196, 1998.

Ricotti, E, Fagioli, F, Garelli, E, Linari, C, Crescenzio, N, Horenstein, AL, Pistamiglio, P, Vai, S, Berger, M, di Montezemolo, LC, Madon, E, Basso, G. c-kit Expressed in soft tissue sarcoma of neuroectodermic origin and its ligand prevents apoptosis of neoplastic cells. Blood 91:2397–2405, 1998.

Risau, W, Flamme, I. Vasculogenesis. Annu Rev Cell Dev Biol 11:73–91, 1995.

Rosenfeld, ME. Cellular mechanisms in the development of atherosclerosis. Diabetes Res Clin Pract 30(suppl):1–11, 1996.

Ryan, JJ, Klein, KA, Neuberger, TJ, Leftwich, JA, Westin, EH, Kauma, S, Fletcher, JA, DeVries, GH, Huff, TF. Role for the stem cell factor/Kit complex in Schwann cell neoplasia and mast cell proliferation associated with neurofibromatosis. J Neurol Res 37:415–432, 1994.

Rygaard, K, Nakamura, T, Spang-Thomsen, M. Expression of the proto-oncogenes c-met and c-kit and their ligands, hepatocyte growth factor/scatter factor and stem cell factor in SCLC cell lines and xenografts. Br J Cancer 67:37–46, 1993.

Samaniego, F, Markham, PD, Gendelman, R, Watanabe, Y, Kao, V, Kowalski, K, Sonnabend, JA, Pintus, A, Gallo, RC, Ensoli, B. Vascular endothelial growth factor and basic fibroblast growth factor present in Kaposi's sarcoma (KS) are induced by inflammatory cytokines and synergize to promote vascular permeability and KS lesion development. Am J Pathol 152:1433–1443, 1998.

Sandlow, JI, Feng, HL, Cohen, MB, Sandra, A. Expression of c-kit and its ligand, stem cell factor, in normal and subfertile human testicular tissue. J Androl 17:403–408, 1996.

Sawada, K, Leko, M, Notoya, A, Tarumi, T, Koizumi, K, Kitayama, S, Nishio, H, Fukada, Y, Yasukouchi, T, Yamaguchi, M, Katoh, S, Koike, T. Role of cytokines in leukemia type growth of myelodysplastic CD34+ cells. Blood 88:319–327, 1996.

Sawai, N, Koike, K, Ito, S, Okumura, N, Kamijo, T, Shiohara, M, Amano, Y, Tsuji, K, Nakahata, T, Oda, M, Okamura, J, Kobayashi, M, Komiyama, A. Aberrant growth of granulocyte-macrophage progenitors in juvenile chronic myelogenous leukemia in serum-free culture. Exp Hematol 2:116–122, 1996.

Scheffner, M, Werness, BA, Huibregtse, JM, Levine, AJ, Howley, PM. The E6 oncoprotein encoded by human papillomavirus types 16 and 18 promotes degradation of p53. cell 63:1129–1136, 1990.

Schmitt, M, Wilhelm, O, Janicke, F, Magdolen, V, Reuning, U, Ohi, H, Moniwa, N, Kobayashi, H, Weidle, U, Graeff, H. Urokinase-type plasminogen activator (uPA) and its receptor (CD87): a new target in tumor invasion and metastasis. J Obstet Gynecol 21:151–165, 1995.

Schulze-Gahmen, U, Brandsen, J, Jones, HD, Morgan, DO, Meijer, L, Vesely, J, Kim, SH. Multiple modes of ligand recognition: crystal structures of cyclin-dependent protein kinase 2 in complex with ATP and two inhibitors, olomoucine and isopentyladenine. Proteins Struct Funct Genet 22:378–391, 1995.

Sekido, Y, Obata, Y, Ueda, R, Hida, T, Suyama, M, Shimokata, K, Ariyoshi, Y, Takahashi, T. Preferential expression of c-kit protooncogene transcripts in small cell lung cancer. Cancer Res 51:2416–2419, 1991.

Senger, DR, Galli, SJ, Dvorak, AM, Peruzzi, CA, Harvey, VS, Dvorak, HF. Tumor cells secrete a vascular permeability factor that promotes accumulation of ascites fluid. Science 219:983–985, 1983.

Shaheen, RM, Davis, DW, Liu, W, Zebrowski, BK, Wilson, MR, Bucana, CD, McConkey, DJ, McMahon, G, Ellis, LM. Antiangiogenic therapy targeting the tyrosine kinase receptor for vascular endothelial growth factor receptor inhibits the growth of colon cancer liver metastasis and induces tumor and endothelial cell apoptosis. Cancer Res 59:5412–5416, 1999.

Shawver, LK, Lipson, KE, Fong, TA, McMahon, G, Plowman, GD, Strawn, LM. Receptor tyrosine kinases as targets for inhibition of angiogenesis. Drug Disc Today 2:50–63, 1997.

Shen, H, Clauss, M, Ryan, J, Schmidt, AM, Tijburg, P, Borden, L, Connolly, D, Stern, D, Kao, J. Characterization of vascular permeability factor/vascular endothelial growth factor receptors on mononuclear phagocytes. Blood 81:2767–2773, 1993.

Shibuya, M. Role of VEGF-FLT receptor system in normal and tumor angiogenesis Adv Cancer Res 67:281–317, 1995.

Shibuya, M, Yamaguchi, S, Yamane, A, Ikeda, T, Tojo, A, Matsushime, H, Sato, M. Nucleotide sequence and expression of a novel human receptor-type tyrosine kinase gene (flt). Oncogene 5:519–524, 1990.

Shui, C, Khan, WB, Leigh, BR, Turner, AM, Wilder, RB, Knox, SJ. Effects of stem cell factor on the growth and radiation survival of tumor cells. Cancer Res 55: 3431–3437, 1995.

Sicheri, F, Moarefi, I, Kuriyan, J. Crystal structure of the src family tyrosine kinase Hck. Nature 385:602–609, 1997.

Soker, S, Takashima, S, Miao, HQ, Neufeld, G. Neuropilin-1 is expressed by endothelial and tumor cells as an isoform-specific receptor for vascular endothelial growth factor cell. 92:735–745, 1998.

Sperling, C, Schwartz, S, Büchner, T, Thiel, E, Ludwig, W. Expression of the stem cell factor receptor C-KIT (CD117) in acute leukemias. Haematologica 82:617–621, 1997.

Stanulla, M, Welte, K, Hadam, MR, T, P. Coexpression of stem cell factor and its receptor c-kit in human malignant glioma cell lines. Act Neuropathol 89:158–165, 1995.

Strawn, L, Shawver, LK. Tyrosine kinase in disease: overview of kinase inhibitors as therapeutic agents and current drugs in clinical trials. Exp Opin Invest Drugs 7:553–573, 1998.

Strohmeyer, T, Hartmann, Peter, S, Munemitsu, S, Ackerman, R, Ullrich, A, Slamon, DJ. Expression of the hst-1 and c-kit protooncogenes in human testicular germ cell tumors. Cancer Res 51:1811–1816, 1991.

Strohmeyer, T, Reese, D, Press, M, Ackermann, R, Hartmann, M, Slamon, D. Expression of the c-kit proto-oncogene and its ligand stem cell factor (SCF) in normal and malignant human testicular tissue. J Urol 153:511–515, 1995.

Sun, L, McMahon, G. Inhibition of tumor angiogenesis by synthetic tyrosine kinase inhibitors. Drug Discovery Today. 5:344–353, 2000.

Sun, L, Tran, N, Liang, C, Tang, F, Rice, A, Schreck, R, Waltz, K, Shawver, LK, McMahon, G, Tang, C. Design, synthesis, and evaluations of substituted 3-[(3- or 4-carboxyethylpyrrol-2-yl)methylidenyl]indolin-2-ones as inhibitors of VEGF, FGF, and PDGF receptor tyrosine kinases. J Med Chem 42:5120–5130, 1999.

Sun, L, Tran, N, Liang, C, Hubbard, S, Tang, F, Lipson, K, Schreck, R, Zhou, Y, McMahon, G, Tang, C. Discovery of substituted 3-[(4,5,6,7-tetrahydro-1*H*-

indol-2-yl)methylene]-1,3-dihydro-indol-2-one growth factor receptor inhibitors for VEGF-R2 (Flk-1/KDR), FGF-R1, and PDGF-R tyrosine kinases. J Med Chem, in press, 2000.

Sun, L, Tran, N, Tang, F, App, H, Hirth, P, McMahon, G, Tang, C. Synthesis and biological evaluations of of 3-substituted indolin-2-ones: a novel class of tyrosine kinase inhibitors that exhibit selectivity toward particular RTKs. J Med Chem 41:2588–2603, 1998.

Tada, M, Diserens, AC, Desbaillets, I, de Tribolet, N. Analysis of cytokine receptor messenger RNA expression in human glioblastoma cells and normal astrocytes by reverse-transcription polymerase chain reaction. J Neurosurgery 80:1063–1073, 1994.

Terman, BI, Dougher-Vermazen, M. Biological properties of VEGF/VPF receptors. Cancer Metast Rev 15:159–163, 1996.

Terman, BI, Carrion, ME, Kovacs, E, Rasmussen, BA, Eddy, RL, Shows, TB. Identification of a new endothelial cell growth factor receptor kinase. Oncogene 6: 1677–1683, 1991.

Terman, BI, Dougher-Vermazen, M, Carrion, ME, Dimitrov, D, Amellino, DC, Gospoderowicz, D, Behlen, P. Identification of the KDR tyrosine kinase as a receptor for vascular endothelial cell growth factor. Bichem Biophys Res Commun 187: 1579–1586, 1992.

Thomas, LH, Warner, JA. The eosinohpil and its role in asthma. Gen Pharmacol 27: 593–597, 1996.

Tischer, E, Gospodarowicz, D, Mitchell, R, Silva, M, Schilling, J, Lau, K, Crisp, T, Fiddes, JC, Abraham, JA. Vascular endothelial growth facor: a new member of the platelet-derived growth factor gene. Biochem Biophys Res Commun 165: 1198–1206, 1989.

Tolnay, E, Kuhnen, C, Voss, B, Wiethege, T, Muller, K-M. Expression and localization of vascular endothelial growth factor and its receptor fit in pulmonary sarcoidosis. Virchows Arch 432:61–65, 1998.

Toyota, M, Hinoda, Y, Takoaka, A, Makiguchi, Y, Takahashi, T, Itoh, F, Imai, K, Yachi, A. Expression of c-kit and kit ligand in human colon carcinoma cells. Tumour Biol 14:295–302, 1993.

Traxler, P. Tyrosine kinase inhibitors in cancer treatment (part II). Exp Opin Ther Pat 12:1599–1625, 1998.

Trinks, U, Buchdunger, E, Furet, P, Kump, W, Mett, H, Meyer, TH, Muller, M, Regenass, U, Rihs, G, Lydon, N, Traxler, P. Dianilinophthalimides: potent and selective, ATP-competitive inhibitors of the EGF-receptor protein tyrosine kinase. J Med Chem 37:1015–1027, 1994.

Tsujimura, T. Role of c-kit receptor tyrosine kinase in the development, survival, and neoplastic transformation of mast cells. Pathol Int 46:933–938, 1996.

Tsujumura, T, Furitsu, T, Morimoto, M, Isozaki, K, Nomura, S, Matsuzawa, Y, Kitamura, Y, Kankura, Y. Ligand-independent activation of c-kit receptor tyrosine in a murine mastocytoma cell line P-815 generated by a point mutation. Blood 9:2619–2626, 1994.

Tsujimura, T, Furitsu, T, Morimoto, M, Kanayama, Y, Nomura, S, Matsuzawa, Y,

Katamura, Y, Kanakura, Y. Substitution of an aspartic acid results in constitutive activation of c-kit receptor tyrosine kinase in a rat tumor mast cell line RBL-2H3. Int Arch Aller Immunol 106:377–385, 1995.

Tsujimura, T, Morimoto, M, Hashimoto, K, Moriyama, Y, Kitayama, H, Matsuzawa, Y, Kitamura, Y, Kanakura, Y. Constitutive activation of c-kit in FMA3 murine mastocytoma cells caused by deletion of seven amino acids at the juxtamembrane domain. Blood 87:273–283, 1996.

Turner, AM, Zsebo, KM, Martin, F, Jacobsen, FW, Bennett, LG, Broudy, VC. Nonhematopoietic tumor cell lines express stem cell factor and display c-kit receptors. Blood 80:374–381, 1992.

Valent, P. Biology, classification and treatment of human mastocytosis. Wein Klin Wochenschr 108:385–397, 1996.

Verfaillie, CM. Chronic myelogenous leukemia: too much or too little growth, or both? Leukemia 12:136–138, 1998.

Vliagoftis, H, Worobec, AS, Metcalf, DD. The protooncogene c-kit and c-kit ligand in human disease. J Allergy Clin Immunol 100:435–440, 1997.

Watanabe, T, Haroaka, S, Shimokama, T. Inflammatory and immunological nature of atherosclerosis. Int J Cardiol 54:S51–S60, 1997.

Wedge, SR, Ogilvie, DJ, Dukes, M, Kendrew, J, Hennegiun, LF, Stokes, ESE, Curry, B. VEGF receptor tyrosine kinase inhibitors as potential anti-tumour agents. 91st Annual Meeting of American Association for Cancer Research, April 1–5, 2000, San Francisco, CA, Abstract 3610.

Werness, BA, Levine, AJ, Howley, PM. Association of human papillomavirus types 16 and 18 E6 proteins with p53. Science 248:76–79, 1990.

Williams, DA. Differences between membrane-bound and secreted isoforms of stem cell factor. CIBA Found Symp 204:57–59, 1997.

Williams, DE, Eisenman, J, Baird, A, Rauch, C, Van Ness, K, March, CJ, Park, LS, Martin, U, Mochizuki, DY, Boswell, HS, Burgess, GS, Cosman, D, Lyman, SD. Identification of a ligand for the c-kit proto-oncogene. Cell 63:167–174, 1990.

Wood, JM, Bold, G, Buchdunger, E, Cozens, R, Ferrari, S, Frei, J, Hofmann, F, Mestan, J, Mett, H, O'Reilly, T, Persohn, E, Rösel, J, Schnell, C, Stover, D, Theuer, A, Towbin, H, Wenger, F, Woods-Cook, K, Menrad, A, Siemeister, G, Schirner, M, Thierauch, KH, Schneider, MR, Drevs, J, Martiny-Baron, G, Totzke, F, Marmé, D. PTK787/ZK222584, a novel and potent inhibitor of vascular endothelial growth factor receptor tyrosine kinases, impairs vascular endothelial growth factor-induced responses and tumor growth after oral administration. Cancer Res 60:2178–2189, 2000.

Xu, W, Harrison, SC, Eck, MJ. Three-dimensional sctructure of the tyrosine kinase c-src. Nature 385:595–602, 1997.

Yamada, Y, Nezu, J, Shimane, M, Hirata, Y. Molecular cloning of a novel vascular endothelial growth factor, VEGF-D. Genome 42:483–488, 1997.

Yamanishi, Y, Maeda, H, Hiyama, K, Ishioka, S, Yamakido, M. Specific growth inhibition of small-cell lung cancer cells by adenovirus vector expressing antisense c-kit transcripts. Jpn J Cancer Res 87:534–542, 1996.

Yarden, Y, Kuang, WJ, Yang-Feng, T, Coussens, L, Munemitsu, S, Dull, TJ, Chen,

E, Schlessinger, J, Francke, U, Ullrich, A. Human proto-oncogene c-kit: a new cell surface receptor tyrosine kinase for an unidentified ligand. EMBO J 11: 3341–3351, 1987.

Yee, NS, Pack, I, Bessmer, P. Role of kit-Ligand in proliferation and suppression of apoptosis in mast cells: basis for radiosensitivity of white-spotting and Steel mutant mice. J Exp Med 179:1777–1787, 1994.

Yuan, Q, Austen, KF, Friend, DS, Heidtman, M, Boyce, JA. Human peripheral blood eosinophils express a functional c-kit receptor for stem cell factor that stimulates very late antigen 4 (VLA-4)–mediated cell adhesion to fibronectin and vascular cell adhesion molecule (VCAM-1). J Exp Med 186:313–323, 1997.

Zakut, R, Perlis, R, Eliyahu, S, Yarden, Y, Givol, D, Lyman, SD, Halaban, R. KIT ligand (mast cell growth factor) inhibits the growth of KIT-expressing melanoma cells. Oncogene 8:2221–2229, 1993.

Zhang, F, Strand, A, Robbins, D, Cobb, MH, Goldsmith, EJ. Atomic structure of the Map kinase ERK2 at 2.3-Ao resolution. Nature 367:704–711, 1994.

Zheng, J, Trafny, EA, Knighton, DR, Xuong, NH, Taylor, SS, Ten Eyck, LF, Sowadsky, JM. 2.2 Å refined crystal structure of the catalytic subunit of c-AMP–dependent protein kinase complexed with MnATP and a peptide inhibitor. Acta Crystallogr 363:595–602, 1993.

REFERENCES TO PATENT LITERATURE

Barker, AJ, Davies, DH. Therapeutic preparations containing quinazoline derivatives. Zeneca, Pat. No. EP-00520722-A1, 1992.

Bilodeau, MT, Hungate, RW, Cunningham, AP, Koester, TJ. Novel angiogenesis inhibitors. Merck, Pat. No. WO-9916755 A1 990408, 1999.

Bold, G, Frei, J, Traxler, P, Altmann, KH, Mett, H, Stover, DR, Wood, J. Preparation of 1-arylamino-4-piridylmethylphthalazines and analogs as VEGF receptor inhibitors. Novartis, Pat. No. WO-9835958, 1998.

Lohman, JJM, Hennequin, LFA, Thomas, AP. Preparation and angiogenic and/or vascular permeability reducing effect of quinazoline derivatives. Zeneca, Pat. No. WO-9722596, 1997.

Tang, PC, Sun, L, McMahon, G, Hirth, KP, Shawver, LK. Indolinone combinatorial libraries and related products and methods for the treatment of disease. SUGEN, Pat. No. WO-09807695, 1998.

Thomas, AP, Hennequin, LFA, Johnstone, C. 4-Anilinoquinazoline derivatives. Zeneca, Pat. No. WO-09732856, 1997a.

Thomas, AP, Hennequin, LFA, Johnstone, C. 4-Anilinoquinazoline derivatives. Zeneca, Pat. No. WO-09742187, 1997b.

Thomas, AP, Hennequin, LFA, Ple, P. Quinoline derivatives inhibiting the effect of growth factors such as VEGF. Zeneca, Pat. No. WO-09813350, 1998.

12
p38 Inhibition

Rebecca J. Gum
Abbott Laboratories, Abbott Park, Illinois

Peter R. Young
DuPont Pharmaceuticals, Wilmington, Delaware

INTRODUCTION

Interleukin-1 (IL-1) and tumor necrosis factor-α (TNF-α) have long been known as key early mediators of the inflammatory response in mammals. Numerous studies have shown that overexpression or administration of these cytokines can cause both systemic and local inflammatory effects (1,2). Furthermore, proteins which block IL-1 and TNF biological activity, such as soluble receptors, monoclonal antibodies, and receptor antagonists, show some efficacy in several established models of acute or chronic inflammatory disease (3). More recently, a soluble TNF receptor (Enbrel) (Enbrel, Immunex, Seattle, WA) and an antibody to TNF (Remicade, Centocor, Malvern, PA) have shown efficacy in clinical trials for the treatment of rheumatoid arthritis and inflammatory bowel disease (4,5). These treatments block IL-1 or TNF by preventing their binding to target cells, but it may be equally attractive to block the biological effects of these cytokines by inhibiting their production or effect on target cells. A potential advantage of this approach could be to inhibit both IL-1 and TNF-α, as evidenced by the known suppression of these and many other proinflammatory and immunostimulatory cytokines by glucocorticoids (6).

p38 MITOGEN-ACTIVATED PROTEIN KINASES

Identification of p38 Mitogen-Activated Protein Kinase as a Target for Inhibition of IL-1 and TNF

Several structurally related pyridinyl imidazoles, which were originally designed to inhibit arachadonic acid metabolism (7), were found to inhibit IL-1 synthesis from lipopolysaccharide (LPS)–stimulated human monocytes and monocytic cell lines (see Fig. 1 for example structures). Further examination revealed that these compounds inhibited the production of both IL-1 isoforms, TNF, and IL-6 (8), and thus they were called cytokine suppressive anti-inflammatory drugs (CSAID) (CSAID™, SmithKline Beecham, King of Prussia, PA) to distinguish their mechanism from nonsteroidal anti-inflammatory drugs (NSAIDs), which inhibit eicosanoid production but have no effect on cytokine production. In contrast, the pyridinyl imidazoles did not show any

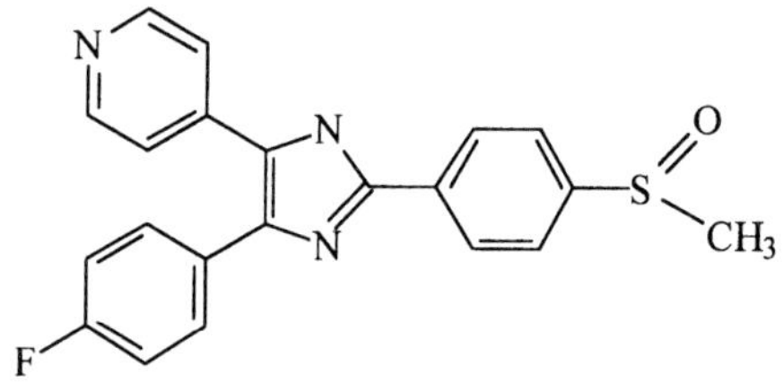

SB203580

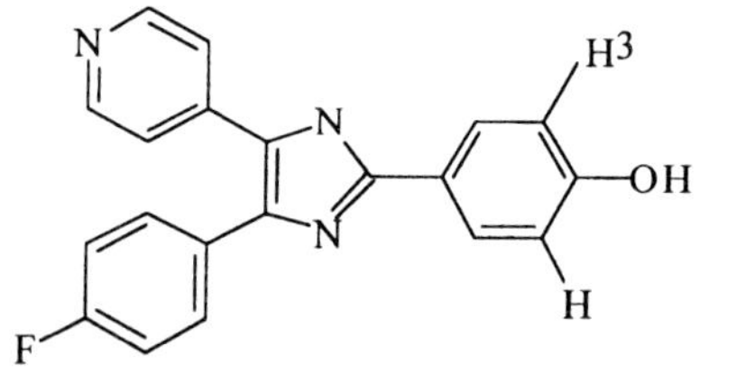

SB202190

Figure 1 Structures of two pyridinyl imidazole inhibitors of p38 MAP kinase.

correlation between cytokine suppression and eicosanoid inhibition, suggesting that they worked through an entirely different mechanism.

In order further to characterize these compounds and determine their mechanism of action, both radiolabeled and radiophotoaffinity analogs of these compounds were synthesized and used to identify, characterize, purify, and sequence the molecular target. The partial peptide sequence obtained allowed the isolation of full-length cDNAs encoding the target, which was called CSBP for *CSAID* *b*inding *p*rotein (9). Its sequence revealed that it was a new member of the mitogen-activated protein (MAP) kinase family of serine-threonine protein kinases. At around the same time, the same kinase was identified through its strong phosphorylation in response to LPS binding to CD14, and it was named p38 MAP kinase (MAPK) (10). This has become the more accepted name and will be used through the rest of this chapter.

Multiple Isoforms of p38 MAP Kinase

p38 MAPK has at least three splice variants. Two of these, CSBP1 and CSBP2 differ in an internal 25 amino acid region which is encoded by two alternatively spliced exons conserved in both mice and humans. CSBP2 is identical to human p38 (10,11). Both forms are ubiquitously expressed and bind to the pyridinyl imidazoles (9), but there are no known differences in activation stimuli or substrate preferences. The third splice variant, Mxi2, was discovered in a yeast two-hybrid screen due to its association with the transcription factor Max (12). This variant lacks 65 amino acids at the carboxyl-terminal end of the protein relative to p38. To date, no function has been associated with Mxi2. Species homologs of p38 have also been identified, and indicate that p38 is conserved from humans to yeast. Indeed, human and mouse p38 are able partially to complement a strain lacking the yeast counterpart, HOG1.

Three homologs of p38 have been identified, as summarized in Table 1. For the purposes of this chapter, we will call them p38β, p38γ, and p38δ. The degree of sequence similarity between p38 and these isoforms is illustrated in Table 1. The closest homolog, p38β, was isolated in two forms differing by the presence (p38β [13]) or absence (p38β2 [14,15]) of an internal region of eight amino acids. It is also called SAPK2b (16) and p38-2 (17). Of the two, the predominant form appears to be p38β2, since p38β seems to be difficult to isolate by reverse transcriptase polymerase chain reaction (RT-PCR) and behaves aberrantly in functional assays (14). Like p38, p38β2 is expressed ubiquitously, whereas the other two homologs, p38γ (18–20) and p38δ (14,16,21), are more restricted in their expression. Other differences between the isoforms are summarized in Table 1 and discussed further below.

Table 1 Members of the p38 MAP Kinase Family

Name	p38	p38β/β2	p38γ	p38δ
Alternative names	SAPK2a CSBP, RK	SAPK2b p38-2	SAPK3 ERK6	SAPK4
% Amino acid identity to p38	100	73	63	61
Size	360	372/364	367	365
Chromosomal location	6p21.2 (167)	—	22q13.3 (168)	6p21.2 (GenBank)
Accession number	L35263/L35264 Z95152	AF001174 AF001008	U66243 Y10487	AF004709 Y10488 Z95152
mRNA expression	Ubiquitous	Ubiquitous	Skeletal muscle	Testes, pancreas, prostate, small intestine
stimulation by PMA in HeLa	No	No	Yes	Yes
Substrates	MAPKAPK2/3 MNK1, MSK1, RSK-B, ATF2, ATF6, MBP, Elk1, SAP-1a MEF2C	MAPKAPK2/3 ATF2, MBP, Elk1 ?	— ATF2, MBP ?	— ATF2, MBP ?
Inhibition by SB203580 or SB202190	Yes	Yes	No	No

Interestingly, at least two of the isoforms, p38 and p38δ, map to the same genetic locus, 6p21.2, as the histocompatability antigen locus. Complete sequencing of this region has revealed the genomic structure for both p38 splice forms and for p38δ.

Comparison of p38 to Other MAP Kinases

MAPK are proline-directed serine-threonine kinases which are highly conserved in evolution in species as diverse as *Homo sapiens, Drosophila melanogaster*, and *Saccharomyces cerevisiae* (22). They are involved in a variety of cellular processes, including cell proliferation, differentiation, and death. Activation of the kinase activity of MAPK occurs by phosphorylation of both a Thr and Tyr residue in a Thr-Xaa-Tyr motif close to the active site (23,24). This phosphorylation is carried out by a MAPKK, which in turn is activated by phosphorylation by a MAPKKK in response to various extracellular or intracellular stimuli (Fig. 2). After activation, the MAPK undergoes a conformational change and can then phosphorylate a number of substrates, including transcription factors, kinases, enzymes, and cytoskeletal proteins (25,26).

Both the identity of the central X amino acid in the TXY motif and the length of the "activation loop" which contains these three residues vary between different MAPK (23). Based on these differences, three different fami-

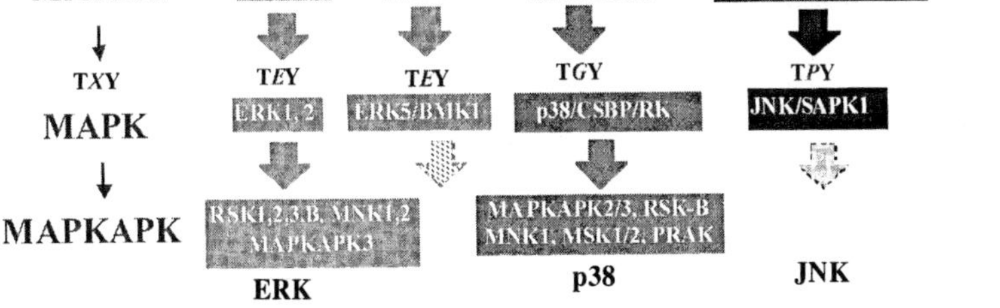

Figure 2 MAP kinase cascades. The generic MAP kinase cascade is shown on the left of the figure, and the specific MAP kinase families and their known members at each level of the kinase cascade on the right.

lies of MAPK have been identified in mammalian cells, the extracellular regulated kinases (ERK) (TEY), p38s (TGY), and the JNK/SAPK (TPY), where the erks have the longest activation loop and the p38s the shortest (23). These MAPK are in turn activated by distinct MAPKK, which themselves are activated by multiple MAPKKK in a more promiscuous fashion (26).

The three families have also been divided by stimuli that activate them. Based on the initial association of the ERK with mitogenic stimuli and the JNK/SAPK and p38 with cellular stresses, they have often been called mitogen-activated kinases (MAPK) and stress-activated kinases (SAPK), respectively. (9,10,23,25–29). However, this is an oversimplification, since many stimuli activate multiple pathways. Furthermore, ERK5 (also called BMK1), which by sequence would appear to be an ERK, is activated by oxidative stress (30,31). Similarly, the p38s are activated by multiple ''mitogenic'' stimuli, as will be further detailed below.

Activation Stimuli for p38 and Its Homologs

p38 can be activated by a diverse number of stimuli that are summarized in Figure 3. The stimuli can be divided into several different classes, including

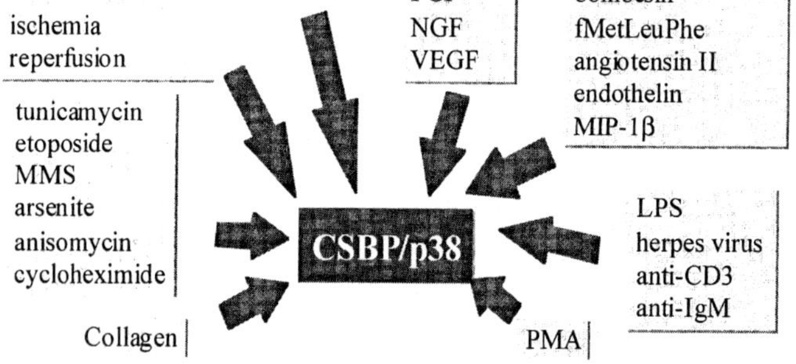

Figure 3 Stimuli reported to activate p38, as in refs. 9, 10, 33, 52, 55, 67, 68, 72, 75, 81, 84, 90, 120, 124, 137, 138, 140, 141, and 146–166.

physiological stress, chemical stress, radiation, inflammatory cytokines (such as members of the TNF and IL-1 families), growth factors, G protein coupled receptor ligands, hormones, and immune receptor activation. Many of these same stimuli also activate the JNK/SAPK pathway, suggesting some conservation in the pathways leading to their activation. Although the ERK do not in general respond to chemical or physiological stress, there are several stimuli for which all three pathways respond (e.g., TNF).

Three major observations can be made. First, the yeast homolog of p38, HOG1, responds to osmotic stress, and in so doing enhances the survival of the host cell (32). Hence, it is likely that physiological stress was one of the most primordial p38 stimuli that has been conserved through evolution. In multicellular organisms, the number of p38 stimuli has expanded to include a wider range of physiological stresses such as ischemia-reperfusion and growth factor withdrawal. In addition, in mammals, p38 is stimulated by cell surface molecules that initiate the host defense to infection, by cytokines that drive recruitment and activation of inflammatory and immune cells, and by hormones that mediate repair. These additional stimuli could be viewed as a way in which p38 has retained its key role in the survival of the host in response to injury and infection and expanded its influence beyond the survival of an individual cell.

Second, some of the p38-dependent secreted products of one stimulus, such as IL-1 and TNF secreted from LPS-activated monocytes, are themselves able to stimulate p38 activity in target cells. This sets up a cascade of p38-dependent steps that result in the expansion of the host response to surrounding cells and tissues.

Third, not all these stimuli work in all cell types. For example, phorbol myristate acetate (PMA) is inactive in HeLa and KB cells, but it does activate p38 and p38β in squamous cell carcinoma cells (33) and acts as a costimulator with calcium ionophores in Jurkat cells (34).

Although less is known about the individual behavior of the p38 homologs, the studies that have been conducted indicate that they are activated by stress stimuli and proinflammatory cytokines in a manner similar to that seen with p38. Some differences have emerged, however. For instance, p38β2 seems to be preferentially activated by bradykinin over p38 in neuroblastoma/glioma hybrid cells (35). In addition, p38γ and p38δ seem to be more susceptible than p38 or p38β2 to activation by some nonstress stimulus such as PMA (14) (see Table 1). Conversely, in 293 cells, p38δ is less susceptible than p38 to activation by IL-1 (21) and p38γ is less inducible by anisomycin (36).

Some of these differences may be due to the MAPKK through which each p38 homolog is activated. p38 can be activated by MKK6, MKK3, and to a lesser extent MKK4 (37–42). However, p38β2 is preferentially activated by MKK6 (15), and there is some evidence that other MKK, perhaps unidentified ones, may be able to activate p38δ in certain cell types (21).

MECHANISM OF ACTION AND SELECTIVITY OF CSAIDs™

Mechanism of Action

As described above, pyridinyl imidazoles were used to isolate CSBP2/p38. It was soon determined that the compounds also inhibit the kinase activity of p38, and that the potency for this inhibition, which reflects binding to the active, phosphorylated form, correlated with the potency for binding to the inactive, nonphosphorylated form. This suggested that the final p38/inhibitor complex may be similar whether the compound binds to the inactive or active form. Although binding of inhibitor to the active form inhibits kinase activity, binding to the inactive form does not block phosphorylation of p38 by MKK3/6 (43). In some reports, the presence of inhibitor can reduce the level of p38 phosphorylation in response to stimulation (44), but preliminary evidence suggests that some p38 phosphorylation may itself be p38 dependent and hence susceptible to inhibition of p38 kinase activity (44a).

Inhibitors Compete with ATP for Binding to p38

A number of biochemical experiments show that the pyridinyl imidazole inhibitors compete with adenosine triphosphate (ATP) for binding to p38 whether in the active or inactive state (43,44). In the inactive state, p38 binds ATP very weakly, and exhibits a K_m of 9.6 mM for an intrinsic ATPase activity. SB203580 binds to this form with a K_D of 15 nM, and binds with even higher affinity in the presence of its protein substrate, MAP kinase-activated protein (MAPKAP) kinase-3 (M. Doyle, unpublished data). In the active form of p38, the K_m for ATP drops to 200 μM with a peptide substrate, and 23 μM when the substrate is the protein ATF2 (43–45). SB203580 competes with ATP with a K_i of 21 nM with a peptide substrate; similar to the affinity seen for the inactive form. Hence, the pyridinyl imidazoles compete better with ATP for binding to the inactive versus the active form of p38.

These biochemical data have been amply confirmed by structural data emerging from cocomplexes of the inactive form of p38 with various inhibitors, such as SB203580, that show that the inhibitors bind in the ATP pocket and overlap the region where ATP might be expected to bind (46–48) (Fig.

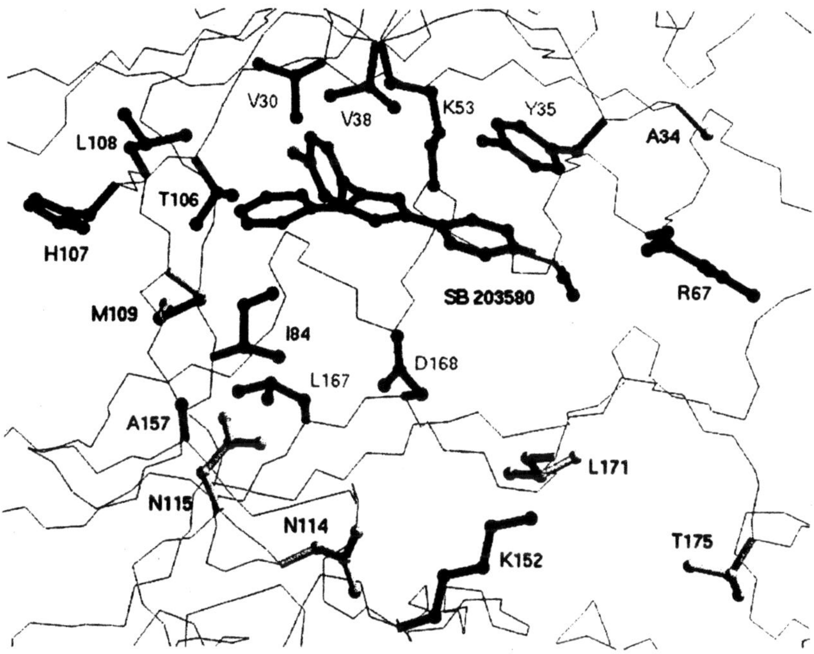

Figure 4 Complex of SB203580 with p38 showing the interaction of SB 203580 with specific side chains in the ATP binding site of p38 (50).

4). However, the inhibitors are located further back in the pocket than ATP and closer to the crossover connection between the amino- and carboxyl-terminal domains that together form the ATP-binding site. Based on comparisons with the structure of inactive and activate ERK, activation of p38 through phosphorylation of tyrosine and threonine should allow the amino- and carboxyl-terminal domains to come together and form a more favorable, catalytically active binding site for ATP. Since the inhibitors bind in a part of the ATP pocket more distant from where these conformational changes occur, this may explain why the inhibitor can bind equally well to both the inactive and active forms of p38, whereas ATP binding is favored in the active conformation.

Selectivity of Inhibitors

SB203580 and SB202190, two commonly used p38 inhibitors, show remarkable selectivity in inhibiting p38 and p38β2 kinase activity but little to no inhibition of many other serine-threonine and tyrosine kinases. This selectivity extends to other closely related MAP kinases such as the ERK JNK/SAPK and even the two p38 homologs p38γ and p38δ (13,14,16,17,49–51). Higher concentrations of the compound appear to inhibit two isoforms of JNK2, JNK2β1 and JNK2β2 (49,52), tumor growth factor (TGF-βR) type I (TGF-βRI) and the tyrosine kinase Lck in vitro (51).

Why p38 Inhibitors Are Selective

The structures of p38 inhibitor complexes have provided many clues as to why the pyridinyl imidazole inhibitors are so selective. Inferences from the structures have been further confirmed by mutagenesis of the inhibitor-binding site. As noted above, the inhibitors overlap the binding site for ATP but appear to occupy a region closer to the crossover connection between the two lobes of the kinase. Consequently, there are some amino acids in the pocket that are required for binding both ATP and inhibitor and some that are specific for binding the inhibitor but with no effect on kinase activity (50).

In particular, one group of residues near the hinge of p38's two lobes form a pocket which uniquely interacts with the fluorophenyl ring of the pyridinyl imidazole inhibitors but which has no interaction with ATP. Mutation of the p38 residue, Thr106, to a larger amino acid such as Met was expected to disrupt the pocket and indeed prevented the binding of inhibitor to p38 (50) (see Fig. 4). Similarly, the Thr106 to Met or Glu mutants were no longer inhibited by SB203580 in kinase assays using protein expressed in bacterial cells (48,51,53,54), although it was inhibited to some extent when expressed in mammalian cells. However, if Thr106 and the adjacent His107 and Leu108 were altered, the mammalian-expressed p38 also completely lost sensitivity to inhibition by SB203580 (50). The differences may be due to the interaction of p38 with other proteins present in the mammalian extract that are missing in the purified bacterially expressed protein. In contrast, substitution of Thr106 with a smaller residue such as Ala had no effect on inhibitor activity (51).

Interestingly, these three residues in p38 and p38β2 (Thr106, His107, and Leu108), which are inhibited by SB203580, are different in p38γ and p38δ, which are not inhibited by SB203580. Substituting these three residues of p38 into p38γ or p38δ resulted in them becoming sensitive to inhibition by SB203580 (50). Furthermore, substitution of these residues into the more distantly related JNK1 or ERK1 also resulted in these MAP kinases becoming sensitive to SB203580 inhibition (50,51,53).

These findings have been taken further, and the presence of Thr or smaller residues at the position equivalent to 106 of p38 has been shown to be somewhat diagnostic for sensitivity to pyridinyl imidazole inhibitors such as SB203580. For example, both TGF-βRI and TGF-βRII kinases and the Src-related tyrosine kinase, Lck have a small residue at this position and show some sensitivity to SB203580, although this sensitivity is still 100-fold less than that seen with p38 (51). Hence, a small amino acid in this position may be shorthand for the presence of a fluorophenyl-binding pocket.

PHYSIOLOGICAL ROLE OF p38

Use of SB203580 to Determine Role of p38

The in vitro data indicating that SB203580 is highly selective for p38 and p38β2 have led to its use in evaluating the role of p38 in cell signaling and in the determination on its in vivo substrates. Some of the effects and molecular targets of p38 activity are summarized in Table 2. Before describing these in more depth, it is worth providing some considerations about SB203580 that impact on our interpretation of the published literature in this area.

Early experiments established that SB203580 inhibited the activation of MAPKAP kinase-2 in cells and the phosphorylation of heat shock protein-27 (HSP27) in a dose-dependent manner, with maximal inhibition seen at concentrations below 10 μM (55). Apart from proving that MAPKAP kinase-2 was a direct substrate of p38 in cells, these results also provided a benchmark dose response which could be used to determine the effect of p38 on other

Table 2 Phenotypic Effects and Molecular Targets of p38 MAP Kinase Signaling

Effect	Target molecules
Gene expression	Transcription factors, chromatin
Transcription	MEF2C, Elk1/SAP-1a, CREB/ATF1, ATF2, ATF6,
mRNA stability	CHOP, IUF1, NFAT; histone H3, HMG14
Translation	Translation factors
Cytoskeletal	eIF-4E
Survival/apoptosis	Cytokines
Differentiation	IL-1, TNF-α, IL-6, IL-8, GM-CSF, IFN-γ, CD23, ANF
Aggregation	Enzymes
	iNOS, COX2, PLA2, MAPKAPK2/3, MSK 1,2,
	MNK 1, PRAK, RSK-B, tyrosine hydroxylase
	Cytoskeletal
	HSP27, LSP1, α B-crystallin, ICAM-1

cellular events, such as transcription and phosphorylation. Hence, interpretation of data from experiments using higher concentrations of SB203580 should consider the possible role of weaker cellular targets of SB203580. For example, SB203580 inhibits the JNK2β isoforms at higher concentrations in vitro, although this has not been verified in cells, since activation of c-jun is not affected by the compound (49,52). It was also previously reported and recently confirmed that SB203580 may inhibit enzymes involved in arachadonic acid metabolism (56,57). Hence, it is probably best to use SB203580 at concentrations below 10 μM. As additional inhibitors become available, it will be important to assess their selectivity prior to use, since even small changes in the structure can lead to reduced selectivity toward other kinases, as was recently reported for one pyridinyl imidazole that inhibited both p38 and ERK (46).

Substrates for p38

A number of p38 substrates are now known. In addition to MAPKAP kinase-2, p38 can phosphorylate and activate the related serine-threonine kinases MAPKAPK-3 (also called 3PK), PRAK, MNK1, MSK1 and RSK-B (58–66). In turn, MAPKAPK-2, MAPKAPK-3, and PRAK phosphorylate HSP27 and LSP-1, which regulate cytoskeletal changes (55,59,60,64,67–70), MNK1 phosphorylates eukaryotic initiation factor 4E (eIF4E), a factor involved in protein translation (61,71), and MSK1 and RSK-B phosphorylate the transcription factor CREB (65,66). MAPKAP kinase-2 (and/or MAPKAP kinase-3, MSK1, PRAK, RSK-B, or MNK1) also phosphorylates tyrosine hydroxylase and alpha B-crystallin. p38 also directly phosphorylates $cPLA_2$ (72–74) and regulates its enzymatic activity. Phosphorylation of all of these substrates is inhibited by SB203580 in cells.

p38 can also phosphorylate several transcription factors in vitro, including ATF2, ATF6, MEF-2C, CHOP, Elk1, and SAP-1a (49,75–80). However, treatment of cells with SB203580 inhibits the phosphorylation of MEF-2C, CHOP, and to some extent Elk-1 and SAP-1a but is without effect on ATF2 (81), suggesting that other MAP kinases may substitute. The inhibitor also suppresses the transcriptional activity of many of these factors. In the case of Elk1 and SAP-1a activation by ultraviolet (UV) radiation, p38 acts in concert with the ERK pathways (77,78).

In other cases, the p38 homologs that are not inhibited by SB203580, p38γ, and p38δ may be involved. Hence, p38β2 shows the same in vitro substrate profile as p38 where it has been tested, including MBP, ATF2, Elk1, SAP-1a, MAPKAPK-2, and MAPKAPK-3 (14,15,17), whereas p38γ and p38δ phosphorylate MBP, ATF2, and Elk1 but are unable to phosphorylate either MAPKAPK-2 or MAPKAPK-3 (14,16,21,36,82). It is likely that unique substrates will eventually be found for p38γ and p38δ which are not shared in common with p38. This bifurcation in p38 homolog signaling is illustrated in Figure 5.

It is now emerging that the selection of MAPK substrates occurs through two regions. One region in part selects the specific amino acid sequence surrounding the serine or threonine to be phosphorylated (83). For the MAP kinases, this includes the adjacent proline. The second region is responsible for specifically binding the substrate and is separated from the active site of the kinase. For p38, interaction with substrate can occur before the kinase is activated and does not require the binding of ATP (59,84). In activated p38, ATP will bind only after substrate is bound (45). Recent studies with chimeras have helped to define this second region. MAPKAPK-2 and MAPKAPK-3 bind to residues in the αD/L8/αE regions of p38 (44a), whereas c-Jun binds to residues in the adjacent αF/L13/αG region of JNK (85) (Fig. 6).

Binding of p38 to members of the MAPKAP kinase family plays a role in its cytoplasmic localization (66,86,87). When unactivated, the complex is predominantly in the nucleus, but on activation of p38, the nuclear export sequence of MAPKAPK-2 is exposed and the complex migrates to the cytoplasm. The interaction is stoichiometric, and inhibition of p38 activity prevents

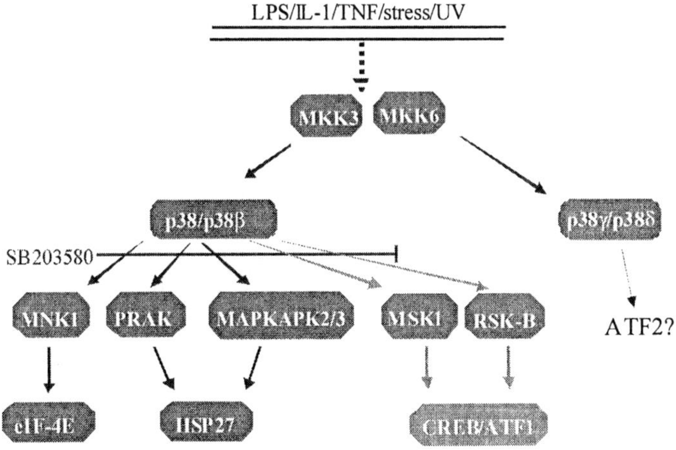

Figure 5 p38 MAP kinase pathways inhibited by SB203580.

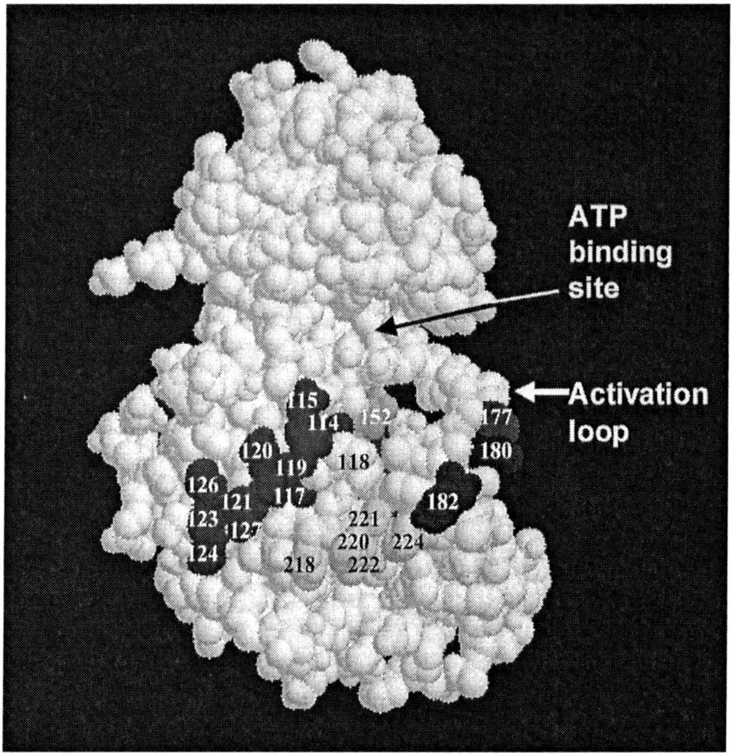

Figure 6 Region of p38 MAP kinase involved in binding to MAPKAP kinase-2. This region (residues 114–124) is shaded. In contrast, c-jun binds to JNK in the region equivalent to p38 MAP kinase residues 218–224. Residues Thr180 and Tyr182 are those phosphorylated by MKK3/MKK6.

migration of the complex to the cytoplasm on cell stimulation. It is likely that this relocalization affects the substrates available for phosphorylation by either kinase.

Role of p38 in Regulation of Gene Expression

One of the results of inhibiting p38 is to affect gene expression. As noted above, several transcription factors are phosphorylated by p38 either directly or indirectly in response to appropriate stimuli such as LPS, UV, radiation, FGF, arsenite, and oxidative stress. However, phosphorylation, even at known sites for activation of the transcription factor, does not appear to guarantee that the transcription factor will activate transcription on every promoter to which it binds, since this may depend on other factors bound to the same promoter (78). Hence, the most revealing studies have been those examining the regulation of natural promoters.

Two key transcription complexes associated with inflammatory responses are regulated by p38, AP-1, and nuclear factor-κB (NF-κB). AP-1 is a complex of c-fos and c-jun, and the stress- and cytokine-activated expression of both is inhibited by SB203580 at the transcriptional level (81,88,89). The target transcription factors of p38 in these cases are MEF2C for the c-jun promoter (76) and Sap-1 and CREB for the c-fos promoter (77,78,90,91). In contrast, p38 has no effect on the activation of NF-κB or degradation of IκB by various cytokines or stress, but it does appear to inhibit several NF-κB–driven promoters, such as human immunodeficiency virus long terminal repeat (HIV LTR) and IL-6 through a mechanism that has not yet been elucidated (92–94).

Other promoters are also regulated by p38 in vivo. These include the IUF1-mediated induction of the insulin promoter in response to high glucose (95), the ATF6-mediated induction of atrial natruretic factor (ANF) expression in response to stress (80), and the Elk1-regulated expression of the immediate-early egr promoter in response to stress (96).

However, regulation of gene expression is not limited to transcription. Recently, evidence was obtained for the regulation of LPS-induced cyclooxygenase-2 (COX2) production in monocytes and IL-1–induced IL-6 production in synovial fibroblasts at the level of mRNA turnover (97,98). Also, the production of IL-1 and TNF in monocytes in response to LPS and TNF-α–induced vascular cell adhesion molecule (VCAM) expression in human umbilical vein endothelial cells (HUVEC) may involve regulation at the translational level (99,100).

Role of p38 in Production of Cytokines and Inflammatory Mediators

The original discovery of p38 was intimately tied to its role in regulating inflammatory cytokine production from LPS-activated monocytes (9). Since that time, additional cytokines have been shown to depend on p38 activity for production in response to various stimuli (see Table 2). These include LPS-induced CD23 (101); LPS-, IL-1-, or TNF-induced IL-6 (93,97,102–104); TNF-induced GM-CSF (93); LPS, PAF, IL-1, TNF, osmotic and temperature stress-induced IL-8 (105–108); and activated T-cell production of IL-2 and interferon-γ (IFN-γ) (34,109).

A number of enzymes responsible for the production of inflammatory mediators or degradative enzymes are also regulated by p38 at the level of production. These include cyclooxygenase-2 (COX2) (98,110,111), inducible nitric oxide (iNOS) (104,111–114), and matrix metalloproteinases (33). In contrast, PLA_2 is directly regulated through phosphorylation by p38 (72).

Interestingly, all of these cytokines and enzymes are associated with the initial inflammatory response, which leads to recruitment of circulating leukocytes to localized areas of infection or injury, systemic responses through acute phase protein synthesis, the generation of a cellular immune response, degradation of tissue, and the generation of pain and swelling. These are the components of many chronic inflammatory diseases, and may help to explain the efficacy of p38 MAP kinase inhibitors in animal models of these diseases, as discussed further below (115).

Role of p38 in Cell Proliferation and Apoptosis

Several studies have suggested a role for p38 in regulating proliferation and apoptosis. For example, SB203580 has been reported to inhibit the proliferation of CD40-stimulated tonsillar B cells and B-cell lines (92) and G-CSF–stimulated BaF3 cells expressing transfected G-CSF receptors (116). In the latter case, proliferation required activation both p38 and ERK kinase pathways. Similarly, IL-2 and IL-7 proliferation is inhibited by SB203580, but at higher concentrations than those needed to inhibit p38 MAP kinase (117). However, an inhibitor of both p38 and JNK pathways, CNI1493, did not inhibit proliferation, suggesting that the effect of SB203580 on proliferation at these higher concentrations may be through a target other than p38 (118). These data are also consistent with the lack of significant effect of the compound in various antigen-induced T-cell proliferation assays (115).

The effects seen on apoptosis appear contradictory depending on the cell type and stimulus. Thus, apoptosis induced by Fas in Jurkat cells (119,120), by UV in U937 cells (121), and by H_2O_2 in HeLa cells (122) is not affected by p38 inhibition. Although p38 inhibition enhances apoptosis in TNF-treated L929 or NIH 3T3 cells (123), it increases survival in Rat-1 fibroblasts or differentiated PC12 neural cells deprived of growth factors. A similar contradiction is seen in cells isolated from natural sources. Thus, apoptosis induced by LPS in eosinophils, by B-cell antigen receptor cross linking in B cells, and anisomycin in cardiac myocytes are not affected by p38 inhibition (124,125), whereas stress-induced apoptosis in neutrophils (126) and glutamate-induced apoptosis in matured cerebellar granular cells (127) are inhibited. This apparent dichotomy may in part be due to different proapoptotic and antiapoptotic activities of the p38 MAP kinase family members (119,128,129). For example, p38α was proapoptotic and p38β hypertrophic in cardiomyocytes (129), so that the effects of the p38 inhibitor, which blocks both of these isoforms, may depend on which p38 isoforms are most activated in a cell by a particular stimulus. p38 may also collaborate with other MAP kinase pathways (130) in regulating cell survival.

Role of p38 in Cell Differentiation

Several cell lines known to differentiate in vitro have been tested for the role of p38 through the use of the inhibitor SB203580. The differentiation of the pheochromocytoma cell line PC12 to neurons by NGF is mediated by a sustained activation of either ERK and p38 MAP kinase combined with a transient activation of the other kinase (131). Inhibition of either the ERK or p38 kinase pathways results in inhibition of NGF-driven differentiation, suggesting that the two kinases work in concert. Similarly, p38 seems to be important in the adipocyte differentiation of 3T3-L1 cells in response to insulin (132), the differentiation of C2C12 myoblasts to myotubes, and the erythropoietin-induced proliferation and/or differentiation of erythroid progenitor cell lines (133). Like PC12 cells, p38 MAP kinase acts in concert with other pathways to induce differentiation. In C2C12 cells, the rapamycin-sensitive p70 S6 kinase pathway was also essential, and in erythroid cells the JNK pathway was required.

However, all of these effects are observations in cell lines, which may have altered signaling properties as a result of long-term culture. Therefore, it is of interest to see if these observations are extended to naturally derived cells or observed in in vivo toxicology studies. One such example is the finding that the p38 pathway plays a role in negative selection of thymocytes in contrast to the positive selection driven by the ERK pathway (134).

Role of p38 in Cell Morphology, Aggregation, and Migration

p38 has been implicated in changes of morphology. As mentioned above, p38β can directly stimulate cardiac myocyte hypertrophy. The relevance of this physiologically is the finding that the hypertrophic response of rat neonatal ventricular myocytes to various agonists such as endothelin-1, phenylephrine, and leukemia-inhibitory factor (LIF) is altered by p38 inhibition (135).

Other changes in the cytoskeleton have been associated with the phosphorylation of HSP27 by the p38 MAP kinase pathway through MAPKAP kinase-2. For example, in HUVEC cells subjected to oxidative stress, F-actin is reorganized from corticol microfilaments into transcytoplasmic stress fibers and vinculin is recruited to focal adhesions. All these changes are blocked by SB203580 (136). Similar p38-dependent changes are seen in cholecystokinin (CCK)–treated rat pancreatic acini (137), and the follicule-stimulating hormone (FSH) induction of rounding and aggregation of immature granulosa cells likely involves a similar mechanism (138).

Some activities of inflammatory cells are also mediated by p38. Thus, macrophage spreading in response to oxidant stress (139), TGF-β1–induced chemoattraction, actin polymerization of human neutrophils (140), and platelet aggregation induced by thromboxane and collagen (141) are all inhibited by SB203580. p38 can also play a role in the regulation of adhesion molecule expression as in the upregulation of ICAM-1 on human pulmonary microvascular endothelial cells in response to LPS or TNF (142).

p38 AS A THERAPEUTIC TARGET

By the time the molecular target of the pyridinyl imidazole inhibitors of inflammatory cytokine production was discovered to be p38 MAP kinase, it was already known that these compounds were effective in several models of inflammatory disease. As more potent and selective p38 inhibitors have been developed, they have helped further to substantiate the key role of this kinase in several models of disease that contain an inflammatory component.

The ability of p38 inhibitors to inhibit inflammatory cytokine production in vitro has been demonstrated in vivo. For example, pretreatment of rats or mice with SB203580 or the related SB220025 inhibited LPS-induced TNF production at doses below 50 mg/kg (115,143). These data have been used to establish the dose range for testing the activity of p38 inhibitors in various animal models.

In the adjuvant arthritis model in the rat induced by a single injection of *Mycobacterium butyricum*, prophylactic treatment with SB203580 dose dependently reduced the hindpaw swelling observed after 16 and 22 days (115). Reversal was also seen in bone mineral density and bone joint damage, and the rise in circulating IL-6 was inhibited. Similarly, improvements in disease severity and reduction of acute phase reactant levels were seen in collagen-induced arthritis in the mouse with therapeutic dosing of either SB203580 or SB220025 (115,143). Part of the effectiveness of CSAID™ in rheumatoid arthritis may be due in part to their effect on angiogenesis, as illustrated by the inhibition of angiogenesis by SB220025 in a murine air pouch granuloma induced by Freund's complete adjuvant and croton oil (143). Another component of CSAID™ action may be its effect on bone remodeling, since SB203580 inhibited osteoclast-mediated bone resorption of fetal rat long bones (115). These data suggest that p38 inhibitors have good potential in treating rheumatoid arthritis.

The copious production of inflammatory cytokines has also been associated with acute endotoxin shock, and like inhibitors of IL-1 and TNF action, SB203580 improved survival in an acute model of endotoxin shock in which mice sensitized with D-(+)-Gal were treated with a lethal dose of endotoxin (115). However, SB203580 appeared to be without activity in more chronic shock models (P. DeMarsh, unpublished studies), which is consistent with the lack of activity of IL-Ira and antibodies to TNF in clinical septicemia.

p38 may also play a significant role in vascular disease, since it is activated in both ischemia and reperfusion. This has been borne out by studies in both stroke and cardiac restenosis models. In a middle cerebral artery occlusion (MCAO) model in hypertensive (SHR) and normotensive rats, a p38 inhibitor SB239063, which is more potent and selective than SB203580, led to significant brain protection and a reduced neurological deficit (144). Similarly, in a perfused rabbit heart model of ischemia-reperfusion, SB203580 decreased myocardial apoptosis and improved postischemic cardiac function (145). In this case, the strongest effects were seen if the compound was added at the beginning of reperfusion when p38 activation was highest. These data suggest that p38 inhibitors may have utility in the treatment of stroke and ischemic heart disease. Preliminary data also suggest that p38 inhibition may suppress neointima formation due to restenosis (E. Ohlstein, personal communication) and therefore may be of use in percutaneous transluminal coronary angioplasty (PTCA).

Early characteristics of the p38 inhibitors in vivo suggest that they are not immunosuppressive (115), but rather inhibit the production of macrophage- and T-cell–derived cytokines and inflammatory mediators that stimulate inflammatory cell infiltration, proliferation, and action. Hence, the compounds may have an advantage over currently available therapies for inflammatory diseases. Immunosuppressive compounds, such as cyclosporine and methotrexate, may increase the susceptibility of the patients to infection, whereas inhibitors of enzymes involved in eicosanoid production, such as NSAID, treat only the acute symptoms and not the underlying degenerative pathology.

An analysis of the effects of p38 inhibition by SB203580 in cell culture might suggest that inhibitors of p38 MAP kinase would have several other activities in vivo that could lead to side effects. However, preliminary studies with more potent and selective compounds have not confirmed these effects in vivo. This might be because the in vitro data with SB203580 does not translate to what is seen in vivo, that SB203580 may bind (and inhibit) additional molecular targets, or that there are differences in pharmacokinetics and distribution of the compounds. To address completely how many cellular processes are regulated by p38 in vivo will likely require a comparison of the activities of p38 inhibitors from at least two different structural classes, since this should presumably rule out effects through a second molecular target.

As several p38 inhibitors are approaching clinical trials, we should soon know whether they bear out the promise of the animal model studies and will provide a new and superior treatment for various inflammatory diseases.

NOMENCLATURE

ATF	activating transcription factor
CHOP	CCAAT enhancer binding protein homologous protein
$cPLA_2$	cytosolic phospholipase A_2
CREB	cAMP response element binding protein
IκB	inhibitor of NF-κB
IUF	insulin upstream factor
LSP	lymphocyte specific protein
MBP	myelin basic protein
MEF	myocyte enhancer factor
MNK	MAP kinase signal-integrating kinase
MSK1	mitogen and stress-activated protein kinase
RSK-B	ribosomal S6 kinase-B
SAP	serum response factor accessory protein

ACKNOWLEDGMENT

We would like to thank John Kyriakis for an update on the current understanding of MAPKKK specificity.

REFERENCES

1. CA Dinarello. The interleukin-1 family: 10 years of discovery. FASEB J 8: 1314–1325, 1994.
2. HJ Gruss, SK Dower. Tumor necrosis factor ligand superfamily: involvement in the pathology of malignant lymphomas. Blood 85:3378–3404, 1995.

3. CA Dinarello. Controlling the production of interleukin-1 and tumor necrosis factor in disease. Nutrition 11:695–697, 1995.
4. KM Murray, SL Dahl. Recombinant human tumor necrosis factor receptor (p75) Fc fusion protein (TNFR:Fc) in rheumatoid arthritis. Ann Pharmacother 31:1335–1338, 1997.
5. RN Maini, FC Breedveld, JR Kalden, JS Smolen, D Davis, JD Macfarlane, C Antoni, B Leeb, MJ Elliott, JN Woody, TF Schaible, M Feldmann. Therapeutic efficacy of multiple intravenous infusions of anti-tumor necrosis factor alpha monoclonal antibody combined with low-dose weekly methotrexate in rheumatoid arthritis (see Comments). Arthritis Rheuma 41:1552–1563, 1998.
6. SW Lee, AP Tsou, H Chan, J Thomas, K Petrie, EM Eugui, AC Allison. Glucocorticoids selectively inhibit the transcription of the interleukin 1 beta gene and decrease the stability of interleukin 1 beta mRNA. Proc Natl Acad Sci USA 85:1204–1208, 1988.
7. I Lantos, PE Bender, KS Razgaitis, BM Sutton, MJ DiMartino, DE Griswold, DT Walz. Anti-inflammatory activity of 5,6-diaryl-2,3-dihydroimidazo-2 [2,1-b]thiazoles, Isomeric 4-pyridyl and 4-substituted phyl derivatives. J Med Chem 27:72–75, 1984.
8. JC Lee, L Rebar, JT Laydon. Effects of SK&F 86002 on cytokine production by human monocytes. Agents Actions 27:277–279, 1989.
9. JC Lee, JT Laydon, PC McDonnell, TF Gallagher, S Kumar, D Green, D McNulty, MJ Blumenthal, JR Heys, SW Landvatter, JE Strickler, MM McLaughlin, IR Siemens, SM Fisher, GP Livi, JR White, JL Adams, PR Young. A protein kinase involved in the regulation of inflammatory cytokine biosynthesis. Nature 372:739–746, 1994.
10. J Han, JD Lee, L Bibbs, RJ Ulevitch. A MAP kinase targeted by endotoxin and hyperosmolarity in mammalian cells. Science 265:808–811, 1994.
11. J Han, B Richter, Z Li, V Kravchenko, RJ Ulevitch. Molecular cloning of human p38 MAP kinase. Biochim Biophys Acta 1265:224–227, 1995.
12. AS Zervos, L Faccio, JP Gatto, JM Kyriakis, R Brent. Mxi2, a mitogen-activated protein kinase that recognizes and phosphorylates Max protein. Proc Natl Acad Sci USA 92:10531–10534, 1995.
13. Y Jiang, C Chen, Z Li, W Guo, JA Gegner, S Lin, J Han. Characterization of the structure and function of a new mitogen-activated protein kinase (p38β). J Biol Chem 271:17920–17926, 1996.
14. S Kumar, PC McDonnell, RJ Gum, AT Hand, JC Lee, PR Young. Novel homologues of CSBP/p38 MAP kinase: activation, substrate specificity and sensitivity to inhibition by pyridinyl imidazoles. Biochem Biophys Res Commun 235: 533–538, 1997.
15. H Enslen, J Raingeaud, RJ Davis. Selective activation of p38 mitogen-activated protein (MAP) kinase isoforms by the MAP kinase kinase MKK3 and MKK6. J Biol Chem 273:1741–1748, 1998.
16. M Goedert, A Cuenda, M Craxton, R Jakes, P Cohen. Activation of the novel stress-activated protein kinase SAPK4 by cytokines and cellular stresses is me-

diated by SKK3 (MKK6); comparison of its substrate specificity with that of other SAP kinases. EMBO J 16:3563–3571, 1997.

17. B Stein, MX Yang, DB Young, R Janknecht, T Hunter, BW Murray, MS Barbosa. p38-2, A novel mitogen-activated protein kinase with distinct properties. J Biol Chem 272:19509–19517, 1997.
18. S Mertens, M Craxton, M Goedert. SAP kinase-3, a new member of the family of mammalian stress-activated protein kinases. FEBS Lett 383:273–276, 1996.
19. Z Li, Y Jiang, RJ Ulevitch, J Han. The primary structure of p38g: a new member of the p38 group of MAP kinases. Biochem Biophys Res Commun 228:334–340, 1996.
20. C Lechner, MA Zahalka, J-F Giot, NPH Moller, A Ullrich. ERK6, a mitogen-activated protein kinase involved in C2C12 myoblast differentiation. Proc Natl Acad Sci USA 93:4355–4359, 1996.
21. Y Jiang, H Gram, M Zhao, L New, J Gu, L Feng, F Di Padova, RJ Ulevitch, J Han. Characterization of the structure and function of the fourth member of p38 group mitogen-activated protein kinases, p38delta. J Biol Chem 272: 30122–30128, 1997.
22. D Kultz. Phylogenetic and functional classification of mitogen- and stress-activated protein kinases. J Mol Evol 46:571–88, 1998.
23. MH Cobb, EJ Goldsmith. How MAP kinases are regulated. J Biol Chem 270: 14843–14846, 1995.
24. SK Hanks, AM Quinn, T Hunter. The protein kinase family: conserved feature and deduced phylogeny of the catalytic domains. Science 241:42–52, 1988.
25. E Cano, LC Mahadevan. Parallel signal processing among mammalian MAPKs. Trends Biochem Sci 20:117–122, 1995.
26. JM Kyriakis, J Avruch. Sounding the alarm: protein kinase cascades activated by stress and inflammation. J Biol Chem 271:24313–24316, 1996.
27. J Blenis. Signal transduction via the MAP kinases: proceed at your own RSK. Proc Natl Acad Sci USA 90:5889–5892, 1993.
28. RJ Davis. The mitogen-activated protein kinase signal transduction pathway. J Biol Chem 268:14553–14556, 1993.
29. P Cohen. The search for physiological substrates of MAP and SAP kinases in mammalian cells. Trends Cell Biol 7:353–361, 1997.
30. G Zhou, ZQ Bao, JE Dixon. Components of a new human protein kinase signal transduction pathway. J Biol Chem 270:12665–12669, 1995.
31. J Abe, M Kusuhara, RJ Ulevitch, BC Berk, JD Lee. Big mitogen-activated protein kinase 1 (BMK1) is a redox-sensitive kinase. J Biol Chem 271:16586–16590, 1996.
32. JL Brewster, VT de, ND Dwyer, E Winter, MC Gustin. An osmosensing signal transduction pathway in yeast. Science 259:1760–1763, 1993.
33. C Simon, H Goepfert, D Boyd. Inhibition of the p38 mitogen-activated protein kinase by SB 203580 blocks PMA-induced Mr 92,000 type IV collagenase secretion and in vitro invasion. Cancer Res 58:1135–1139, 1998.

34. S Matsuda, T Moriguchi, S Koyasu, E Nishida. T lymphocyte activation signals for interleukin-2 production involve activation of MKK6-p38 and MKK7-SAPK/JNK signaling pathways sensitive to cyclosporin A. J Biol Chem 273: 12378–12382, 1998.
35. MA Wilk Blaszczak, B Stein, S Xu, MS Barbosa, MH Cobb, F Belardetti. The mitogen-activated protein kinase p38-2 is necessary for the inhibition of N-type calcium current by bradykinin. J Neurosci 18:112–118, 1998.
36. A Cuenda, P Cohen, V Buee Scherrer, M Goedert. Activation of stress-activated protein kinase-3 (SAPK3) by cytokines and cellular stresses is mediated via SAPKK3 (MKK6); comparison of the specificities of SAPK3 and SAPK2 (RK/p38). EMBO J 16:295–305, 1997.
37. B Derijard, J Raingeaud, T Barrett, I-H Wu, J Han, RJ Ulevitch, RJ Davis. Independent human MAP kinase signal transduction pathways defined by MEK and MKK Isoforms. Science 267:682–685, 1995.
38. A Lin, A Minden, H Martinetto, FX Claret, C Lange Carter, F Mercurio, GL Johnson, M Karin. Identification of a dual specificity kinase that activates the Jun kinases and p38-Mpk2. Science 268:286–290, 1995.
39. J Han, JD Lee, Y Jiang, Z Li, L Feng, RJ Ulevitch. Characterization of the structure and function of a novel MAP kinase kinase (MKK6). J Biol Chem 271:2886–2891, 1996.
40. T Moriguchi, N Kuroyanagi, K Yamaguchi, Y Gotoh, K Irie, T Kano, K Shirakabe, Y Muro, H Shibuya, K Matsumoto, E Nishida, M Hagiwara. A novel kinase cascade mediated by mitogen-activated protein kinase kinase 6 and MKK3. J Biol Chem 271:13675–13679, 1996.
41. B Stein, H Brady, MX Yang, DB Young, MS Barbosa. Cloning and characterization of MEK6, a novel member of the mitogen-activated protein kinase kinase cascade. J Biol Chem 271:11427–11433, 1996.
42. A Cuenda, G Alonso, N Morrice, M Jones, R Meier, P Cohen, AR Nebreda. Purification and cDNA cloning of SAPKK3, the major activator of RK/p38 in stress- and cytokine-stimulated monocytes and epithelial cells. EMBO J 15: 4156–4164, 1996.
43. PR Young, MM McLaughlin, S Kumar, S Kassis, ML Doyle, D McNulty, TF Gallagher, S Fisher, PC McDonnell, SA Carr, MJ Huddleston, G Seibel, TG Porter, GP Livi, JL Adams, JC Lee. Pyridinyl imidazole inhibitors of p38 mitogen-activated protein kinase bind in the ATP site. J Biol Chem 272: 12116–12121, 1997.
44. B Frantz, T Klatt, M Pang, J Parsons, A Rolando, H Williams, MJ Tocci, SJ Okeefe, EA Oneill. The activation state of P38 mitogen-activated protein kinase determines the efficiency of ATP competition for pyridinylimidazole inhibitor binding. Biochemistry 37:13846–13853, 1998.
44a. RJ Gum, PR Young. Identification of two distinct regions of p38 MAP kinase required for substrate binding and phosphorylation. Biochem Biophys Res Commun 266:284–289, 1999.
45. PV LoGrasso, B Frantz, AM Rolando, SJ O'Keefe, JD Hermes, EA O'Neill.

Kinetic mechanism for p38 MAP kinase. Biochemistry 36:10422–10427, 1997.
46. Z Wang, BJ Canagarajah, JC Boehm, S Kassis, MH Cobb, PR Young, JL Adams, EJ Goldsmith. How protein kinases can be specific. Structure 6:1117–1128, 1998.
47. L Tong, S Pav, DM White, W Rogers, KM Crane, CL Cywin, ML Brown, CA Pargellis. A highly specific inhibitor of human p38 MAP kinase binds in the ATP pocket. Nat Struct Biol 4:311–316, 1997.
48. KP Wilson, PG McCaffrey, K Hsiao, S Pazhanisamy, V Galullo, GW Bemis, MJ Fitzgibbon, PR Caron, MA Murcko, MSS Su. The structural basis for the specificity of pyridinylimidazole inhibitors of p38 MAP kinase. Chem Biol 4: 423–431, 1997.
49. AJ Whitmarsh, S-H Yang, MSS Su, AD Sharrocks, RJ Davis. Role of p38 and JNK mitogen-activated protein kinases in the activation of ternary complex factors. Mol Cell Biol 17:2360–2371, 1997.
50. RJ Gum, MM McLaughlin, S Kumar, Z Wang, MJ Bower, JC Lee, JL Adams, GP Livi, EJ Goldsmith, PR Young. Acquisition of sensitivity of stress-activated protein kinases to the p38 inhibitor, SB 203580, by alteration of one or more amino acids within the ATP binding pocket. J Biol Chem 273:15605–15610, 1998.
51. PA Eyers, M Craxton, N Morrice, P Cohen, M Goedert. Conversion of SB 203580–insensitive MAP kinase family members to drug-sensitive forms by a single amino acid substitution. Chem Biol 5:321–328, 1998.
52. A Clerk, PH Sugden. The p38-MAPK inhibitor, SB203580, inhibits cardiac stress-activated protein kinases/c-Jun N-terminal kinases (SAPKs/JNKs). FEBS Lett 426:93–96, 1998.
53. T Fox, JT Coll, XL Xie, PJ Ford, UA Germann, MD Porter, S Pazhanisamy, MA Fleming, V Galullo, MSS Su, KP Wilson. A single amino acid substitution makes Erk2 susceptible to pyridinyl imidazole inhibitors of P38 MAP kinase. Protein Sci 7:2249–2255, 1998.
54. J Lisnock, A Tebben, B Frantz, EA O'Neill, G Croft, SJ O'Keefe, B Li, C Hacker, S de Laszlo, A Smith, B Libby, N Liverton, J Hermes, P LoGrasso. Molecular basis for p38 protein kinase inhibitor specificity. Biochemistry 37: 16573–16581, 1998.
55. A Cuenda, J Rouse, YN Doza, R Meier, P Cohen, TF Gallagher, PR Young, JC Lee. SB 203580 is a specific inhibitor of a MAP kinase homologue which is stimulated by cellular stresses and interleukin-1. FEBS Lett 364:229–233, 1995.
56. AG Borsch Haubold, S Pasquet, SP Watson. Direct inhibition of cyclooxygenase-1 and-2 by the kinase inhibitors SB 203580 and PD 98059. J Biol Chem 273:28766–28772, 1998.
57. DE Griswold, PR Young. Pharmacology of cytokine suppressive anti-inflammatory drug binding protein (CSBP), a novel stress induced kinase. Pharmacol Commun 7, 1996.

58. D Stokoe, DG Campbell, S Nakielny, H Hidaka, SJ Leevers, CJ Marshall, P Cohen. MAPKAP kinase-2: a novel protein kinase activated by mitogen activated protein kinase. EMBO J 11:3985–3992, 1992.
59. MM McLaughlin, S Kumar, PC McDonnell, S Van Horn, JC Lee, GP Livi, PR Young. Identification of mitogen-activated protein (MAP) kinase-activated protein kinase-3, a novel substrate of CSBP p38 MAP kinase. J Biol Chem 271:8488–8492, 1996.
60. S Ludwig, K Engel, A Hoffmeyer, G Sithanandam, B Neufeld, D Palm, M Gaestel, UR Rapp. 3pK, A novel mitogen-activated protein (MAP) kinase–activated protein kinase, is targeted by three MAP kinase pathways. Mol Cell Biol 16, 1996.
61. AJ Waskiewicz, A Flynn, CG Proud, JA Cooper. Mitogen-activated protein kinases activate the serine/threonine kinases Mnk1 and Mnk2. EMBO J 16: 1909–1920, 1997.
62. R Fukunaga, T Hunter. MNK1, a new MAP kinase-activated protein kinase, isolated by a novel expression screening method for identifying protein kinase substrates. EMBO J 16:1921–1933, 1997.
63. H Ni, XS Wang, K Diener, Z Yao, MAPKAPK5, a novel mitogen-activated protein kinase (MAPK)–activated protein kinase, is a substrate of the extracellular-regulated kinase (ERK) and p38 kinase. Biochem Biophys Res Commun 243:492–496, 1998.
64. L New, Y Jiang, M Zhao, K Liu, W Zhu, LJ Flood, Y Kato, GCN Parry, J Han. PRAK, a novel protein kinase regulated by the p38 MAP kinase. EMBO J 17:3372–3384, 1998.
65. M Deak, AD Clifton, JM Lucocq, DR Alessi. Mitogen and Stress Activated protein kinase (MSK1), a novel two-kinase domain enzyme that is directly activated by MAPK and SAPK2/p38, and which may mediate the activation of CREB. EMBO J 17:4426–4441, 1998.
66. B Pierrat, JD Correia, JL Mary, M Tomaszuber, W Lesslauer. Rsk-B, a novel ribosomal S6 kinase family member, is a CREB kinase under dominant control of p38-alpha mitogen-activated protein kinase (p38-Alpha(MAPK)). J Biol Chem 273:29661–29671, 1998.
67. J Rouse, P Cohen, S Trigon, M Morange, A Alonso-Llamazares, D Zamanillo, T Hunt, AR Nebreda. Identification of a novel protein kinase cascade stimulated by chemical stress and heat shock which activates MAP kinase–activated protein MAPKAP kinase-2 and induces phosphorylation of the small heat shock proteins. Cell 78:1027–1037, 1994.
68. NW Freshney, L Rawlinson, F Guesdon, E Jones, S Cowley, J Hsuan, J Saklatvala. Interleukin-1 activates a novel protein kinase cascade that results in the phosphorylation of Hsp27. Cell 78:1039–1049, 1994.
69. J Landry, J Huot. Modulation of actin dynamics during stress and physiological stimulation by a signaling pathway involving p38 MAP kinase and heat-shock protein 27. Biochem Cell Biol 73:703–707, 1995.
70. C-K Huang, L Zhan, Y Ai, J Jongstra. LSP1 is the major substrate for mitogen-

activated protein kinase-activated protein kinase 2 in human neutrophils. J Biol Chem 272:17–19, 1997.
71. SJ Morley, L McKendrick. Involvement of stress-activated protein kinase and p38/RK mitogen-activated protein kinase signaling pathways in the enhanced phosphorylation of initiation factor 4E in NIH 3T3 cells. J Biol Chem 272: 17887–17893, 1997.
72. RM Kramer, EF Roberts, SL Um, AG Borsch Haubold, SP Watson, MJ Fisher, JA Jakubowski. p38 mitogen-activated protein kinase phosphorylates cytosolic phospholipase A2 ($cPLA_2$) in thrombin-stimulated platelets. Evidence that proline-directed phosphorylation is not required for mobilization of arachidonic acid by $cPLA_2$. J Biol Chem 271:27723–27729, 1996.
73. AG Borsch Haubold, F Bartoli, J Asselin, T Dudler, RM Kramer, R Apitz Castro, SP Watson, MH Gelb. Identification of the phosphorylation sites of cytosolic phospholipase A2 in agonist-stimulated human platelets and HeLa cells. J Biol Chem 273:4449–4458, 1998.
74. AG Borsch Haubold, RM Kramer, SP Watson. Phosphorylation and activation of cytosolic phospholipase A2 by 38-kDa mitogen-activated protein kinase in collagen-stimulated human platelets. Eur J Biochem 245:751–759, 1997.
75. XZ Wang, D Ron. Stress-induced phosphorylation and activation of the transcription factor CHOP (GADD153) by p38 MAP Kinase. Science 272:1347–1349, 1996.
76. J Han, Y Jiang, Z Li, VV Kravchenko, RJ Ulevitch. Activation of the transcription factor MEF2C by the MAP kinase p38 in inflammation. Nature 386:296–299, 1997.
77. MA Price, FH Cruzalegui, R Treisman. The p38 and ERK MAP kinase pathways cooperate to activate ternary complex factors and c-fos transcription in response to UV light. EMBO J 15:6552–6563, 1996.
78. R Janknecht, T Hunter. Convergence of MAP kinase pathways on the ternary complex factor Sap-1a. EMBO J 16:1620–1627, 1997.
79. J Raingeaud, AJ Whitmarsh, T Barrett, B Derijard, RJ Davis. MKK3- and MKK6-regulated gene expression is mediated by the p38 mitogen-activated protein kinase signal transduction pathway. Mol Cell Biol 16:1247–1255, 1996.
80. DJ Thuerauf, ND Arnold, D Zechner, DS Hanford, KM DeMartin, PM McDonough, R Prywes, CC Glembotski. p38 Mitogen-activated protein kinase mediates the transcriptional induction of the atrial natriuretic factor gene through a serum response element. A potential role for the transcription factor ATF6. J Biol Chem 273:20636–20643, 1998.
81. CA Hazzalin, E Cano, A Cuenda, MJ Barratt, P Cohen, LC Mahadevan. p38/RK is essential for stress-induced nuclear responses: Jnk/Sapks and c-Jun/Atf-2 phosphorylation are insufficient. Curr Biol 6:1028–1031, 1996.
82. XS Wang, K Diener, CL Manthey, S Wang, B Rosenzweig, J Bray, J Delaney, CN Cole, PY Chan Hui, N Mantlo, HS Lichenstein, M Zukowski, Z Yao. Mo-

lecular cloning and characterization of a novel p38 mitogen-activated protein kinase. J Biol Chem 272:23668–23674, 1997.
83. Y Jiang, Z Li, EM Schwarz, A Lin, K Guan, RJ Ulevitch, J Han. Structure-function studies of p38 mitogen-activated protein kinase. Loop 12 influences substrate specificity and autophosphorylation, but not upstream kinase selection. J Biol Chem 272:11096–11102, 1997.
84. E Krump, JS Sanghera, SL Pelech, W Furuya, S Grinstein. Chemotactic peptide N-formyl-met-leu-phe activation of p38 mitogen-activated protein kinase (MAPK) and MAPK-activated protein kinase-2 in human neutrophils. J Biol Chem 272:937–944, 1997.
85. T Kallunki, B Su, I Tsigelny, HK Sluss, B Derijard, G Moore, R Davis, M Karin. JNK2 contains a specificity-determining region responsible for efficient c-jun binding and phosphorylation. Genes Dev 8:2996–3007, 1994.
86. R Ben-Levy, S Hooper, R Wilson, HF Paterson, CJ Marshall. Nuclear export of the stress-activated protein kinase p38 mediated by its substrate MAPKAP kinase-2. Curr Biol 8:1049–1057, 1998.
87. K Engel, A Kotlyarov, M Gaestel. Leptomycin B-sensitive nuclear export of MAPKAP kinase 2 is regulated by phosphorylation. EMBO J 17:3363–3371, 1998.
88. CA Hazzalin, A Cuenda, E Cano, P Cohen, LC Mahadevan. Effects of the inhibition of p38/RK MAP kinase on induction of five fos and jun genes by diverse stimuli. Oncogene 15:2321–2331, 1997.
89. M Fujihara, K Ikebuchi, TL Maekawa, S Wakamoto, C Ogiso, T Ito, TA Takahashi, T Suzuki, S Sekiguchi. Lipopolysaccharide-induced desensitization of junB gene expression in a mouse macrophage-like cell line, P388D1. J Immunol 161:3659–3665, 1998.
90. Y Tan, J Rouse, A Zhang, S Cariati, P Cohen, MJ Comb. FGF and stress regulate CREB and ATF-1 via a pathway involving p38 MAP kinase and MAPKAP kinase-2. EMBO J 15:4629–4642, 1996.
91. M Iordanov, K Bender, T Ade, W Schmid, C Sachsenmaier, K Engel, M Gaestel, HJ Rahmsdorf, P Herrlich. CREB is activated by UVC through a p38/HOG-1–dependent protein kinase. EMBO J 16:1009–1022, 1997.
92. A Craxton, G Shu, JD Graves, J Saklatvala, EG Krebs, EA Clark. p38 MAPK is required for CD40-induced gene expression and proliferation in B lymphocytes. J Immunol 161:3225–3236, 1998.
93. R Beyaert, A Cuenda, W Vanden Berghe, S Plaisance, JC Lee, G Haegeman, P Cohen, W Fiers. The p38/RK mitogen-activated protein kinase pathway regulates interleukin-6 synthesis response to tumor necrosis factor. EMBO J 15:1914–1923, 1996.
94. S Kumar, MJ Orsini, JC Lee, PC McDonnell, C Debouck, PR Young. Activation of the HIV-1 long terminal repeat by cytokines and environmental stress requires an active CSBP/p38 MAP kinase. J Biol Chem 271:30864–30869, 1996.
95. WM McFarlane, SB Smith, RFL James, AD Clifton, YN Doza, P Cohen, K

Docherty. The p38/reactivating kinase mitogen-activated protein kinase cascade mediates the activation of the transcription factor insulin upstream factor 1 and insulin gene transcription by high glucose in pancreatic β-cells. J Biol Chem 272:20936–20944, 1997.

96. CP Lim, N Jain, X Cao. Stress-induced immediate-early gene, egr-1, involves activation of p38/JNK1. Oncogene 16:2915–2926, 1998.
97. K Miyazawa, A Mori, H Miyata, M Akahane, Y Ajisawa, H Okudaira. Regulation of interleukin-1beta–induced interleukin-6 gene expression in human fibroblast-like synoviocytes by p38 mitogen-activated protein kinase. J Biol Chem 273:24832–24838, 1998.
98. JLE Dean, M Brook, AR Clark, J Saklatvala. p38 Mitogen-activated protein kinase regulates cyclooxygenase-2 mRNA stability and transcription in lipopolysaccharide-treated human monocytes. J Biol Chem 274:264–269, 1999.
99. W Prichett, A Hand, J Sheilds, D Dunnington. Mechanism of action of bicyclic imidazoles defines a translational regulatory pathway on tumor necrosis factor α. J Inflamm. 45:97–105, 1995.
100. A Pietersma, BC Tilly, M Gaestel, N de Jong, JC Lee, JF Koster, W Sluiter. p38 Mitogen activated protein kinase regulates endothelial VCAM-1 expression at the post-transcriptional level. Biochem Biophys Res Commun 230:44–48, 1997.
101. LA Marshall, MJ Hansbury, BJ Bolognese, RJ Gum, PR Young, RJ Mayer. Inhibitors of the p38 mitogen-activated kinase modulate interleukin-4 induction of low affinity IgE receptor (CD23) in human monocytes. J Immunol 161: 6005–6013, 1998.
102. P De Cesaris, D Starace, A Riccioli, F Padula, A Filippini, E Ziparo. Tumor necrosis factor-alpha induces interleukin-6 production and integrin ligand expression by distinct transduction pathways. J Biol Chem 273:7566–7571, 1998.
103. JG Bode, T Peters-Regehr, F Schliess, D Haussinger. Activation of mitogen-activated protein kinases and IL-6 release in response to lipopolysaccharides in Kupffer cells in modulated by anisoosmolarity. J Hepatol 28:795–802, 1998.
104. NR Bhat, P Zhang, JC Lee, EL Hogan. Extracellular signal- regulated kinase and p38 subgroups of mitogen-activated protein kinases regulate inducible nitric oxide synthase and tumor necrosis factor-alpha gene expression in endotoxin-stimulated primary glial cultures. J Neurosci 18:1633–1641, 1998.
105. CL Manthey, SW Wang, SD Kinney, Z Yao. SB202190, a selective inhibitor of p38 mitogen-activated protein kinase, is a powerful regulator of LPS-induced mRNAs in monocytes (in process citation). J Leukoc Biol 64:409–417, 1998.
106. L Shapiro, CA Dinarello. Osmotic regulation of cytokine synthesis in vitro. Proc Natl Acad Sci USA 92:12230–12234, 1995.
107. K Matsumoto, S Hashimoto, Y Gon, T Nakayama, T Horie. Proinflammatory cytokine-induced and chemical mediator-induced IL-8 expression in human bronchial epithelial cells through p38 mitogen-activated protein kinase–dependent pathway. J Allergy Clin Immunol 101:825–831, 1998.

108. Y Gon, S Hashimoto, K Matsumoto, T Nakayama, I Takeshita, T Horie. Cooling and rewarming-induced IL-8 expression in human bronchial epithelial cells through p38 MAP kinase–dependent pathway. Biochem Biophys Res Commun 249:156–160, 1998.
109. M Rincon, H Enslen, J Raingeaud, M Recht, T Zapton, MS-S Su, LA Penix, RJ Davis, RA Flavell. Interferon γ expression by Th1 effector T cells mediated by the p38 MAP kinase signaling pathway. EMBO J 17:2817–2829, 1998.
110. CO Reiser, T Lanz, F Hofmann, G Hofer, HD Rupprecht, M Goppelt Struebe. Lysophosphatidic acid–mediated signal-transduction pathways involved in the induction of the early-response genes prostaglandin G/H synthase-2 and Egr-1: a critical role for the mitogen-activated protein kinase p38 and for Rho proteins. Biochem J 330:1107–1114, 1998.
111. Z Guan, LD Baier, AR Morrison. p38 Mitogen-activated protein kinase downregulates nitric oxide and up-regulates prostaglandin E2 biosynthesis stimulated by interleukin-1beta. J Biol Chem 272:8083–8089, 1997.
112. H Pyo, I Jou, S Jung, S Hong, EH Joe. Mitogen-activated protein kinases activated by lipopolysaccharide and beta-amyloid in cultured rat microglia. Neuroreport 9:871–874, 1998.
113. J Da Silva, B Pierrat, JL Mary, W Lesslauer. Blockade of p38 mitogen-activated protein kinase pathway inhibits inducible nitric-oxide synthase expression in mouse astrocytes. J Biol Chem 272:28373–28380, 1997.
114. AM Badger, MN Cook, MW Lark, TM Newmann-Tarr, BA Swift, AH Nelson, FC Barone, S Kumar. SB 203580 inhibits p38 mitogen-activated protein kinase, nitric oxide production, and inducible nitric oxide synthase in bovine cartilage-derived chondrocytes. J Immunol 161:467–473, 1998.
115. AM Badger, JN Bradbeer, B Votta, JC Lee, JL Adams, DE Griswold. Pharmacological profile of SB 203580, a selective inhibitor of cytokine suppressive binding protein/p38 kinase, in animal models of arthritis, bone resorption, endotoxin shock and immune function. J Pharmacol Exp Ther 279:1453–1461, 1996.
116. O Rausch, CJ Marshall. Cooperation of p38 and extracellular signal-regulated kinase mitogen activated protein kinase pathways during granulocyte colony-stimulating factor–induced cell proliferation. J Biol Chem 274:4096–4105, 1999.
117. JB Crawley, L Rawlinson, FV Lali, TH Page, J Saklatvala, BM Foxwell. T cell proliferation in response to interleukins 2 and 7 requires p38MAP kinase activation. J Biol Chem 272:15023–15027, 1997.
118. FV Lali, AE Hunt, J Lord, B Nelson, K Tracey, T Miyazaki, T Taniguchi, BMJ Foxwell. The activation of MAPK pathways is not essential for interleukin 2 driven lymphocyte proliferation. Eur Cytokine Netw 9:456, 1998.
119. S Huang, Y Jiang, Z Li, E Nishida, P Mathias, S Lin, RJ Ulevitch, GR Nemerow, J Han. Apoptosis signaling pathway in T cells is composed of ICE/Ced-3 family proteases and MAP kinase kinase 6b. Immunity 6:739–749, 1997.
120. P Juo, CJ Kuo, SE Reynolds, RF Konz, J Raingeaud, RJ Davis, HP Biemann,

J Blenis. Fas activation of the p38 mitogen-activated protein kinase signalling pathway requires ICE/CED-3 family proteases (published erratum appears in Mol Cell Biol 1997 Mar; 17(3):1757). Mol Cell Biol 17:24–35, 1997.

121. CC Franklin, S Srikanth, AS Kraft. Conditional expression of mitogen-activated protein kinase phosphatase-1, MKP-1, is cytoprotective against UV-induced apoptosis. Proc Natl Acad Sci USA 95:3014–3019, 1998.

122. X Wang, JL Martindale, Y Liu, NJ Holbrook. The cellular response to oxidative stress: influences of mitogen-activated protein kinase signalling pathways on cell survival. Biochem J 333:291–300, 1998.

123. A Roulston, C Reinhard, P Amiri, LT Williams. Early activation of c-Jun N-terminal kinase and p38 kinase regulate cell survival in response to tumor necrosis factor alpha. J Biol Chem 273:10232–10239, 1998.

124. RA Salmon, IN Foltz, PR Young, JW Schrader. The p38 mitogen-activated protein kinase is activated by ligation of the T or B lymphocyte antigen receptors, Fas or CD40, but suppression of kinase activity does not inhibit apoptosis induced by antigen receptors. J Immunol 159:5309–5317, 1997.

125. D Zechner, R Craig, DS Hanford, PM McDonough, RA Sabbadini, CC Glembotski. MKK6 activates myocardial cell NF-kappaB and inhibits apoptosis in a p38 mitogen-activated protein kinase-dependent manner. J Biol Chem 273:8232–8239, 1998.

126. SC Frasch, JA Nick, VA Fadok, DL Bratton, GS Worthen, PM Henson. p38 mitogen-activated protein kinase-dependent and -independent intracellular signal transduction pathways leading to apoptosis in human neutrophils. J Biol Chem 273:8389–8397, 1998.

127. T Ishizuka, H Kawasome, N Terada, K Takeda, P Gerwins, GM Keller, GL Johnson, EW Gelfand. Stem cell factor augments Fc epsilon RI-mediated TNF-alpha production and stimulates MAP kinases via a different pathway in MC/9 mast cells. J Immunol 161:3624–3630, 1998.

128. S Nemoto, J Xiang, S Huang, A Lin. Induction of apoptosis by SB202190 through inhibition of p38beta mitogen-activated protein kinase. J Biol Chem 273:16415–16420, 1998.

129. Y Wang, S Huang, VP Sah, J Ross, Jr., JH Brown, J Han, KR Chien. Cardiac muscle cell hypertrophy and apoptosis induced by distinct members of the p38 mitogen-activated protein kinase family. J Biol Chem 273:2161–2168, 1998.

130. B Brenner, U Koppenhoefer, C Weinstock, O Linderkamp, F Lang, E Gulbins. Fas- or ceramide-induced apoptosis is mediated by a Rac1-regulated activation of Jun N-terminal kinase/p38 kinases and GADD153. J Biol Chem 272:22173–22181, 1997.

131. T Morooka, E Nishida. Requirement of p38 Mitogen-activated protein kinase for neuronal differentiation in PC12 Cells. J Biol Chem 273:24285–24288, 1998.

132. JA Engelman, MP Lisanti, PE Scherer. Specific inhibitors of p38 mitogen-activated protein kinase block 3T3-L1 adipogenesis. J Biol Chem 273:32111–32120, 1998.

133. Y Nagata, N Takahashi, RJ Davis, K Todokoro. Activation of p38 MAP kinase and JNK but not ERK is required for erythropoietin-induced erythroid differentiation (in process citation). Blood 92:1859–1869, 1998.
134. T Sugawara, T Moriguchi, E Nishida, Y Takahama. Differential roles of Erk and p38 MAP kinase pathways in positive and negative selection of T lymphocytes. Immunity 9:565–574, 1998.
135. S Nemoto, Z Sheng, A Lin. Opposing effects of Jun kinase and p38 mitogen-activated protein kinases on cardiomyocyte hypertrophy. Mol Cell Biol 18: 3518–3526, 1998.
136. J Huot, F Houle, F Marceau, J Landry. Oxidative stress-induced actin reorganization mediated by the p38 mitogen-activated protein kinase/heat shock protein 27 pathway in vascular endothelial cells. Circ Res 80:383–392, 1997.
137. C Schafer, SE Ross, MJ Bragado, GE Groblewski, SA Ernst, JA Williams. A role for the p38 mitogen-activated protein kinase/Hsp 27 pathway in cholecystokinin-induced changes in the actin cytoskeleton in rat pancreatic acini (in process Citation). J Biol Chem 273:24173–24180, 1998.
138. ET Maizels, J Cottom, JC Jones, M Hunzicker-Dunn. Follicle stimulating hormone (FSH) activates the p38 mitogen-activated protein kinase pathway, inducing small heat shock protein phosphorylation and cell rounding in immature rat ovarian granulosa cells. Endocrinology 139:3353–3366, 1998.
139. M Ogura, M Kitamura. Oxidant stress incites spreading of macrophages via extracellular signal-regulated kinases and p38 mitogen-activated protein kinase. J Immunol 161:3569–3574, 1998.
140. M Hannigan, L Zhan, Y Ai, CK Huang. The role of p38 MAP kinase in TGFβ1-induced signal transduction in human neutrophils. Biochem Biophys Res Commun. 246:55–58, 1998.
141. J Saklatvala, L Rawlinson, RJ Waller, S Sarsfield, JC Lee, LF Morton, MJ Barnes, RW Farndale. Role for p38 mitogen-activated protein kinase in platelet aggregation caused by collagen or a thromboxane analogue. J Biol Chem 271: 6586–6589, 1996.
142. DY Tamura, EE Moore, JL Johnson, G Zallen, J Aiboshi, CC Silliman. p38 mitogen-activated protein kinase inhibition attenuates intercellular adhesion molecule-1 up-regulation on human pulmonary microvascular endothelial cells. Surgery 124:403–407; discussion 408, 1998.
143. JR Jackson, B Bolognese, L Hillegass, S Kassis, J Adams, DE Griswold, JD Winkler. Pharmacological effects of SB 220025, a selective inhibitor of p38 mitogen-activated protein kinase, in angiogenesis and chronic inflammatory disease models. J Pharmacol Exp Ther 284:687–692, 1998.
144. FC Barone, GZ Feuerstein, RF White, EA Irving, AA Parsons, SJ Hadingham, J Roberts, AJ Hunter, GE Archer, S Kumar, JC Lee, BR Smith, JL Adams. Selective Inhibition of p38 mitogen activated kinase reduces brain injury and neurological deficits in rat focal stroke models. J Cereb Blood Flow Metab 19(Suppl. 1):S613, 1999.
145. XL Ma, S Kumar, F Gao, CS Louden, BL Lopez, TA Christopher, C Wang,

JC Lee, GZ Feuerstein, T-L Yue. Inhibition of p38 mitogen-activated protein kinase decreases cardiomyocyte apoptosis and improves cardiac function after myocardial ischemia and reperfusion. Circulation 99:1685–1691, 1999.

146. J Raingeaud, S Gupta, JS Rogers, M Dickens, J Han, RJ Ulevitch, RJ Davis. Pro-inflammatory cytokines and environmental stress cause p38 mitogen-activated protein kinase activation by dual phosphorylation on tyrosine and threonine. J Biol Chem 270:7420–7426, 1995.
147. CA Hazzalin, R Le Panse, E Cano, LC Mahadevan. Anisomycin selectively desensitizes signalling components involved in stress kinase activation and fos and jun induction. Mol Cell Biol 18:1844–1854, 1998.
148. JL Kummer, PK Rao, KA Heidenreich. Apoptosis induced by withdrawal of trophic factors is mediated by p38 mitogen-activated protein kinase. J Biol Chem 272:20490–20494, 1997.
149. WM Macfarlane, SB Smith, RF James, AD Clifton, YN Doza, P Cohen, K Docherty. The p38/reactivating kinase mitogen-activated protein kinase cascade mediates the activation of the transcription factor insulin upstream factor 1 and insulin gene transcription by high glucose in pancreatic beta-cells. J Biol Chem 272:20936–20944, 1997.
150. MA Bogoyevitch, J Gillespie Brown, AJ Ketterman, SJ Fuller, R Ben Levy, A Ashworth, CJ Marshall, PH Sugden. Stimulation of the stress-activated mitogen-activated protein kinase subfamilies in perfused heart. p38/RK mitogen-activated protein kinases and c-Jun N-terminal kinases are activated by ischemia/reperfusion. Circ Res 79:162–173, 1996.
151. T Yin, G Sandhu, CD Wolfgang, A Burrier, RL Webb, DF Rigel, T Hai, J Whelan. Tissue-specific pattern of stress kinase activation in ischemic/reperfused heart and kidney. J Biol Chem 272:19943–19950, 1997.
152. CL Sutherland, AW Heath, SL Pelech, PR Young, MR Gold. Differential activation of the ERK, JNK, and p38 mitogen activated kinases by CD40 and the B cell antigen receptor. J Immunol 157:3381–3390, 1996.
153. RM Kramer, EF Roberts, BA Strifler, EM Johnstone. Thrombin induces activation of p38 MAP kinase in human platelets. J Biol Chem 270:27395–27398, 1995.
154. J Yamauchi, M Nagao, Y Kaziro, H Itoh. Activation of p38 mitogen-activated protein kinase by signaling through G protein–coupled receptors. Involvement of Gbetagamma and Galphaq/11 subunits. J Biol Chem 272:27771–27777, 1997.
155. IN Foltz, JC Lee, PR Young, JW Schrader. Hemopoietic growth factors with the exception of interleukin-4 activate the p38 mitogen-activated protein kinase pathway. J Biol Chem 272:3296–3301, 1997.
156. S Rousseau, F Houle, J Landry, J Huot. p38 Map kinase activation by vascular endothelial growth factor mediates actin reorganization and cell migration in human endothelial cells. Oncogene 15:2169–2177, 1997.
157. J Xing, JM Kornhauser, Z Xia, EA Thiele, ME Greenberg. Nerve growth factor activates extracellular signal-regulated kinase and p38 mitogen-activated pro-

tein kinase pathways to stimulate CREB serine 133 phosphorylation. Mol Cell Biol 18:1946–1955, 1998.
158. H Schultz, T Rogalla, K Engel, JC Lee, M Gaestel. The protein kinase inhibitor SB203580 uncouples PMA-induced differentiation of HL-60 cells from phosphorylation of Hsp27. Cell Stress Chaperones 2:41–49, 1997.
159. M Ushio-Fukai, RW Alexander, M Akers, KK Griendling. p38 Mitogen-activated protein kinase is a critical component of the redox-sensitive signaling pathways activated by angiotensin II. Role in vascular smooth muscle cell hypertrophy. J Biol Chem 273:15022–15029, 1998.
160. H Kawasaki, T Morooka, S Shimohama, J Kimura, T Hirano, Y Gotoh, E Nishida. Activation and involvement of p38 mitogen-activated protein kinase in glutamate-induced apoptosis in rat cerebellar granule cells. J Biol Chem 272: 18518–18521, 1997.
161. A Clerk, A Michael, PH Sugden. Stimulation of the p38 mitogen-activated protein kinase pathway in neonatal rat ventricular myocytes by the G protein–coupled receptor agonists, endothelin-1 and phenylephrine: a role in cardiac myocyte hypertrophy? J Cell Biol 142:523–535, 1998.
162. A Clerk, A Michael, PH Sugden. Stimulation of multiple mitogen-activated protein kinase sub-families by oxidative stress and phosphorylation of the small heat shock protein, HSP25/27, in neonatal ventricular myocytes (in process citation). Biochem J 333:581–589, 1998.
163. A Lazou, PH Sugden, A Clerk. Activation of mitogen-activated protein kinases (p38-MAPKs, SAPKs/JNKs and ERKs) by the G-protein–coupled receptor agonist phenylephrine in the perfused rat heart. Biochem J 332:459–465, 1998.
164. JF Larrivee, DR Bachvarov, F Houle, J Landry, J Huot, F Marceau. Role of the mitogen-activated protein kinases in the expression of the kinin B1 receptors induced by tissue injury. J Immunol 160:1419–1426, 1998.
165. LD Shrode, EA Rubie, JR Woodgett, S Grinstein. Cytosolic alkalinization increases stress-activated protein kinase/c-Jun NH2-terminal kinase (SAPK/JNK) activity and p38 mitogen-activated protein kinase activity by a calcium-independent mechanism. J Biol Chem 272:13653–13659, 1997.
166. J Tao, JS Sanghera, SL Pelech, G Wong, JG Levy. Stimulation of stress-activated protein kinase and p38 HOG1 kinase in murine keratinocytes following photodynamic therapy with benzoporphyrin derivative. J Biol Chem 271: 27107–27115, 1996.
167. PC McDonnell, AG DiLella, JC Lee, PR Young. Localization of the human stress responsive MAP kinase-like CSAIDs binding protein (CSBP) gene to chromosome 6q21.3/21.2. Genomics 29:301–302, 1995.
168. M Goedert, J Hasegawa, M Craxton, MA Leversha, S Clegg. Assignment of the human stress-activated protein kinase-3 gene (SAPK3) to chromosome 22q13.3 by fluorescence in situ hybridization. Genomics 41:501–502, 1997.

Index